SERIES EDITOR: ALAN SMITH

Modular Maths for Edexcel

A2
Pure Mathematics

Second Edition

- JOHN SYKES
- CATHERINE BERRY, VAL HANRAHAN, PETER SECKER

Hodder Murray
A MEMBER OF THE HODDER HEADLINE GROUP

C3 C4

Acknowledgements

We are grateful to the following companies, institutions and individuals who have given permission to reproduce copyright material in this book. Every effort has been made to trace and acknowledge ownership of copyright. The publishers will be glad to make suitable arrangements with any copyright holders whom it has not been possible to contact.

Photos
page 14, © David Simson
page 193, © Guinness Records

OCR, AQA and Edexcel accept no responsibility whatsoever for the accuracy or method of working in the answers given.

Orders: please contact Bookpoint Ltd, 130 Milton Park, Abingdon, Oxon OX14 4SB. Telephone: (44) 01235 827720, Fax: (44) 01235 400454. Lines are open from 9.00–5.00, Monday to Saturday, with a 24-hour message answering service. Visit our website at www.hoddereducation.co.uk

British Library Cataloguing in Publication Data
A catalogue record for this title is available from The British Library.

ISBN-10: 0-340-88530-0
ISBN-13: 978-0-340-88530-7

First published 2001
Second edition published 2005
Impression number   10 9 8 7 6 5 4 3 2
Year   2010 2009 2008 2007 2006

Copyright in this format © 2000, 2005 John Sykes

This work includes material adapted from the MEI Structured Mathematics series.

All rights reserved. Apart from any use permitted under UK copyright law, no part of this publication may be reproduced or transmitted in any form or by any means, electronic or mechanical, including photocopying, recording, or any information storage and retrieval system, without permission in writing from the publisher or under licence from the Copyright Licensing Agency Limited. Further details of such licences (for reprographic reproduction) may be obtained from the Copyright Licensing Agency Limited, 90 Tottenham Court Road, London W1T 4LP.

Papers used in this book are natural, renewable and recyclable products. They are made from wood grown in sustainable forests. The logging and manufacturing processes conform to the environmental regulations of the country of origin.

Cover photo from The Image Bank/Getty Images

Typeset by Pantek Arts Ltd, Maidstone, Kent.
Printed in Great Britain for Hodder Murray, a member of the Hodder Headline Group, 338 Euston Road, London NW1 3BH by Martins the Printers Ltd

Edexcel Advanced Mathematics

The Edexcel course is based on units in the four strands of Pure Mathematics, Mechanics, Statistics and Decision Mathematics. The first unit in each of these strands is designated AS and so is Pure Mathematics: Core 2; all others are A2.

The units may be aggregated as follows:

3 units	AS Mathematics
6 units	A Level Mathematics
9 units	A Level Mathematics + AS Further Mathematics
12 units	A Level Mathematics + A Level Further Mathematics

Core 1 and 2 are compulsory for AS Mathematics, and Core 3 and 4 must also be included in a full A Level award.

Examinations are offered by Edexcel twice a year, in January (most units) and in June (all units). All units are assessed by examination only; there is no longer any coursework in the scheme.

Candidates are not permitted to use electronic calculators in the Core 1 examination. In all other examinations candidates may use any legal calculator of their choice, including graphical calculators.

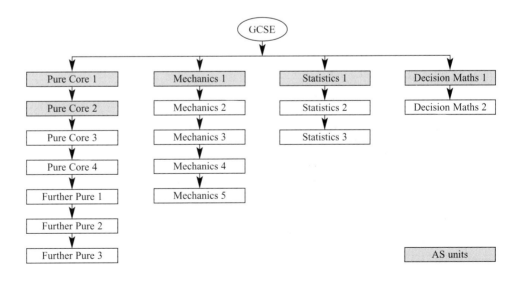

INTRODUCTION

This is the second book written to support the Pure Mathematics units in the Edexcel Advanced Level Mathematics scheme. Much of the material is new and there has been a substantial re-write of previous material to provide complete coverage of the Edexcel Pure Mathematics Core 3 and Core 4 units.

Each book follows the order and structure of the Edexcel examination specifications. Throughout the series the emphasis is on understanding and applying a wide variety of mathematical skills. The style of writing provides clear explanations of the methods and techniques. There is a large number of examples illustrating the problem-solving processes. There are many questions from Edexcel question papers.

The Core 3 section of this book has five chapters relating to the five sections of the syllabus. The first chapter looks at algebraic fractions and the algebra of functions. Chapter 2 builds upon the trigonometry seen in Chapter 9 of Core 2. In Chapter 3 the reader learns about logarithmic and exponential functions. In Chapter 4 differentiation and integration methods are developed, with particular emphasis on differentiation techniques. Chapter 5 covers numerical methods.

In the Core 4 section there are six chapters. Chapter 6 is mainly about partial fractions. Chapter 7 covers the coordinate geometry of functions defined parametrically, with Chapter 8 explaining the general binomial expansion. Chapter 9 has more differentiation methods, building upon those of Core 3, whilst Chapter 10 develops a significant number of integration methods. Chapter 11 is about vectors.

I would like to thank the many people in the preparation and checking of material. Further thanks go to my colleagues at Sedbergh School.

John Sykes

Contents

Core 3

1 Algebra and functions — 2

Rational expressions — 2
Proof — 5
Equations involving algebraic fractions — 7
Algebraic division — 11
Mathematical mappings — 14
Composite functions — 20
Inverse functions — 25
The modulus function — 32
Combinations of transformations — 36

2 Trigonometry — 51

Trigonometric functions — 51
Reciprocal trigonometric functions — 53
Inverse trigonometric functions — 56
Fundamental trigonometric identity — 59
Compound-angle formulae — 63
Double-angle formulae — 67
Half-angle formulae — 69
$a\cos\theta + b\sin\theta$ in the form $r\cos(\theta \pm \alpha)$ or $r\sin(\theta \pm \alpha)$ — 71

3 Exponentials and logarithms — 81

- The function e^x — 81
- The function $\ln x$ — 85
- Equations of the form $e^{ax+b} = p$ and $\ln(ax+b) = q$ — 88

4 Differentiation — 97

- Derivative of e^x — 100
- Derivative of $\ln x$ — 101
- The chain rule — 105
- The product rule — 113
- The quotient rule — 116
- The derivative $\dfrac{dx}{dy}$ — 123
- Differentiating trigonometric functions — 129

5 Numerical methods — 140

- Numerical solutions of equations — 140
- Change of sign method — 141
- Iterative methods to solve $f(x) = 0$ — 147

Core 4

6 Algebra — 162

- Rational functions — 162
- Partial fractions — 163
- Partial fractions and differentiation — 170

7 Coordinate geometry in the (x, y) plane — 175

- Parametric equations of curves — 175
- Finding the parametric equation by eliminating the parameter — 175

	Parametric equations of some standard curves	180
	Integrating functions defined parametrically	184

8 Sequences and series — 193

Expansion of $(ax + b)^n$ when n is a positive integer	194
The general binomial expansion	195

9 Differentiation — 209

Differentiating functions defined implicitly	209
Differentiating functions defined parametrically	215
Exponential growth and decay, and the derivative of a^x	222
Differential equations	225

10 Integration — 234

Review of integration	234
Integral of e^x	239
Integral of $\frac{1}{x}$	242
Further integration	245
Integration by substitution	253
Integration by parts	262
Using partial fractions in integration	271
Using trigonometric identities in integration	275
Finding volumes of revolution by integration	279
Volumes of revolution with equations in parametric form	285
Solving differential equations	287
Choosing an appropriate method of integration	296
Numerical integration of functions	298

11 Vectors — 310

Vectors in two dimensions — 310
Vectors in three dimensions — 323
Coordinate geometry using vectors — 326
The scalar product — 342

Answers — 358

Pure Mathematics: Core 3 — 358
Pure Mathematics: Core 4 — 385

Index — 410

Core 3

C3

Chapter one

ALGEBRA AND FUNCTIONS

She must know all the needs of a rational being.

Sir Owen Seaman

RATIONAL EXPRESSIONS

If $f(x)$ and $g(x)$ are polynomials, the expression $\frac{f(x)}{g(x)}$ is an *algebraic fraction* or *rational function*. It may also be called a *rational expression*. There are many occasions in mathematics when a problem reduces to the manipulation of algebraic fractions, and the rules for this are exactly the same as those for numerical fractions.

SIMPLIFYING FRACTIONS

To simplify a fraction, you look for a factor common to both the numerator (top line) and the denominator (bottom line) and cancel it.

For example, in arithmetic

$$\frac{15}{20} = \frac{5 \times 3}{5 \times 4} = \frac{3}{4}$$

and in algebra

$$\frac{6a}{9a^2} = \frac{2 \times 3 \times a}{3 \times 3 \times a \times a} = \frac{2}{3a}.$$

Note

You must *factorise* both the numerator and denominator before cancelling, since it is only possible to cancel a *common factor*. In some cases this involves putting brackets in. For example:

$$\frac{2a+4}{a^2-4} = \frac{2(a+2)}{(a+2)(a-2)} = \frac{2}{(a-2)}$$

RATIONAL EXPRESSIONS

MULTIPLYING AND DIVIDING FRACTIONS

Multiplying fractions involves cancelling any factors common to the numerator and denominator. For example:

$$\frac{10a}{3b^2} \times \frac{9ab}{25} = \frac{6a^2}{5b}$$

As with simplifying, it is often necessary to factorise any algebraic expressions first:

$$\frac{a^2 + 3a + 2}{9} \times \frac{12}{a+1} = \frac{(a+1)(a+2)}{3 \times 3} \times \frac{3 \times 4}{(a+1)}$$

$$= \frac{(a+2)}{3} \times \frac{4}{1}$$

$$= \frac{4(a+2)}{3}$$

Remember that when one fraction is divided by another, you change ÷ to × and invert the fraction which follows the ÷ symbol.

$$\frac{12}{x^2-1} \div \frac{4}{x+1} = \frac{12}{(x+1)(x-1)} \times \frac{(x+1)}{4}$$

$$= \frac{3}{(x-1)}$$

ADDITION AND SUBTRACTION OF FRACTIONS

To add or subtract two fractions they must be replaced by equivalent fractions, both of which have the same denominator.

For example:

$$\frac{2}{3} + \frac{1}{4} = \frac{8}{12} + \frac{3}{12} = \frac{11}{12}$$

Similarly, in algebra:

$$\frac{2x}{3} + \frac{x}{4} = \frac{8x}{12} + \frac{3x}{12} = \frac{11x}{12}$$

and $\quad \dfrac{2}{3x} + \dfrac{1}{4x} = \dfrac{8}{12x} + \dfrac{3}{12x} = \dfrac{11}{12x}$

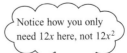
Notice how you only need 12x here, not $12x^2$

You must take particular care when the subtraction of fractions introduces a sign change. For example:

$$\frac{4x-3}{6} - \frac{2x+1}{4} = \frac{2(4x-3) - 3(2x+1)}{12}$$

$$= \frac{8x - 6 - 6x - 3}{12}$$

$$= \frac{2x - 9}{12}$$

Note

In addition and subtraction, the new denominator is the *lowest common multiple* of the original denominators. When two denominators have no *common factor*, their product gives the new denominator. For example:

$$\frac{2}{y+3} + \frac{3}{y-2} = \frac{2(y-2) + 3(y+3)}{(y+3)(y-2)}$$

$$= \frac{2y - 4 + 3y + 9}{(y+3)(y-2)}$$

$$= \frac{5y + 5}{(y+3)(y-2)}$$

$$= \frac{5(y+1)}{(y+3)(y-2)}$$

It may be necessary to factorise denominators in order to identify common factors, as shown here:

$$\frac{2b}{a^2 - b^2} - \frac{3}{a+b} = \frac{2b}{(a+b)(a-b)} - \frac{3}{(a+b)}$$

$(a+b)$ is a common factor

$$= \frac{2b - 3(a-b)}{(a+b)(a-b)}$$

$$= \frac{5b - 3a}{(a+b)(a-b)}$$

PROOF

You may have to be able to show how to obtain a given result. If one expression is equivalent to another the ≡ symbol is used. This means that one side of the expression is a rearrangement of the other side. You came across this idea in Chapter 9 of *Pure Mathematics: Core 1 and Core 2*. In each case we start with the left-hand side (LHS) of the equation and derive the right-hand side (RHS).

EXAMPLE 1.1

Prove that $\dfrac{x^2-1}{x+1} \times \dfrac{x}{x-1} \equiv x$.

Solution

$$\begin{aligned}
\text{LHS} &= \dfrac{x^2-1}{x+1} \times \dfrac{x}{x-1} \\
&= \dfrac{(x+1)(x-1)}{x+1} \times \dfrac{x}{x-1} \\
&= x \\
&= \text{RHS}
\end{aligned}$$

Hence the result is proven.

EXAMPLE 1.2

Prove that $\dfrac{(x+1)(x+2)}{(x-3)(x+4)} - \dfrac{2}{x-3} \equiv \dfrac{(x+3)(x-2)}{(x-3)(x+4)}$.

Solution

$$\begin{aligned}
\text{LHS} &= \dfrac{(x+1)(x+2)}{(x-3)(x+4)} - \dfrac{2}{x-3} \\
&= \dfrac{x^2+3x+2-2(x+4)}{(x-3)(x+4)} \\
&= \dfrac{x^2+x-6}{(x-3)(x+4)} \\
&= \dfrac{(x+3)(x-2)}{(x-3)(x+4)} \\
&= \text{RHS}
\end{aligned}$$

Hence the result is proven.

Exercise 1A

Simplify the expressions in questions 1–10.

1. $\dfrac{6a}{b} \times \dfrac{a}{9b^2}$

2. $\dfrac{5xy}{3} \div 15xy^2$

3. $\dfrac{x^2 - 9}{x^2 - 9x + 18}$

4. $\dfrac{5x - 1}{x + 3} \times \dfrac{x^2 + 6x + 9}{5x^2 + 4x - 1}$

5. $\dfrac{4x^2 - 25}{4x^2 + 20x + 25}$

6. $\dfrac{a^2 + a - 12}{5} \times \dfrac{3}{4a - 12}$

7. $\dfrac{4x^2 - 9}{x^2 + 2x + 1} \div \dfrac{2x - 3}{x^2 + x}$

8. $\dfrac{2p + 4}{5} \div (p^2 - 4)$

9. $\dfrac{a^2 - b^2}{2a^2 + ab - b^2}$

10. $\dfrac{x^2 + 8x + 16}{x^2 + 6x + 9} \times \dfrac{x^2 + 2x - 3}{x^2 + 4x}$

In questions 11–24 write each of the expressions as a single fraction in its simplest form.

11. $\dfrac{1}{4x} + \dfrac{1}{5x}$

12. $\dfrac{x}{3} - \dfrac{(x + 1)}{4}$

13. $\dfrac{a}{a + 1} + \dfrac{1}{a - 1}$

14. $\dfrac{2}{x - 3} + \dfrac{3}{x - 2}$

15. $\dfrac{x}{x^2 - 4} - \dfrac{1}{x + 2}$

16. $\dfrac{p^2}{p^2 - 1} - \dfrac{p^2}{p^2 + 1}$

17. $\dfrac{2}{a + 1} - \dfrac{a}{a^2 + 1}$

18. $\dfrac{2y}{(y + 2)^2} - \dfrac{4}{y + 4}$

19. $x + \dfrac{1}{x + 1}$

20. $\dfrac{2}{b^2 + 2b + 1} - \dfrac{3}{b + 1}$

21. $\dfrac{2}{3(x - 1)} + \dfrac{3}{2(x + 1)}$

22. $\dfrac{6}{5(x + 2)} - \dfrac{2x}{(x + 2)^2}$

23. $\dfrac{2}{a + 2} - \dfrac{a - 2}{2a^2 + a - 6}$

24. $\dfrac{1}{x - 2} + \dfrac{1}{x} + \dfrac{1}{x + 2}$

In questions 25–30 prove the given results.

25. $\dfrac{x^2 + 2x + 1}{x^2 + 3x + 2} \equiv \dfrac{x + 1}{x + 2}$

26. $\dfrac{1}{x + 1} - \dfrac{1}{x + 2} \equiv \dfrac{1}{(x + 1)(x + 2)}$

27. $\dfrac{(x + 1)(x + 3)}{(x - 1)(x - 3)} \div \dfrac{x^2 - 9}{x^2 - 1} \equiv \dfrac{(x + 1)^2}{(x - 3)^2}$

28. $\dfrac{3(x + 2)}{(x + 1)(x - 4)} - \dfrac{3}{x - 4} \equiv \dfrac{3}{(x + 1)(x - 4)}$

29. $\dfrac{x^2 + 8x + 12}{x^2 + 7x + 12} \times \dfrac{x^2 + 6x + 8}{x^2 + 9x + 18} \equiv \left(\dfrac{x + 2}{x + 3}\right)^2$

30. $\dfrac{(x + 1)(x + 4)}{(x - 2)(x + 3)} + \dfrac{2}{x - 2} \equiv \dfrac{(x + 2)(x + 5)}{(x - 2)(x + 3)}$

Equations involving algebraic fractions

The easiest way to solve an equation involving fractions is usually to multiply both sides by an expression which will cancel out the fractions.

EXAMPLE 1.3

Solve $\dfrac{x}{3} + \dfrac{2x}{5} = 4$.

Solution Multiplying by 15 (the lowest common multiple of 3 and 5) gives

$$15 \times \frac{x}{3} + 15 \times \frac{2x}{5} = 15 \times 4$$

$$\Rightarrow \quad 5x + 6x = 60$$

$$\Rightarrow \quad 11x = 60$$

$$\Rightarrow \quad x = \frac{60}{11}.$$

Notice that all three terms must be multiplied by 15

A similar method applies when denominators are algebraic.

EXAMPLE 1.4

Solve $\dfrac{5}{x} - \dfrac{4}{x+1} = 1$.

Solution Multiplying by $x(x+1)$ (the least common multiple of x and $x+1$) gives

$$\frac{5x(x+1)}{x} - \frac{4x(x+1)}{x+1} = x(x+1)$$

$$\Rightarrow \quad 5(x+1) - 4x = x(x+1)$$

$$\Rightarrow \quad 5x + 5 - 4x = x^2 + x$$

$$\Rightarrow \quad x^2 = 5$$

$$\Rightarrow \quad x = \pm\sqrt{5}.$$

In Example 1.4, the lowest common multiple of the denominators is their product, but this is not always the case.

EXAMPLE 1.5

Solve $\dfrac{1}{(x-3)(x-1)} + \dfrac{1}{x(x-1)} = -\dfrac{1}{x(x-3)}$.

Solution Here you only need to multiply by $x(x-3)(x-1)$ to eliminate all the fractions. This gives

$$\frac{x(x-3)(x-1)}{(x-3)(x-1)} + \frac{x(x-3)(x-1)}{x(x-1)} = \frac{-x(x-3)(x-1)}{x(x-3)}$$

$$\Rightarrow x + (x-3) = -(x-1)$$
$$\Rightarrow 2x - 3 = -x + 1$$
$$\Rightarrow 3x = 4$$
$$\Rightarrow x = \tfrac{4}{3}.$$

Fractional algebraic equations arise in a number of situations, including, as in the following example, problems connecting distance, speed and time. The relationship $\text{time} = \dfrac{\text{distance}}{\text{speed}}$ is useful here.

EXAMPLE 1.6

Each day I travel 10 km from home to work. One day, because of road works, my average speed was 5 km h^{-1} slower than usual, and my journey took an extra 10 minutes.

Taking x km h^{-1} as my usual speed:
(a) write down an expression in x which represents my usual time in hours;
(b) write down an expression in x which represents my time when I travel 5 km h^{-1} slower than usual;
(c) use these expressions to form an equation in x and solve it.
(d) How long did my journey usually take?

Solution (a) Time = $\dfrac{\text{distance}}{\text{speed}} \Rightarrow$ usual time = $\dfrac{10}{x}$.

(b) I now travel at $(x-5)$ km h^{-1}, so the longer time = $\dfrac{10}{x-5}$.

(c) The difference in these times is 10 minutes, or $\tfrac{1}{6}$ hour, so

$$\frac{10}{x-5} - \frac{10}{x} = \frac{1}{6}.$$

Multiplying by $6x(x-5)$ gives

$$\frac{60x(x-5)}{(x-5)} - \frac{60x(x-5)}{x} = \frac{6x(x-5)}{6}$$

$$\Rightarrow \quad 60x - 60(x-5) = x(x-5)$$

$$\Rightarrow \quad 60x - 60x + 300 = x^2 - 5x$$

$$\Rightarrow \quad x^2 - 5x - 300 = 0$$

$$\Rightarrow \quad (x-20)(x+15) = 0$$

$$\Rightarrow \quad x = 20 \quad \text{or} \quad x = -15.$$

(d) Reject $x = -15$, since x km h^{-1} is a speed.

Usual speed = 20 km h^{-1}

Usual time = $\frac{10}{x}$ hours = 30 minutes

EXERCISE 1B

1 Solve the following equations.

(a) $\dfrac{2x}{7} - \dfrac{x}{4} = 3$

(b) $\dfrac{5}{4x} + \dfrac{3}{2x} = \dfrac{11}{16}$

(c) $\dfrac{2}{x} - \dfrac{5}{2x-1} = 0$

(d) $x - 3 = \dfrac{x+2}{x-2}$

(e) $\dfrac{1}{x} + x + 1 = \dfrac{13}{3}$

(f) $\dfrac{2x}{x+1} - \dfrac{1}{x-1} = 1$

(g) $\dfrac{x}{x-1} - \dfrac{x-1}{x} = 2$

2 I have £6 to spend on crisps for a party. When I get to the shop I find that the price has been reduced by 1 penny per packet, and I can buy one packet more than I expected. Taking x pence as the original cost of a packet of crisps:
(a) write down an expression in x which represents the number of packets that I expected to buy;
(b) write down an expression in x which represents the number of packets bought at the reduced price;
(c) form an equation in x and solve it to find the original cost.

3 The distance from Manchester to Oxford is 270 km. One day, roadworks on the M6 meant that my average speed was 10 km h^{-1} less than I had anticipated, and so I arrived 18 minutes later than planned. Taking x km h^{-1} as the anticipated average speed:

(a) write down an expression in x for the anticipated and actual times of the journey;
(b) form an equation in x and solve it;
(c) find the time of my arrival in Oxford if I left home at 10 am.

4 Each time somebody leaves the firm of Honeys, he or she is taken out for a meal by the rest of the staff. On one such occasion the bill came to £272, and each member of staff remaining with the firm paid an extra £1 to cover the cost of the meal for the one who was leaving. Taking £x as the cost of the meal, write down an equation in x and solve it.

How many staff were left working for Honeys?

5 A Swiss Roll cake is 21 cm long. When I cut it into slices, I can get two extra slices if I reduce the thickness of each slice by $\frac{1}{4}$ cm. Taking x as the number of thicker slices, write down an equation in x and solve it.

6 Two electrical resistances may be connected in series or in parallel. In series, the equivalent single resistance is the sum of the two resistances, but in parallel, the two resistances R_1 and R_2 are equivalent to a single resistance R where

$$\frac{1}{R_1} + \frac{1}{R_2} = \frac{1}{R}.$$

(a) Find the single resistance which is equivalent to resistances of 3 and 4 ohms connected in parallel.
(b) What resistance must be added in parallel to a resistance of 6 ohms to give a resistance of 2.4 ohms?
(c) What is the effect of connecting two equal resistances in parallel?

7 Express $x^3 + 1$ as the product of a linear and a quadratic factor.
Hence simplify

$$\frac{x^3 + 1}{x^2 + 3x + 2}.$$

8 Solve

$$1 + \frac{6}{x} = \frac{6}{x - 1}.$$

9 Express as a single fraction in its simplest form

$$\frac{x + 1}{x^2 - 4} \times \frac{x + 2}{x^2 - 1}.$$

10 Express as a single fraction in its simplest form

$$\frac{2(x + 1)(x - 2)}{(x - 5)(x + 4)} - \frac{x - 1}{x - 5}.$$

Algebraic division

You learnt in Chapter 6 of *Pure Mathematics: Core 1 and Core 2* how to divide by functions of the type $x + a$ or $x - a$.

For example to divide $6x^3 + 13x^3 + 2x - 5$ by $x + 1$ using the 'long division' method gives:

$$
\begin{array}{r}
6x^2 + 7x - 5 \\
x + 1 \overline{) 6x^3 + 13x^2 + 2x - 5} \\
\underline{6x^3 + 6x^2 } \\
7x^2 + 2x \\
\underline{7x^2 + 7x } \\
-5x - 5 \\
\underline{-5x - 5} \\
0
\end{array}
$$

i.e. $\dfrac{6x^3 + 13x^2 + 2x - 5}{x + 1} = 6x^2 + 7x - 5$.

We may use the same method to divide by functions of the type $ax + b$.

EXAMPLE 1.7

Divide $6x^3 + 13x^2 + 2x - 5$ by $2x - 1$.

Solution

$$
\begin{array}{r}
3x^2 + 8x + 5 \\
2x - 1 \overline{) 6x^3 + 13x^2 + 2x - 5} \\
\underline{6x^3 - 3x^2 } \\
16x^2 + 2x \\
\underline{16x^2 - 8x } \\
10x - 5 \\
\underline{10x - 5} \\
0
\end{array}
$$

i.e. $\dfrac{6x^3 + 13x^2 + 2x - 5}{2x - 1} = 3x^2 + 8x + 5$.

In Example 1.7 there is no remainder, therefore $2x - 1$ is a factor of $6x^3 + 13x^2 + 2x - 5$.

EXAMPLE 1.8

Find the quotient and remainder when $4x^3 + 2x^2 - 4x - 1$ is divided by $2x + 3$.

Solution

$$
\begin{array}{r}
2x^2 - 2x + 1 \\
2x+3{\overline{\smash{\big)}\,4x^3 + 2x^2 - 4x - 1}} \\
\underline{4x^3 + 6x^2} \\
-4x^2 - 4x \\
\underline{-4x^2 - 6x} \\
2x - 1 \\
\underline{2x + 3} \\
-4
\end{array}
$$

The quotient is $2x^2 - 2x + 1$ and the remainder is -4.

This will remind you of the remainder theorem where

$$p(x) = Q(x)\,(ax + b) + R$$

where $Q(x)$ is the Quotient and R is the Remainder.

It follows that putting $ax + b = 0$ gives the general rule $R = p\left(-\dfrac{b}{a}\right)$.

EXAMPLE 1.9

Show that $x^3 - x^2 + x + 3$ has a factor $x + 1$. Hence divide $x^3 - x^2 + x + 3$ by $x^2 - 1$.

Solution Let $f(x) = x^3 - x^2 + x + 3$

then $f(-1) = -1 - 1 - 1 + 3 = 0 \quad \therefore x + 1$ is a factor

$$\frac{x^3 - x^2 + x + 3}{x^2 - 1} = \frac{\cancel{(x+1)}(x^2 - 2x + 3)}{\cancel{(x+1)}(x - 1)}$$

(This could be found by division if you cannot do the factorisation by inspection.)

Dividing gives:
$$
\begin{array}{r}
x - 3 \\
x+1{\overline{\smash{\big)}\,x^2 - 2x + 3}} \\
\underline{x^2 + x} \\
-3x + 3 \\
\underline{-3x - 3} \\
6
\end{array}
$$

The final answer is: $\dfrac{x^3 - x^2 + x + 3}{x^2 - 1} = x - 3 + \dfrac{6}{x + 1}$

EXERCISE 1C

1 In the following, find the quotient and the remainder.
   (a) $x^2 - 3x + 3$ divided by $x - 2$
   (b) $2x^3 + 3x^2 - 4x + 6$ divided by $x + 3$
   (c) $2x^3 + 5x^2 - 13x + 1$ divided by $2x - 1$
   (d) $4x^3 - 6x^2 - 2x + 5$ divided by $2x + 1$
   (e) $3x^3 + 4x^2 - x - 4$ divided by $3x - 2$
   (f) $10x^3 - 9x^2 - 29x - 10$ divided by $5x + 3$
   (g) $x^4 - x^3 - x^2 + 8x + 4$ divided by $x + 2$
   (h) $4x^4 + 4x^3 - 23x^2 + 14x + 3$ divided by $2x - 3$
   (i) $2x^4 - x^3 + 7x^2 + 3x - 2$ divided by $x^2 + x + 1$
   (j) $6x^4 - x^3 - 18x^2 + 20x - 10$ divided by $2x^2 - 3x + 4$

2 $f(x) = \dfrac{2x^2 - 3x - 1}{x - 3}$

   (a) Express $f(x)$ in the form $ax + b + \dfrac{c}{x-3}$.
   (b) Find $f'(x)$.
   (c) Show that $f'(4) = -6$.

3 You are given that
   $$f(x) = \dfrac{3x^2 + 2x - 3}{x + 2}.$$
   Prove that
   $$f(x) = 3x - 4 + \dfrac{5}{x+2}.$$
   Explain why $f(x) \to 3x - 4$ as x becomes large.

4 You are given that $p(x) = 4x^3 - 4x^2 + 5x + 120$.
   (a) Divide $p(x)$ by $2x + 5$.
   (b) Express $p(x)$ in the form $Q(x) \times (2x + 5) + R$.
   (c) Find R using the remainder theorem and so verify that your remainder in part (a) was correct.

5 (a) Given that $p(x) = 2x^3 - 5x^2 - 9x + 18$ and $q(x) = x^2 - x - 6$ show that $x - 3$ is a factor of both $p(x)$ and $q(x)$.
   (b) Simplify $\dfrac{p(x)}{q(x)}$.
   (c) Solve $\dfrac{p(x)}{q(x)} = 0$.

Mathematical mappings

Why fly to Geneva in January?

Several people arriving at Geneva airport from London were asked the main purpose of their visit. Their answers were recorded:

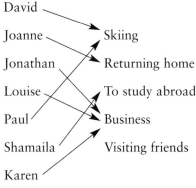

This is an example of a *mapping*.

A mapping is any rule which associates two sets of items. In this example, each of the names on the left is an *object*, or *input*, and each of the reasons on the right is an *image*, or *output*.

For a mapping to make sense or to have any practical application, the inputs and outputs must each form a natural collection or set. The set of possible inputs (in this case, all of the people who flew to Geneva from London in January) is called the *domain* of the mapping. The set of possible outputs (in this case, the set of all possible reasons for flying to Geneva including 'visiting friends') is called the *co-domain* of the mapping.

The seven people questioned in this example gave a set of four reasons, or outputs. These form the *range* of the mapping for this particular set of inputs. The range of any mapping forms part or all of its co-domain.

Notice that Jonathan, Louise and Karen are all visiting Geneva on business: each person gave only one reason for the trip, but the same reason was given by several people. This mapping is said to be many-to-one. A mapping can also be one-to-one, one-to-many, or many-to-many. The relationship between the people and their passport numbers will be one-to-one. The relationship between the people and their items of luggage is likely to be one-to-many, and that between the people and the countries they have visited in the last 10 years will be many-to-many.

MATHEMATICAL MAPPINGS

Mappings expressed using algebra

In mathematics, many (but not all) mappings can be expressed using algebra. Here are some examples of mathematical mappings.

(a)   Domain: integers   Range: real numbers
      Objects   Images
      −1 ⟶ 3
      0 ⟶ 5
      1 ⟶ 7
      2 ⟶ 9
      3 ⟶ 11

General rule: $x \longrightarrow 2x + 5$

(b)   Domain: integers   Range: real numbers
      Objects   Images
                                 1.9
      2   2.1
                                 2.33
                                 2.52
      3   2.99
                                 π

General rule: Rounded whole numbers ↦ Unrounded numbers

(c)   Domain: real numbers   Range: real numbers, $y: -1 \leq y \leq 1$
      Objects   Images
      0
      45   0
      90   0.707
      135   1
      180

General rule: $x° \longrightarrow \sin x°$

(d)   Domain: quadratic equations with real roots   Range: real numbers
      Objects   Images
      $x^2 - 4x + 3 = 0$ 0
      $x^2 - x = 0$ 1
      $x^2 - 3x + 2 = 0$ 2
                                 3

General rule: $ax^2 + bx + c = 0 \longrightarrow x = \dfrac{-b - \sqrt{b^2 - 4ac}}{2a}$
                                                   $x = \dfrac{-b + \sqrt{b^2 - 4ac}}{2a}$

Functions

Mappings which are one-to-one or many-to-one are of particular importance, since in these cases there is only one possible image for any object. Mappings of these types are called *functions*. For example, $x \mapsto x^2$ and $x \mapsto \cos x°$ are both functions, because in each case for any value of x there is only one possible answer. The mapping of rounded whole numbers on to unrounded numbers is not a function, since, for example, the rounded number 5 could mean any number between 4.5 and 5.5.

There are several different but equivalent ways of writing down a function. For example, the function which maps x on to x^2 can be written in any of the following ways.

- $y = x^2$
- $f(x) = x^2$
- $f: x \mapsto x^2$ *(Read this as 'f maps x on to x^2')*

It is often helpful to represent a function graphically, as in the following example, which also illustrates the importance of knowing the domain.

EXAMPLE 1.10

Sketch the graph of $y = 3x + 2$ when the domain of x is:

(a) $x \in \mathbb{R}$
(b) $x \in \mathbb{R}^+$
(c) $x \in \mathbb{N}$,

where $\mathbb{R}$ is the set of all real numbers, $\mathbb{R}^+$ is the set of positive real numbers and $\mathbb{N}$ is the set of natural numbers $\{1, 2, 3 \ldots\}$.

Solution
(a) When the domain is $\mathbb{R}$, all values of y are possible. The range is therefore $\mathbb{R}$, also.
(b) When x is restricted to positive values, all the values of y are greater than 2, so the range is $y > 2$. On the diagram the open circle shows that $(0, 2)$ is not part of the line.
(c) In this case the range is the set of points $\{2, 5, 8, \ldots\}$. These are clearly all of the form $3x + 2$ where x is a natural number $(1, 2, 3, \ldots)$. This set can be written neatly as $\{3x + 2 : x \in \mathbb{N}\}$.

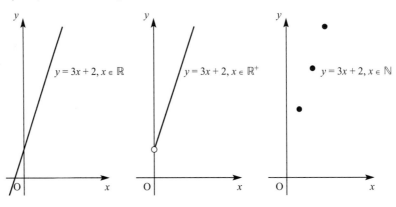

FIGURE 1.1

When you draw the graph of a mapping, the x-coordinate of each point is an input value, the y-coordinate is the corresponding output value. The table below shows this for the mapping $x \mapsto x^2$, or $y = x^2$, and figure 1.2 shows the resulting points on a graph.

Input (x)	Output (y)	Point plotted
−2	4	(−2, 4)
−1	1	(−1, 1)
0	0	(0, 0)
1	1	(1, 1)
2	4	(2, 4)

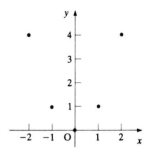

FIGURE 1.2

If the mapping is a function, there is one and only one value of y for every value of x in the domain. Consequently the graph of a function is a curve or line going from left to right, with no doubling back.

Figure 1.3 illustrates some different types of mapping. The graphs in (a) and (b) illustrate functions, those in (c) and (d) do not.

(a) *One-to-one*

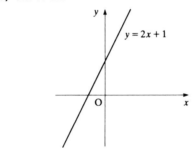

(b) *Many-to-one*

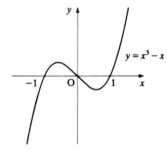

(c) *One-to-many*

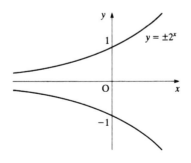

(d) *Many-to-many*

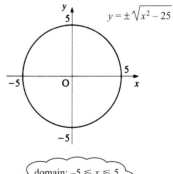

domain: −5 ≤ x ≤ 5
range: −5 ≤ x ≤ 5

FIGURE 1.3

EXAMPLE 1.11

Sketch the graph of $y = f(x)$ where $f(x) = \frac{1}{x}$, $x > 0$. State the type of mapping and whether or not $y = f(x)$ is a function. State the range.

Solution The graph is as shown in figure 1.4.

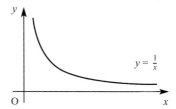

FIGURE 1.4

The mapping is one-to-one since each value of x maps on to a unique value of y. Since the mapping is one-to-one, $f(x)$ is a function.
The range is $y > 0$.

SUMMARY

You have seen that mappings are:

- one-to-one if each one value of x maps onto exactly one value of y;
- many-to-one if two, or more, values of x map onto exactly one value of y;
- one-to-many if one value of x maps onto two, or more, values of y;
- many-to-many if two, or more, values of x map onto two, or more, values of y.

A mapping is said to be a *function* if it is either one-to-one or many-to-one.

The *domain* of the function is the set of x values that are allowed and the *range* is the corresponding set of y values.

EXERCISE 1D

1 Describe each of the following mappings as either one-to-one, many-to-one, one-to-many or many-to-many, and say whether it represents a function.

(a)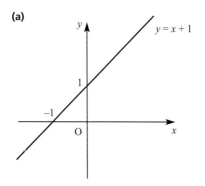

(b)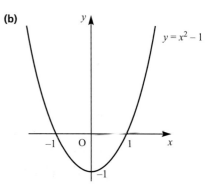

(c)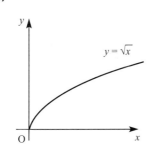

(d)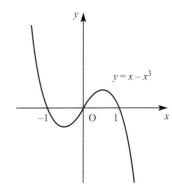

(e)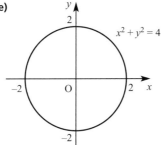

(f)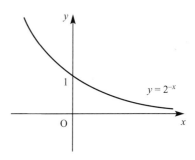

(g)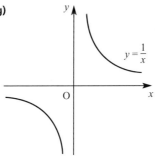

(h)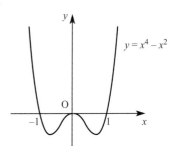

2 Draw the following graphs and state whether the mapping is either one-to-one, many-to-one, one-to-many or many-to-many. For each mapping that is a function, assume the domain is as large as possible and then state range.

(a) $y = 2x + 3$

(b) $y = \dfrac{1}{x^2}$

(c) $y = \pm\sqrt{x}$

(d) $y = 1 - \tfrac{1}{2}x$

(e) $x^2 + y^2 = 9$

(f) $y = \dfrac{1}{x}$

(g) $y = x^2 - 4$

(h) $y = (x - 1)x(x + 1)$

(i) $y = 3^x$

(j) $\dfrac{x^2}{4} + y^2 = 1$

3 (a) A function is defined by $f(x) = 2x - 5$. Write down the values of
   (i) $f(0)$ (ii) $f(7)$ (iii) $f(-3)$.

(b) A function is defined by g:(polygons) ⟼ (number of sides). Find
   (i) g(triangle) (ii) g(pentagon) (iii) g(decagon).

(c) The function t maps Celsius temperatures on to Fahrenheit temperatures. It is defined by t: $C \longmapsto \frac{9C}{5} + 32$. Find
   (i) $t(0)$ (ii) $t(28)$ (iii) $t(-10)$
   (iv) the value of C when $t(C) = C$.

4 Find the range of each of the following functions. (You may find it helpful to draw the graph first.)

(a) $f(x) = 2 - 3x$ $x \geqslant 0$
(b) $f(\theta) = \sin\theta$ $0° \leqslant \theta \leqslant 180°$
(c) $y = x^2 + 2$ $x \in \{0, 1, 2, 3, 4\}$
(d) $y = \tan\theta$ $0° < \theta < 90°$
(e) $f: x \longmapsto 3x - 5$ $x \in \mathbb{R}$
(f) $f: x \longmapsto 2^x$ $x \in \{-1, 0, 1, 2\}$
(g) $y = \cos x$ $-\frac{\pi}{2} \leqslant x \leqslant \frac{\pi}{2}$
(h) $f: \theta \longmapsto \frac{1}{\cos\theta}$ $\theta \in \mathbb{R}$
(i) $f(x) = \frac{1}{1 + x^2}$ $x \in \mathbb{R}$
(j) $f(x) = \sqrt{x - 3} + 3$ $x \geqslant 3$

5 The mapping f is defined by $f(x) = x^2$ $0 \leqslant x \leqslant 3$
                                    $f(x) = 3x$ $3 \leqslant x \leqslant 10$.

The mapping g is defined by $g(x) = x^2$ $0 \leqslant x \leqslant 2$
                              $g(x) = 3x$ $2 \leqslant x \leqslant 10$.

Explain why f is a function and g is not.

COMPOSITE FUNCTIONS

It is possible to combine functions in several different ways, and you have already met some of these. For example, if $f(x) = x^2$ and $g(x) = 2x$, then you could write:

$$f(x) + g(x) = x^2 + 2x.$$

In this example, two functions are added.

Similarly if $f(x) = x$ and $g(x) = \sin x$, then:

$$f(x).g(x) = x\sin x.$$

In this example, two functions are multiplied.

Sometimes you need to apply one function and then apply another to the answer. You are then creating a *composite function* or a *function of a function*.

COMPOSITE FUNCTIONS

EXAMPLE 1.12

A new mother is bathing her baby for the first time. She takes the temperature of the bath water with a thermometer which reads in Celsius, but then has to convert the temperature to degrees Fahrenheit to apply the rule that her own mother taught her:

> At one o five
> He'll cook alive
> But ninety four
> Is rather raw.

Write down the two functions that are involved, and apply them to readings of:
(a) $30\,°C$ (b) $38\,°C$ (c) $45\,°C$.

Solution The first function converts the Celsius temperature C into a Fahrenheit temperature, F:

$$F = \frac{9C}{5} + 32$$

The second function maps Fahrenheit temperatures onto the state of the bath:

$$\begin{aligned}F &\leqslant 94 &&\text{Too cold}\\ 94 &< F < 105 &&\text{All right}\\ F &\geqslant 105 &&\text{Too hot}\end{aligned}$$

This gives

(a) $30\,°C \longmapsto 86\,°F \longmapsto$ too cold
(b) $38\,°C \longmapsto 100.4\,°F \longmapsto$ all right
(c) $45\,°C \longmapsto 113\,°C \longmapsto$ too hot.

In this case the composite function would be (to the nearest degree):

$$\begin{aligned}C &\leqslant 34\,°C &&\text{too cold}\\ 35\,°C \leqslant C &\leqslant 40\,°C &&\text{all right}\\ C &\geqslant 41\,°C &&\text{too hot.}\end{aligned}$$

In algebraic terms, a composite function is constructed as

Input $x \xmapsto{f}$ Output $f(x)$

Input $f(x) \xmapsto{g}$ Output $g[f(x)]$ or $gf(x)$.

> Read this as 'g of f of x'

Thus the composite function $gf(x)$ should be performed from right to left: start with x then apply f and then g.

NOTATION

To apply f twice in succession you would write $f^2(x)$, not $ff(x)$. Similarly $g^3(x)$ means three applications of g. In order to apply a function repeatedly the set of values of the range must be contained within the set of values for the domain.

ORDER OF FUNCTIONS

If f is the rule 'square the input value' and g is the rule 'add 1', then gf is given by

$$x \xrightarrow{f}_{\text{square}} x^2 \xrightarrow{g}_{\text{add 1}} x^2 + 1.$$

So $gf(x) = x^2 + 1.$

Notice that gf(x) is not the same as fg(x), since for fg(x) you must apply g first. In the example above, this would give:

$$x \xrightarrow{g}_{\text{add 1}} (x + 1) \xrightarrow{f}_{\text{square}} (x + 1)^2$$

and so $fg(x) = (x + 1)^2.$

Clearly this is *not* the same result.

EXAMPLE 1.13

Given that $f(x) = 2x$, $g(x) = x^2$ and $h(x) = \frac{1}{x}$, find:

(a) fg(x)   (b) gf(x)   (c) gh(x)
(d) $f^2(x)$ (e) fgh(x)   (f) hfg(x)

Solution

(a) $fg(x) = f[g(x)]$
         $= f(x^2)$
         $= 2x^2$

(b) $gf(x) = g[f(x)]$
         $= g(2x)$
         $= (2x)^2$
         $= 4x^2$

(c) $gh(x) = g[h(x)]$
         $= g(\frac{1}{x})$
         $= \frac{1}{x^2}$

(d) $f^2(x) = f[f(x)]$
         $= f(2x)$
         $= 2(2x)$
         $= 4x$

(e) $fgh(x) = f[gh(x)]$
          $= f\left(\frac{1}{x^2}\right)$ using (c)
          $= \frac{2}{x^2}$

(f) $hfg(x) = h[fg(x)]$
          $= h(2x^2)$ using (a)
          $= \frac{1}{2x^2}$

EXAMPLE 1.14

Functions f and g are defined by

$f: x \mapsto x - 2, \quad x \in \mathbb{R}$
$g: x \mapsto x^2 + 1, \quad x \in \mathbb{R}.$

(a) Find gf in terms of x, stating its domain and range.
(b) Solve $fg(x) = 15.$

EXERCISE 1E

Solution (a) gf means g(f(x)), i.e. substitute f(x) into g,
so
$$gf(x) = (x-2)^2 + 1$$
$$= x^2 - 4x + 5$$
with domain $x \in \mathbb{R}$ and range $y \geq 1$.

Minimum value of gf(x) is when $x = 2$

(b) To solve fg(x) = 15 first find the function fg(x).
$$fg(x) = (x^2 + 1) - 2 = x^2 - 1$$
so
$$x^2 - 1 = 15$$
$$x^2 = 16$$
$$x = \pm 4$$

EXERCISE 1E

1 The functions f, g and h are defined by $f(x) = x^3$, $g(x) = 2x$ and $h(x) = x + 2$. Find each of the following in terms of x.
   (a) fg   (b) gf   (c) fh
   (d) hf   (e) fgh   (f) ghf
   (g) g^2 (h) $(fh)^2$ (i) h^2

2 The functions f and g are defined by
   $$f: x \mapsto x + 4, \quad x \in \mathbb{R}$$
   $$g: x \mapsto x^2 - 2, \quad x \in \mathbb{R}.$$
   (a) Find gf in terms of x stating its domain and range.
   (b) Solve fg(x) = 27.

3 The functions f and g are defined by
   $$f: x \mapsto x + 1, \quad x \in \mathbb{R}$$
   $$g: x \mapsto \sqrt{x}, \quad x > 0.$$
   (a) Find gf(x), stating its domain and range.
   (b) Solve gf(x) = 3.
   (c) Solve fg(x) = 3.

4 The functions f and g are defined by
   $$f: x \mapsto 2x + 3, \quad x \in \mathbb{R}$$
   $$g: x \mapsto \frac{1}{x - 1}, \quad x \neq 1.$$
   (a) Find the composite function gf in its simplest form, stating its domain.
   (b) Find the values of x for which
   $$fg(x) = gf(x),$$
   giving your answers to 3 significant figures.

5 The functions f, g and h are defined by

$$f: x \mapsto 2x + 3, \quad x \in \mathbb{R}$$
$$g: x \mapsto \frac{3}{2x}, \quad x \neq 0$$
$$h: x \mapsto \frac{x-3}{2}, \quad x \in \mathbb{R}.$$

(a) Find the composite function gh in terms of x.
(b) Find the composite function fgh in terms of x.
(c) Which pair of functions form a composite function equal to x?

6 Express the following functions in terms of f and g where

$$f: x \mapsto \sqrt{x} \text{ and } g: x \mapsto x + 4.$$

(a) $x \mapsto \sqrt{x+4}$ (b) $x \mapsto x + 8$
(c) $x \mapsto \sqrt{x+8}$ (d) $x \mapsto \sqrt{x} + 4$

7 The functions f and g are defined by

$$f: x \mapsto x - 2, \quad x > 0$$
$$g: x \mapsto x^2 + 3, \quad x > 0.$$

(a) Find, in terms of x, the composite function gf.
(b) Sketch $y = gf(x)$ for the appropriate domain.
(c) State the range of gf.
(d) Describe the mapping that takes x onto gf(x).
(e) Is gf a function?

8 The functions f and g are defined by

$$f: x \mapsto x^2 + 2x + 3 \quad x \in \mathbb{R}$$
$$g: x \mapsto px + q \quad x \in \mathbb{R}.$$

(a) Given that fg(1) = 6 show that

$$p^2 + 2pq + q^2 + 2p + 2q - 3 = 0.$$

(b) Given that gf(–2) = 7 find another expression in terms of p and q. Solve the simultaneous equations in (a) and (b) to find p and q.

9 The function f is defined for all real values of x by $f(x) = (x+3)^{\frac{1}{3}}$.
(a) Find f(24).
(b) Find f^2(122).
(c) Find f^3(–1334).
(d) Find x if f(x) = 4.

10 The functions f and g are defined by

$$f: x \mapsto 2x^2 + 2x + 3 \quad x \in \mathbb{R}$$
$$g: x \mapsto x - 1 \quad x \in \mathbb{R}.$$

EXERCISE 1E

(a) Find fg as a function in terms of x.

(b) By the method of completing the square express fg in the form

$$p(x + q)^2 + r$$

where the values of the constants p, q, and r are to be stated.

(c) State the range of fg.

INVERSE FUNCTIONS

Look at the mapping $x \mapsto x + 2$ with domain and range the set of integers.

Domain	Range
...	...
...	...
-1	-1
0	0
1	1
2	2
...	3
...	4
x	$x + 2$

The mapping is clearly a function, since for every input there is one and only one output, the number that is two greater than that input.

This mapping can also be seen in reverse. In that case, each number maps onto the number two less than itself: $x \mapsto x - 2$. The reverse mapping is also a function because for any input there is one and only one output. The reverse mapping is called the *inverse function*, f^{-1}.

Function: f: $x \mapsto x + 2$ $x \in \mathbb{Z}$ where $\mathbb{Z} = \{0, \pm 1, \pm 2, \pm 3, ...\}$.

Inverse function: f^{-1}: $x \mapsto x - 2$ $x \in \mathbb{Z}$.

For a mapping to be a function which also has an inverse function, every object in the domain must have one and only one image in the range, and vice versa. This can only be the case if the mapping is one-to-one.

So the condition for a function f to have an inverse function is that, over the given domain and range, f represents a one-to-one mapping. This is a common situation, and many inverse functions are self-evident as in the following examples, for all of which the domain and range are the real numbers:

f: $x \mapsto x - 1$; f^{-1}: $x \mapsto x + 1$

g: $x \mapsto 2x$; g^{-1}: $x \mapsto \frac{1}{2}x$

h: $x \mapsto x^3$; h^{-1}: $x \mapsto \sqrt[3]{x}$.

You can decide whether an algebraic mapping is a function, and whether it has an inverse function, by looking at its graph. The curve or line representing a one-to-one mapping does not double back on itself, has no turning points and covers the full domain and range. Figure 1.5 illustrates the functions given above.

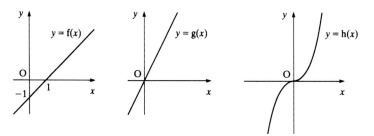

FIGURE 1.5

Now look at $f(x) = x^2$ for $x \in \mathbb{R}$ (figure 1.6). You can see that there are two distinct input values giving the same output: for example $f(2) = f(-2) = 4$. When you want to reverse the effect of the function, you have a mapping which for a single input of 4 gives two outputs, -2 and $+2$. Such a mapping is not a function.

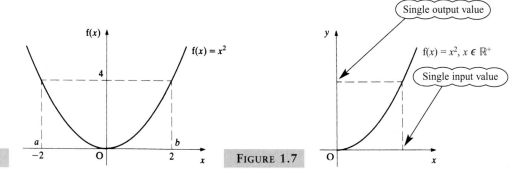

FIGURE 1.6

FIGURE 1.7

If the domain of $f(x) = x^2$ is restricted to $\mathbb{R}^+$ (the set of positive real numbers), you have the situation shown in figure 1.7. This shows that the function which is now defined is one-to-one. The inverse function is given by $f^{-1}(x) = \sqrt{x}$, since the sign $\sqrt{}$ means 'the positive square root of'.

It is often helpful to restrict the domain of a function so that its inverse is also a function. When you use the inv sin (i.e. $\sin^{-1}$ or arcsin) key on your calculator the answer is restricted to the range $-90°$ to $90°$, and is described as the *principal value*. Although there are infinitely many roots of the equation $\sin x = 0.5$ (…$-330°$, $-210°$, $30°$, $150°$, …), only one of these, $30°$, lies in the restricted range and this is the value your calculator will give you.

THE GRAPH OF A FUNCTION AND ITS INVERSE

You have just looked at the graph of $f(x) = x^2$ for $x \in \mathbb{R}^+$. The reverse process of x^2 will be $\pm\sqrt{x}$. But for the function to be one-to-one the positive root only is taken.

Drawing $y = x^2$ and $y = \sqrt{x}$ on the same diagram gives figure 1.8.
Drawing $h(x) = x^3$ and $h^{-1}(x) = \sqrt[3]{x}$ on the same diagram gives figure 1.9.

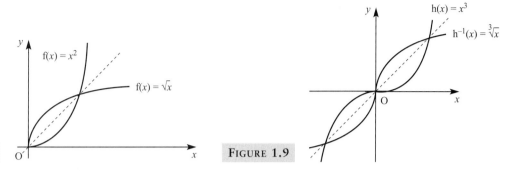

FIGURE 1.8

FIGURE 1.9

You have probably realised by now that the graph of the inverse function is the same shape as that of the function, but reflected in the line $y = x$. To see why this is so, think of a function $f(x)$ mapping a on to b; (a, b) is clearly a point on the graph of $f(x)$. The inverse function $f^{-1}(x)$, maps b on to a and so (b, a) is a point on the graph of $f^{-1}(x)$.

The point (b, a) is the reflection of the point (a, b) in the line $y = x$. This is shown for a number of points in figure 1.10.

This result can be used to obtain a sketch of the inverse function without having to find its equation, provided that the sketch of the original function uses the same scale on both axes.

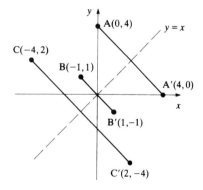

FIGURE 1.10

FINDING THE ALGEBRAIC FORM OF THE INVERSE FUNCTION

To find the algebraic form of the inverse of a function $f(x)$, you should start by changing notation and writing it in the form $y = \ldots$.

Since the graph of the inverse function is the reflection of the graph of the original function in the line $y = x$, it follows that you may find its equation by interchanging y and x in the equation of the original function. You will then need to make y the subject of your new equation. This procedure is illustrated in Example 1.15.

EXAMPLE 1.15

Find $f^{-1}(x)$ when $f(x) = 2x + 1$.

Solution The function $f(x)$ is given by $\qquad y = 2x + 1$
Interchanging x and y gives $\qquad x = 2y + 1$
Re-arranging to make y the subject: $\quad y = \dfrac{x - 1}{2}$

So $f^{-1}(x) = \dfrac{x-1}{2}$.

Sometimes the domain of the function f will not include the whole of $\mathbb{R}$. When any real numbers are excluded from the domain of f, it follows that they will be excluded from the range of f^{-1}, and vice versa:

$\qquad$ domain of f $\equiv$ range of f^{-1}
$\qquad$ range of f $\equiv$ domain of f^{-1}.

EXAMPLE 1.16

Find $f^{-1}(x)$ when $f(x) = 2x - 3$ and the domain of f is $x \geq 4$.

Solution

	Domain	Range
Function: $y = 2x - 3$	$x \geq 4$	$y \geq 5$
Inverse function: $x = 2y - 3$	$x \geq 5$	$y \geq 4$

Rearranging the inverse function to make y the subject, $y = \dfrac{x + 3}{2}$.

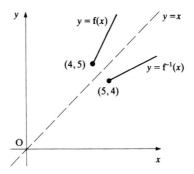

FIGURE 1.11

The full definition of the inverse function is therefore:

$$f^{-1}(x) = \frac{x + 3}{2} \text{ for } x \geq 5.$$

You can see in figure 1.11 that the inverse function is the reflection of a restricted part of the line $y = f(x)$ in the line $y = x$.

INVERSE FUNCTIONS

EXAMPLE 1.17

(a) Find $f^{-1}(x)$ when $f(x) = x^2 + 2$, $x \geq 0$.

(b) Find **(i)** $f(7)$ **(ii)** $f^{-1}f(7)$.
What do you notice?

Solution

(a)
	Domain	Range
Function: $y = x^2 + 2$	$x \geq 0$	$y \geq 2$
Inverse function: $x = y^2 + 2$	$x \geq 2$	$y \geq 0$

Rearranging the inverse function to make y its subject:

$$y^2 = x - 2.$$

This gives $y = \pm\sqrt{x-2}$, but since we know the range of the inverse function to be $y \geq 0$ we can write:

$$y = +\sqrt{x-2} \quad \text{or just} \quad y = \sqrt{x-2}.$$

The full definition of the inverse function is therefore:

$$f^{-1}(x) = \sqrt{x-2} \text{ for } x \geq 2.$$

The function and its inverse function are shown in figure 1.12.

(b) (i) $f(7) = 7^2 + 2 = 51$
(ii) $f^{-1}f(7) = \sqrt{51 - 2} = 7$

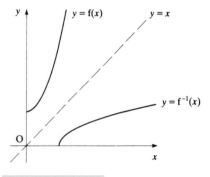

FIGURE 1.12

Notice that applying the function followed by its inverse has brought us back to the original input value.

Note

Part (b) of Example 1.17 illustrates an important general result. For any function $f(x)$ with an inverse $f^{-1}(x)$,

$$f^{-1}f(x) = x.$$

Similarly

$$ff^{-1}(x) = x.$$

The effects of a function and its inverse can be thought of as cancelling each other out.

EXERCISE 1F

1. Find the inverses of the following functions.
   (a) $f(x) = 2x + 7$
   (b) $f(x) = 4 - x$
   (c) $f(x) = \dfrac{4}{2-x}$
   (d) $f(x) = x^2 - 3 \quad x \geq 0$

2. The function f is defined by $f(x) = (x-2)^2 + 3$ for $x \geq 2$.
   (a) Sketch the graph of $f(x)$.
   (b) On the same axes, sketch the graph of $f^{-1}(x)$ without finding its equation.

3. The functions f, g and h are defined by
   $$f(x) = \dfrac{3}{x-4} \quad g(x) = x^2 \quad h(x) = \sqrt{2-x}.$$
   (a) For each function, state any real values of x for which it is not defined.
   (b) Find the inverse functions f^{-1} and h^{-1}.
   (c) Explain why g^{-1} does not exist when the domain of g is $\mathbb{R}$.
   (d) Suggest a suitable domain for g so that g^{-1} does exist.
   (e) Is the domain for the composite function fg the same as for the composite function gf? Give reasons for your answer.

4. A function f is defined by:
   $$f: x \mapsto \tfrac{1}{x} \quad x \in \mathbb{R}, x \neq 0.$$
   Find:
   (a) $f^2(x)$
   (b) $f^3(x)$
   (c) $f^{-1}(x)$
   (d) $f^{999}(x)$.

5. (a) Show that $x^2 + 4x + 7 = (x+2)^2 + a$, where a is to be determined.
   (b) Sketch the graph of $y = x^2 + 4x + 7$, giving the equation of its axis of symmetry and the coordinates of its vertex.

   The function f is defined by $f: x \mapsto x^2 + 4x + 7$ and has as its domain the set of all real numbers.

   (c) Find the range of f.
   (d) Explain, with reference to your sketch, why f has no inverse with its given domain. Suggest a domain for f for which it has an inverse.
   [MEI]

6. The function f is defined by $f: x \mapsto 4x^3 + 3 \quad x \in \mathbb{R}$.
   Give the corresponding definition of f^{-1}.
   State the relationship between the graphs of f and f^{-1}.
   [OCR]

7 The functions f and g are defined by

$$f: x \mapsto 4x - 2, \quad x \in \mathbb{R}$$
$$g: x \mapsto x^2, \quad x \geq 0.$$

(a) Find, in terms of x, the functions f^{-1} and g^{-1}.
(b) On the same diagram sketch $y = f^{-1}(x)$ and $y = g^{-1}(x)$.
(c) How many solutions are there to the equation $f^{-1}(x) = g^{-1}(x)$?
(d) Find the exact values of the solutions to $f^{-1}(x) = g^{-1}(x)$.

8 The function f is given by

$$f: x \mapsto x^2 - 6x + 10, \quad x \in \mathbb{R}.$$

(a) Express f in the form of $(x + a)^2 + b$, stating the values of a and b.
(b) State the coordinate of the minimum turning point on the graph of $y = f(x)$.
(c) Sketch the graph of $y = f(x)$.
(d) State a domain for which $y = f(x)$ is a one-to-one fuction. For this domain state the range.
(e) For your domain sketch the inverse function, f^{-1}, and describe its geometrical relationship with $y = f(x)$.

9 A function is defined by

$$f: x \mapsto 4 - x^2, \quad x \geq 0.$$

(a) Sketch the graph of $y = f(x)$.
(b) State the range of f.
(c) Find f^{-1} in terms of x and state its domain.
(d) Solve $f^{-1}(x) = 3$.

10 The functions f and g are defined as follows:

$$f: x \mapsto x + 2 \quad x \in \mathbb{R}$$
$$g: x \mapsto x^2 + 1 \quad x > 0.$$

(a) Show that $gf(x) = 0$ has no real roots.
(b) State the domain of g^{-1}.
(c) Find, in terms of x, an expression for $g^{-1}(x)$.
(d) Sketch, on a single diagram, the graph of $y = g(x)$ and $y = g^{-1}(x)$.

THE MODULUS FUNCTION

THE GRAPH OF $y = |f(x)|$

The modulus of a number is its absolute value and it disregards the sign of the number. The modulus of x is written as $|x|$.

So $\quad |5| = 5$

$\quad\quad |-5| = 5.$

The graph of $y = |x|$ has all y values non-negative and equal to the magnitude of the value of x. This is illustrated in figure 1.13.

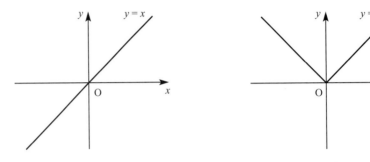

FIGURE 1.13

You see that the geometric effect is to reflect the part of the line below the x-axis to above the x-axis.

In general, the graph of $y = |f(x)|$ is found by drawing $y = f(x)$ and reflecting any part of it that lies below the x-axis to above the x-axis.

EXAMPLE 1.18

Sketch the graph of $y = |2x + 3|$.

Solution

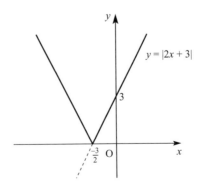

FIGURE 1.14

The Modulus Function

Pure Mathematics: Core 3

EXAMPLE 1.19

Sketch the graph of $y = |4 - x^2|$.

Solution

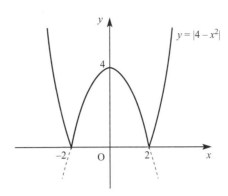

FIGURE 1.15

THE GRAPH OF $y = f(|x|)$

The function $f(|x|)$ has the same values whether the value of x is positive or negative. Since $|-x| = |x|$ the function $f(|x|)$ is said to be an even one which means that the graph of $y = f(|x|)$ will be symmetrical about the y-axis.

To sketch the graph of $y = f(|x|)$ reflect the section of the graph of $y = f(x)$ for $x \geq 0$ in the y-axis.

EXAMPLE 1.20

Sketch the graph of $y = 2|x| - 3$.

Solution

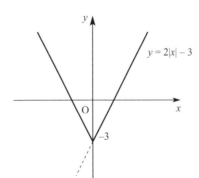

FIGURE 1.16

EXAMPLE 1.21

Figure 1.17 shows a sketch of the graph of $y = f(x)$.

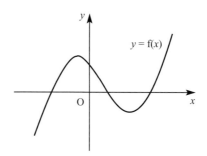

FIGURE 1.17

Sketch the graph of $y = f(|x|)$.

Solution Reflect the section of $y = f(x)$ for which $x \geqslant 0$ in the y-axis to give the graph of $y = f(|x|)$.

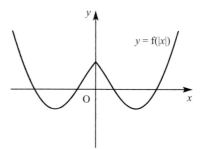

FIGURE 1.18

EXERCISE 1G

1 For each of the following sketches of $y = f(x)$ sketch the graph of $y = |f(x)|$.

(a)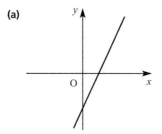

(b)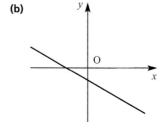

(c)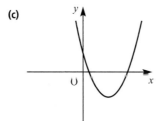

(d)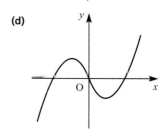

(e) 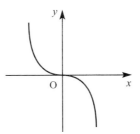　　(f)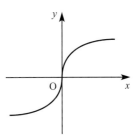

2. For each of the sketches of $y = f(x)$ in question 1 sketch the graph of $y = f(|x|)$.

3. Sketch the following graphs.
   (a) $y = |x + 2|$
   (b) $y = |3x - 2|$
   (c) $y = |x| + 2$
   (d) $y = |x^2 - 2|$
   (e) $y = |x^3|$
   (f) $y = |x^3| - 1$
   (g) $y = \left|\frac{1}{x}\right|$
   (h) $y = |3 - x|$
   (i) $y = |9 - x^2|$
   (j) $y = |\sin x|$

4. The diagrams show sketches of $y = |f(x)|$. For each one draw possible sketches for $y = f(x)$.

   (a) 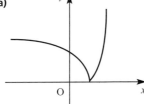　　(b)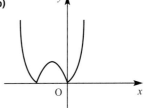

5. The diagrams show sketches of $y = f(|x|)$. For each one draw possible sketches for $y = f(x)$.

   (a) 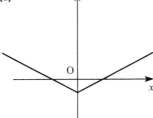　　(b)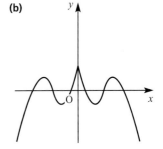

Combinations of transformations

In Chapter 1 of *Pure Mathematics: Core 1 and Core 2* you learnt how to transform the graphs of simple functions. Here are the rules:

$y = f(x)$ mapped on to $y = f(x - a) + b$ $\Rightarrow$	translation with vector $\begin{pmatrix} a \\ b \end{pmatrix}$
$y = f(x)$ mapped on to $y = f(ax)$ $\Rightarrow$	stretch scale factor $\frac{1}{a}$ parallel to the x-axis, y-axis invariant
$y = f(x)$ mapped on to $y = af(x)$ $\Rightarrow$	stretch scale factor a parallel to the y-axis, x-axis invariant
$y = f(x)$ mapped on to $y = f(-x)$ $\Rightarrow$	reflection in the y-axis
$y = f(x)$ mapped onto $y = -f(x)$ $\Rightarrow$	reflection in the x-axis

To illustrate such transformations and remind ourselves of the processes look at the following examples.

Start with the graph of $y = f(x)$ where $f(x) = x^3$:

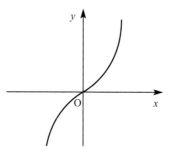

FIGURE 1.19

Then examples of translations are:

$y = (x - 2)^3$ is a translation $\begin{pmatrix} 2 \\ 0 \end{pmatrix}$.

$y = x^3 - 2$ is a translation $\begin{pmatrix} 0 \\ -2 \end{pmatrix}$.

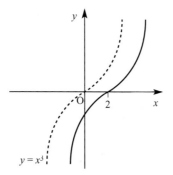
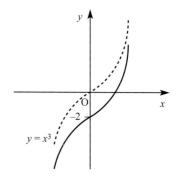

FIGURE 1.20

Examples of stretches are:

$y = (2x)^3$ is a stretch, scale factor $\frac{1}{2}$, parallel to the x-axis.

$y = 2x^3$ is a stretch, scale factor 2, parallel to the y-axis.

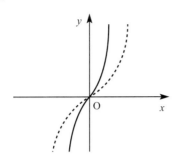
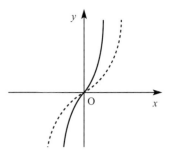

FIGURE 1.21

Examples of reflections are:

$y = (-x)^3$ is a reflection in the y-axis.

$y = -x^3$ is a reflection in the x-axis.

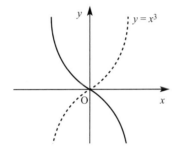
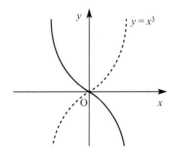

FIGURE 1.22

In this chapter you will combine two or more of these transformations together.

EXAMPLE 1.22

Sketch the graph of $y = (x - 1)^2 + 3$.

Solution Consider this as a transformation of $y = x^2$.

x has been replaced by $x - 1$, giving a translation 1 unit to the right.
3 is added to the function $(x - 1)^2$, giving a translation 3 units up.

Hence $y = (x - 1)^2 + 3$ is a translation of $y = x^2$ by the vector $\begin{pmatrix} 1 \\ 3 \end{pmatrix}$. This is shown in figure 1.23.

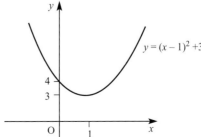

FIGURE 1.23

Note

If the question had been expressed as 'Sketch the graph of $y = x^2 - 2x + 4$', you would have needed to start by completing the square to express the equation in the form with x appearing only once.

EXAMPLE 1.23

Find the values of a, p and q when $y = 2x^2 + 4x - 1$ is written in the form $y = a[(x + p)^2 + q]$.

Show how the graph of $y = 2x^2 + 4x - 1$ can be obtained from the graph of $y = x^2$ by successive transformations, and list the transformations in the order in which they are applied.

Solution First take out 2 as a factor:

$$2[x^2 + 2x - \tfrac{1}{2}].$$

Now complete the square on the contents of the square brackets:

$$2[(x + 1)^2 - \tfrac{3}{2}].$$

So, the equation of the curve can be written as $y = 2[(x + 1)^2 - 1\tfrac{1}{2}]$.

To sketch the graph, start with the curve $y = x^2$.

The curve $y = x^2$ becomes $y = (x + 1)^2 - 1\tfrac{1}{2}$ by applying the translation $\begin{pmatrix} -1 \\ -1\tfrac{1}{2} \end{pmatrix}$.

The curve $y = (x + 1)^2 - 1\tfrac{1}{2}$ becomes $y = 2[(x + 1)^2 - 1\tfrac{1}{2}]$ by applying a stretch of scale factor 2 parallel to the y-axis.

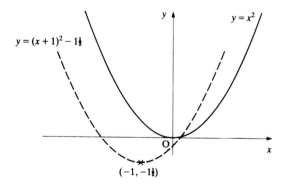
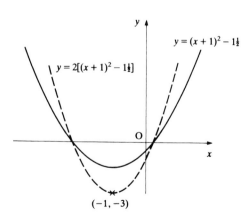

The translation $\begin{pmatrix} -1 \\ -1\tfrac{1}{2} \end{pmatrix}$

The stretch of scale factor 2 parallel to the y axis

FIGURE 1.24

Note

Notice on figure 1.24 how the stretch doubles the *y*-coordinate of every point on the curve, including the turning point.

Points on the *x*-axis have a *y*-coordinate of 0, so are unchanged.

EXAMPLE 1.24

Starting with the curve $y = \cos x$, show how transformations can be used to sketch the curves

(a) $y = 2\cos 3x$ (b) $y = 3 + \cos\frac{x}{2}$ (c) $y = \cos(2x - 60°)$.

Solution

(a) The curve with equation $y = \cos 3x$ is obtained from the curve with equation $y = \cos x$ by a stretch of scale factor $\frac{1}{3}$ parallel to the *x*-axis. There will therefore be one complete wavelength of the curve in 120° (instead of 360°).

The curve of $y = 2\cos 3x$ is obtained from that of $y = \cos 3x$ by a stretch of scale factor 2 parallel to the *y*-axis. The curve therefore oscillates between $y = 2$ and $y = -2$ (instead of $y = 1$ and $y = -1$). This is shown in figure 1.25.

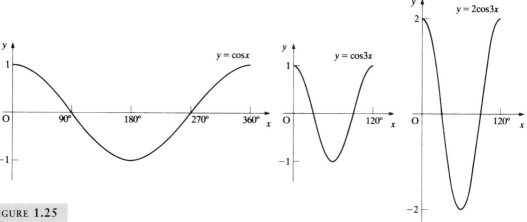

FIGURE 1.25

(b) The curve of $y = \cos\frac{x}{2}$ is obtained from that of $y = \cos x$ by a stretch of scale factor 2 in the *x*-direction. There will therefore be one complete oscillation of the curve in 720° (instead of 360°).

The curve of $y = 3 + \cos\frac{x}{2}$ is obtained from that of $y = \cos\frac{x}{2}$ by a translation $\begin{pmatrix} 0 \\ 3 \end{pmatrix}$.

The curve therefore oscillates between $y = 4$ and $y = 2$ (see figure 1.26).

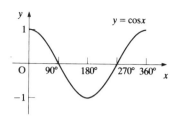
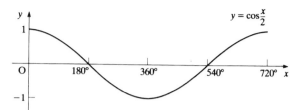

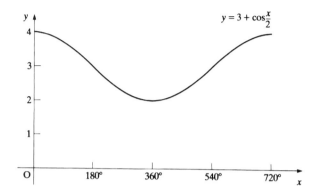

FIGURE 1.26

(c) The curve of $y = \cos(x - 60°)$ is obtained from that of $y = \cos x$ by a translation of $\begin{pmatrix} 60° \\ 0 \end{pmatrix}$.

The curve of $y = \cos(2x - 60°)$ is obtained from that of $y = \cos(x - 60°)$ by a stretch of scale factor $\frac{1}{2}$ parallel to the x-axis (see figure 1.27).

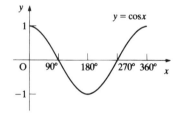 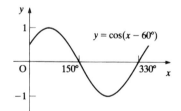 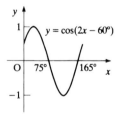

FIGURE 1.27

EXAMPLE 1.25

Starting with $f(x) = \dfrac{1}{x}$, explain the transformation that maps $f(x)$ on to $f(2 - x) + 1$. Hence sketch the graph of $y = f(2 - x) + 1$.

Solution Firstly, you have to think of $f(2 - x)$ as $f(-[x - 2])$.

$x - 2$ is a translation of 2 units to the right.

$[x - 2]$ has been replaced by $-[x - 2]$, which is reflection in the y axis.

The function then has 1 added to it, which is a translation of 1 unit upwards.

The graph looks like this:

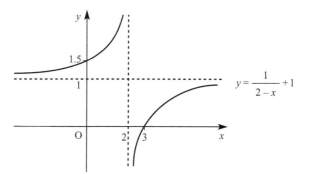

FIGURE 1.28

You may be given questions where the function is not given but the graph of the function f(x) is drawn for you. Such a question is illustrated in Example 1.26.

EXAMPLE 1.26

Figure 1.29 shows the graph of the function $y = f(x)$.

Sketch the graph of $y = 2f(3x) + 1$.

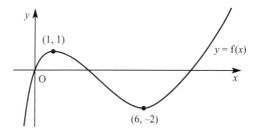

FIGURE 1.29

Solution x has been replaced by $3x$ giving a stretch scale factor $\frac{1}{3}$ parallel to the x-axis. $f(3x)$ is then multiplied by 2 giving a stretch, scale factor 2, parallel to the y-axis. Finally, 1 is added to $2f(3x)$ causing a translation of 1 unit upwards.

The graph is shown in figure 1.30.

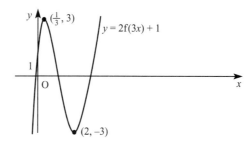

FIGURE 1.30

Note You must be careful to perform the transformations in the correct order. It is always a good idea to check your results using a graphical calculator or a graph drawing package on a computer whenever possible.

EXERCISE 1H

1. Starting with the graph of $y = x^2$, state the transformations in the order in which they are used to obtain the given equations. Sketch the graph of each equation.
   (a) $y = (x + 3)^2 + 1$
   (b) $y = -(x - 1)^2 + 2$
   (c) $y = 2(x - 4)^2$
   (d) $y = (2x)^2 - 4$
   (e) $y = 3x^2 - 5$

2. Starting with $y = \sin x$, state transformations which can be used to sketch the following curves. Specify the transformations in the order in which they are used, and where there is more than one stage in the sketching of the curve, state each stage.
   (a) $y = \sin(x - 90°)$
   (b) $y = \sin 3x$
   (c) $2y = \sin x$
   (d) $y = \sin \frac{x}{2}$
   (e) $y = 2 + \sin 3x$

3. Starting with $y = \cos x$, state transformations which can be used to sketch the following curves. Specify the transformations in the order in which they are used, and where there is more than one stage in the sketching of the curve, state each stage.
   (a) $y = \cos(x + 60°)$
   (b) $3y = \cos x$
   (c) $y = \cos x + 1$
   (d) $y = \cos 2(x + 90°)$

4. For each of the following curves
   (a) sketch the curve;
   (b) identify the curve as being the same as one of the following:

   $y = \pm \sin x$, $y = \pm \cos x$,   or $y = \pm \tan x$.

   (i) $y = \sin(x + 360°)$
   (ii) $y = \sin(x + 90°)$
   (iii) $y = \tan(x - 180°)$
   (iv) $y = \cos(x - 90°)$
   (v) $y = \cos(x + 180°)$

5. (a) Express $2x^2 - 4x + 5$ in the form $a(x + b)^2 + c$ where a, b and c are constants to be determined.
   (b) Sketch the graph of $y = 2x^2 - 4x + 5$.

6. The circle with equation $x^2 + y^2 = 1$ is shown here. It is stretched with scale factor 3 parallel to the x-axis and with scale factor 2 parallel to the y-axis. Sketch both curves on the same graph, and write down the equation of the new curve. (It is an ellipse.)

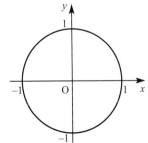

7 The sketch below shows the curve with equation $y = 2 - 6x - 3x^2$ and its axis of symmetry, $x = -1$.

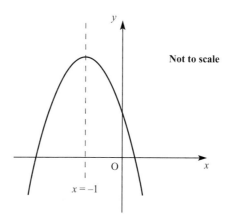

Not to scale

(a) Give the coordinates of the vertex and the value of y when $x = 0$.
(b) Find the values of the constants a, b such that $2 - 6x - 3x^2 = a(x + 1)^2 + b$.
(c) Copy the given sketch and draw in the reflection of the curve with equation $y = 2 - 6x - 3x^2$ in the line $y = 2$.
(d) Write down the equation of the new curve and give the coordinates of its vertex.

[MEI]

8 (a) Sketch the graph of $y = 3^x$.
(b) On the same diagram sketch the graph of $y = 3^{x-1}$. Describe two transformations that map $y = 3^x$ on to $y = 3^{x-1}$.
(c) On a separate diagram sketch the graph of $y = 3^{x-1} - 1$. Describe the transformations needed to map $y = 3^{x-1} - 1$ on to $y = 3^x$.

9 The diagram shows the graph of $y = f(x)$.

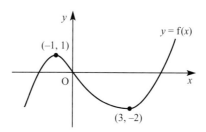

Sketch the following graphs, using a separate set of axes for each graph, and indicate the coordinates of the turning points.
(a) $y = f(x - 2) + 1$
(b) $y = 2f(x) - 1$
(c) $y = -f(3x)$
(d) $y = -f(-x)$
(e) $y = 3f(2x)$

10 The diagram shows the graph of $y = f(x)$.

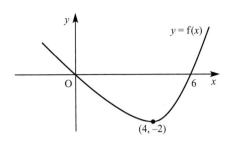

Sketch the following graphs, using a separate set of axes for each graph.
(a) $y = f(x + 1) - 2$
(b) $y = 1 - f(x)$
(c) $y = f(-2x)$
(d) $y = -2f(x)$
(e) $y = 2f(3x)$

EXERCISE 1I   Examination-style questions

1 Express as a single fraction
$$\frac{x+3}{x^2 - 6x + 5} + \frac{2}{x - 1}.$$

2 Solve
$$\frac{2}{x} = 2 + \frac{3}{x + 1}.$$

3 Simplify
$$\frac{x^3 - 1}{x^2 - 2x + 1}.$$

4 Show that
$$\frac{r + 1}{r + 2} - \frac{r}{r + 1} \equiv \frac{1}{(r + 1)(r + 2)}.$$
[Edexcel]

5 Express as a fraction in its simplest form
$$\frac{5(x - 3)(x + 1)}{(x - 12)(x + 3)} - \frac{3(x + 1)}{x - 12}.$$
[Edexcel]

6 $f(x) = \dfrac{2}{x - 1} - \dfrac{6}{(x - 1)(2x + 1)}$, $x > 1$

(a) Prove that $f(x) = \dfrac{4}{2x + 1}$.
(b) Find the range of f.
(c) Find f^{-1}
(d) Find the range of f^{-1}.
[Edexcel]

Exercise 1I

7 Show that
$$\frac{x^2 + 2x - 1}{x - 1} = x + 3 + \frac{2}{x - 1}.$$

8 You are given that $f(x) = \frac{x^2 - 5x - 13}{x + 2}$.

Express $f(x)$ in the form $x + A + \frac{B}{x + 2}$ where A and B are constants to be determined.
Show that $f'(x) = 1 - \frac{1}{(x + 2)^2}$.

9 You are given that $p(x) = 6x^3 - 3x^2 - 10x + 9$.
 (a) Divide $p(x)$ by $2x + 3$.
 (b) Express $p(x)$ in the form $Q(x) \times (2x + 3) + R$.
 (c) Verify that your value of R is correct by using the remainder theorem.

10 You are given that $f(x) = \frac{3x - 1}{x - 1}$.

Express $f(x)$ in the form $a + \frac{b}{x + 1}$ where a and b are constants to be determined.
Sketch the graph of $y = f(x)$.

11 The functions f and g are defined by

$$f: x \mapsto x - 3, \quad x \in \mathbb{R}$$
$$g: x \mapsto 2x^2, \quad x \in \mathbb{R}$$

 (a) Find the range of g.
 (b) Solve $gf(x) = 50$.
 (c) Sketch the graph of $y = gf(x)$.
 (d) Sketch the graph of $y = |f(x)|$ and hence solve $|f(x)| < 4$.

12 The functions f and g are defined by

$$f: x \mapsto 2x + 1, \quad x \in \mathbb{R}$$
$$g: x \mapsto \sqrt{x}, \quad x \geq 0.$$

 (a) State the range of g.
 (b) Find the inverses f^{-1} and g^{-1}.
 (c) On the same diagram sketch $y = f^{-1}(x)$ and $y = g^{-1}(x)$.
 (d) State, with reasons, the number of solutions to $f^{-1}(x) = g^{-1}(x)$.
 (e) Evaluate $gf(5)$ to 3 decimal places.

13 (a) On the same axes draw the graphs of $y = |3x + 1|$ and $y = |x - 1|$.
 (b) Use your diagram to solve $|3x + 1| = |x - 1|$.

14 (a) Express $x^2 - 4x + 5$ in the form $(x - a)^2 + b$, where a and b are integers to be defined.
  (b) Sketch the graph of $y = x^2 - 4x + 5$.
  (c) Define a domain for which $f(x) = x^2 - 4x + 5$ is a one-to-one function.
  (d) Find the inverse function f^{-1}.
  (e) For your domain, sketch f and f^{-1} on the same diagram showing the geometrical connection between them.
  (f) From your graph find the value of x for which $f(x) = f^{-1}(x)$.

15 The functions f and g are defined by
$$f: x \mapsto 2x - 1, \quad x \in \mathbb{R}$$
$$g: x \mapsto \frac{1}{x + 1}, \quad x \neq -1.$$
  (a) Find the inverse function g^{-1}.
  (b) Find and simplify $fg(x)$.
  (c) Sketch, on the same diagram, the graphs of $y = f(x)$ and $y = g(x)$.
  (d) From your sketch find the approximate solutions to $f(x) = g(x)$.
  (e) Solve $f(x) = g(x)$ algebraically, giving your answer to 3 significant figures.

16 The graph shows one wavelength of the curve of $y = A + B\sin(2x)$.

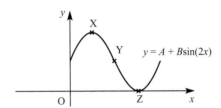

  (a) Given that the point X has coordinates $\left(\frac{\pi}{4}, 4\right)$ state the coordinates of the central point Y and the minimum point Z and find the values of A and B.
  (b) If $f(x) = A + B\sin(2x)$, sketch, on separate diagrams, the graphs of
    (i) $y = f\left(x + \frac{\pi}{2}\right)$,
    (ii) $y = |f(x) - 2|$.

17 The diagram shows the graph of $y = f(x)$.

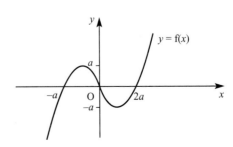

On separate diagrams sketch

(a) $y = f(-x)$,
(b) $y = -f(x)$,
(c) $y = f(x - a)$,
(d) $y = f(x) - a$,
(e) $y = |f(x)|$.

For what values of x is $f(x) > 0$?

18

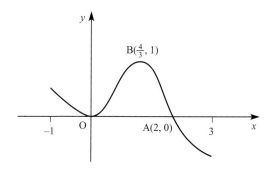

The diagram shows a sketch of the curve with equation $y = f(x)$, $-1 \leq x \leq 3$. The curve touches the x-axis at the origin O, has a maximum at the point $(\frac{4}{3}, 1)$ and crosses the x-axis at the point A(2, 0).

On separate diagrams, show a sketch of the curve with equation

(a) $y = f(x + 1)$,
(b) $y = |f(x)|$,
(c) $y = f(|x|)$,

marking on each sketch the coordinates of points at which the curve

(i) has a turning point,
(ii) meets the x-axis.

[Edexcel]

19 The function f is an odd function defined on the interval $[-2, 2]$. Given that

$$f(x) = -x, \quad 0 \leq x \leq 1,$$
$$f(x) = x - 2, \quad 1 \leq x \leq 2,$$

(a) sketch the graph of f for $-2 \leq x \leq 2$,
(b) find the values of x for which $f(x) = \frac{1}{2}$.

[Edexcel]

20 (a) Using the same scales and axes, sketch the graph of $y = |2x|$ and $y = |x - a|$, where $a > 0$.
(b) Write down the coordinates of the points where the graph of $y = |x - a|$ meets the axes.
(c) Show that the point with coordinates $(-a, 2a)$ lies on both graphs.
(d) Find the coordinates, in terms of a, of a second point which lies on both graphs.

[Edexcel]

21 $f(x) = 5\sin 3x$, $0° \leq x \leq 180°$
(a) Sketch the graph of $y = f(x)$, indicating the value of x at each point where the graph intersects the x-axis.
(b) Write down the coordinates of all the minimum and maximum points of $f(x)$.
(c) Calculate all the values of x for which $f(x) = 2.5$.

[Edexcel]

22 (a) Express $2x^2 + 5x + 1$ in the form $a(x + b)^2 + c$, where a, b and c are constants to be determined.
(b) Describe the transformations that map $y = x^2$ on to $y = 2x^2 + 5x + 1$ and sketch the graphs of both functions on the same axes.

23 (a) Sketch the graph of $y = \dfrac{1}{x}$.
(b) Describe the transformations that map $y = \dfrac{1}{x}$ on to $y = 1 - \dfrac{2}{x + 1}$.
(c) Sketch the graph of $y = 1 - \dfrac{2}{x + 1}$.

24 The function f is defined by $f(x) = \sqrt[3]{2 - x}$.
(a) Find f^{-1}.
(b) On the same diagram sketch the graphs of $y = f(x)$ and $y = f^{-1}(x)$.
(c) What is the geometrical relationship between the two graphs?

25 The diagram shows the graph of $y = f(x)$.

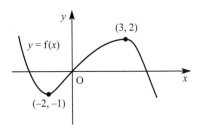

Sketch the following, on separate diagrams, labelling the positions of the turning points.
(a) $y = f(x - 1) - 2$
(b) $y = -2f(x)$
(c) $y = f(-2x)$
(d) $y = 2f(2 - x)$

KEY POINTS

1. **Rational functions**

   A rational expression or rational function is a fraction of the form

   $$\frac{f(x)}{g(x)}$$

   where $f(x)$ and $g(x)$ are functions of x.

2. When simplifying fractions look for common factors to cancel.

   For example: $\dfrac{2x+3}{2x^2+x-3} = \dfrac{2x+3}{(2x+3)(x-1)} = \dfrac{1}{x-1}$

3. When adding or subtracting fractions look for common denominators.

   For example: $\dfrac{1}{x^2-1} + \dfrac{1}{x+1} = \dfrac{1+(x-1)}{(x+1)(x-1)} = \dfrac{x}{(x+1)(x-1)}$

4. **Algebraic division**
   Use the long division method or the remainder theorem.

5. **Functions**
   The **domain** is the set of x-values.

6. The **range** is the set of y-values.

7. A **one-to-one** mapping means each value of x maps on to a unique value of y, e.g. $y = x^3$.

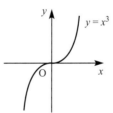

8. **Composite functions**
   fg means fg(x) or f(g(x)),
   i.e. replace x in f(x) by the function g(x).

9. **Inverse functions**
   - For a function to have an inverse it must be one-to-one.
   - Geometrically the inverse function f^{-1} is the reflection of the function f in the line $y = x$.
   - The equation f^{-1} is found by swapping x and y around in the equation $y = f(x)$ and making y the subject.
   - $f^{-1}f(x) = ff^{-1}(x) = x$

10 Modulus function

$y = |f(x)|$ is found by reflecting any parts below the x-axis to above the x-axis.

$y = f(|x|)$ is found by reflecting any parts to the right of the y-axis to the left of the y-axis.

11 Transformations of f(x)

- $f(x - a) + b$ is a translation with vector $\begin{pmatrix} a \\ b \end{pmatrix}$.
- $f(ax)$ is a stretch, scale factor $\frac{1}{a}$, parallel to the x-axis, y-axis invariant.
- $af(x)$ is a stretch, scale factor a, parallel to the y-axis, x-axis invariant.
- $f(-x)$ is reflection in the y-axis.
- $-f(x)$ is a reflection in the x-axis.

Chapter two

TRIGONOMETRY

The young light-hearted Masters of the Waves.

Matthew Arnold

TRIGONOMETRIC FUNCTIONS

In Chapter 9 of *Pure Mathematics: Core 1 and Core 2* you studied the trigonometric functions sine (sinθ), cosine (cosθ) and tangent (tanθ). You will also remember their graphs which are drawn again for you in figure 2.1.

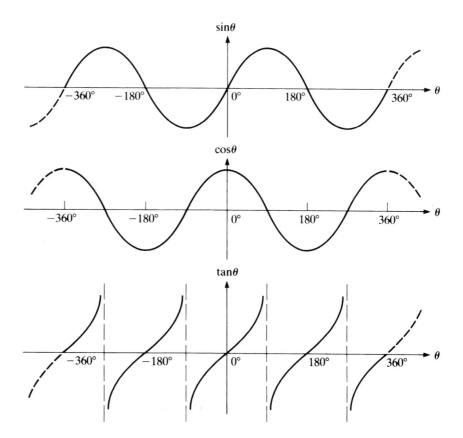

FIGURE 2.1

You will also recall the identities

$$\tan\theta \equiv \frac{\sin\theta}{\cos\theta}$$

and

$$\sin^2\theta + \cos^2\theta \equiv 1.$$

For example, we could solve equations of the type $\sin\theta = \tfrac{1}{2}$ for $0 \leq \theta \leq 2\pi$:

$\sin\theta = \tfrac{1}{2}$

$\theta = \tfrac{\pi}{6}, \tfrac{5\pi}{6}$  From the symmetry of the graph

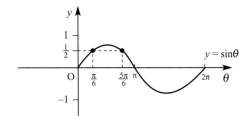

FIGURE 2.2

You will also recall that π radians is equivalent to $180°$.

We could solve problems if the type $6\sin^2\theta - \cos\theta - 4 = 0$

for $0° \leq \theta \leq 360°$ by using the second identity (the fundamental trigonometric identity). Thus:

$$6\sin^2\theta - \cos\theta - 4 = 0$$
$$\therefore 6(1 - \cos^2\theta) - \cos\theta - 4 = 0$$
$$\therefore 6\cos^2\theta + \cos\theta - 2 = 0$$
$$\therefore (3\cos\theta + 2)(2\cos\theta - 1) = 0$$
$$\therefore \cos\theta = -\tfrac{2}{3} \text{ or } \cos\theta = \tfrac{1}{2}$$
$$\therefore \theta = 131.8°, 228.2°, 60°, 300°$$

We will now develop our trigonmetric knowledge and skills by looking at the reciprocal functions and a number of other identities.

Reciprocal trigonometric functions

As well as the three main trigonometric functions, there are three more functions which are commonly used. These are the reciprocals of $\sin\theta$, $\cos\theta$ and $\tan\theta$ and they are called $\operatorname{cosec}\theta$ (short for cosecant), $\sec\theta$ (secant) and $\cot\theta$ (cotangent). These functions are defined by

$$\operatorname{cosec}\theta = \frac{1}{\sin\theta}, \quad \sec\theta = \frac{1}{\cos\theta}, \quad \cot\theta = \frac{1}{\tan\theta}.$$

The graphs of the functions can be found by taking the corresponding sine, cosine and tangent graph and calculating the reciprocals of each point on the graph. The results are shown in figure 2.3.

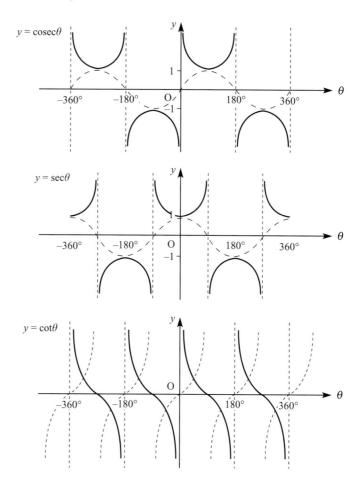

FIGURE 2.3

Note

To help you remember which reciprocal function relates to which of sine, cosine and tangent, take the third letter; so, for example, $\operatorname{cosec}\theta$ has third letter s and it relates to $\sin\theta$.

EXAMPLE 2.1

Solve $\sec\theta = 2$ for $-2\pi \leq \theta \leq 2\pi$.

Solution You could solve this by using the graph of $y = \sec\theta$ but it is easier to write it in terms of $\cos\theta$.

So $\sec\theta = 2$.

Taking the reciprocal of both sides gives

$$\cos\theta = \tfrac{1}{2}$$

which has principal solution

$$\theta = \tfrac{\pi}{3}.$$

Using the graph of $y = \cos\theta$ (see figure 2.4) gives the set of solutions from $-2\pi \leq \theta \leq 2\pi$ as

$$\theta = \pm\tfrac{\pi}{3}, \pm\tfrac{5\pi}{3}.$$

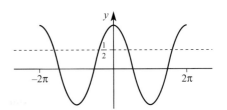

FIGURE 2.4

EXAMPLE 2.2

Solve $3\cot^2\theta - 4\cot\theta + 1 = 0$ for $0° \leq \theta \leq 360°$.

Solution Factorise:

$$3\cot^2\theta - 4\cot\theta + 1 = 0$$
$$(3\cot\theta - 1)(\cot\theta - 1) = 0$$
$$\therefore \quad 3\cot\theta - 1 = 0 \quad \text{or} \quad \cot\theta - 1 = 0$$
$$\text{so} \quad \cot\theta = \tfrac{1}{3} \quad \text{or} \quad \cot\theta = 1$$
$$\therefore \quad \tan\theta = 3 \quad \text{or} \quad \tan\theta = 1$$

and using the graph of $y = \tan\theta$ gives $\theta = 71.6°, 251.6°, 45°$ or $225°$.

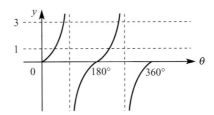

FIGURE 2.5

EXERCISE 2A

EXAMPLE 2.3

Solve $\cot 3x = \sqrt{3}$ for $0° \leq x \leq 360°$.

Solution $\cot 3x = \sqrt{3}$

Take the reciprocal of both sides

$$\tan 3x = \frac{1}{\sqrt{3}}$$

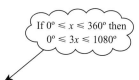
If $0° \leq x \leq 360°$ then $0° \leq 3x \leq 1080°$

$\therefore\ 3x = 30°, 210°, 390°, 570°, 750°, 930°$
$\therefore\ x = 10°, 70°, 130°, 190°, 250°, 310°$.

COTθ IN MORE DETAIL

Since $\tan\theta \equiv \frac{\sin\theta}{\cos\theta}$, if you take the reciprocal of both sides you have

$$\cot\theta \equiv \frac{\cos\theta}{\sin\theta}.$$

EXAMPLE 2.4

Solve $2\cos\theta = \cot\theta$ for $0 \leq \theta \leq 2\pi$.

Solution
$$2\cos\theta = \cot\theta$$
$$2\cos\theta = \frac{\cos\theta}{\sin\theta}$$
$$2\cos\theta\sin\theta - \cos\theta = 0$$
$$\cos\theta(2\sin\theta - 1) = 0$$
$$\cos\theta = 0 \ \text{ or } \ \sin\theta = \tfrac{1}{2}$$
$$\theta = \tfrac{\pi}{2}, \tfrac{3\pi}{2}, \tfrac{\pi}{6} \text{ or } \tfrac{5\pi}{6}.$$

EXERCISE 2A

1 Solve the following.
   (a) $\text{cosec}\,x = 2$ for $0° \leq x \leq 360°$
   (b) $\sec\theta = 5$ for $0° \leq \theta \leq 360°$
   (c) $\cot x = 1$ for $0 \leq x \leq 2\pi$
   (d) $\text{cosec}\,x = -\dfrac{2}{\sqrt{3}}$ for $-2\pi \leq x \leq 2\pi$
   (e) $\cot\theta = 0.5$ for $0° \leq \theta \leq 360°$

2 Solve the following.
  (a) $\sec 2x = 2$ for $0° \leqslant x \leqslant 360°$
  (b) $\operatorname{cosec} 3\theta = -1$ for $-2\pi \leqslant \theta \leqslant 2\pi$
  (c) $\cot 2x = -0.5$ for $-180° \leqslant x \leqslant 180°$
  (d) $\operatorname{cosec} 4\theta = \dfrac{2}{\sqrt{3}}$ for $-\pi \leqslant \theta \leqslant \pi$
  (e) $\cot 2x = \dfrac{1}{\sqrt{3}}$ for $-\pi \leqslant \theta \leqslant 3\pi$

3 Solve the following quadratics.
  (a) $\sec^2 x - 3\sec x + 2 = 0$ for $0° \leqslant x \leqslant 360°$
  (b) $\operatorname{cosec}^2 x - 2\operatorname{cosec} x - 3 = 0$ for $0° \leqslant x \leqslant 360°$
  (c) $2\cot^2 \theta - 3\cot \theta + 1 = 0$ for $0° \leqslant \theta \leqslant 360°$
  (d) $2\sec^2 2\theta + \sec 2\theta - 1 = 0$ for $0 \leqslant \theta \leqslant 2\pi$
  (e) $3\cot^2 3\theta - 10\cot 3\theta + 3 = 0$ for $0° \leqslant \theta \leqslant 180°$

4 Sketch the graphs of the following functions over the given values.
  (a) $y = \sec x$ for $0° \leqslant x \leqslant 360°$
  (b) $y = \operatorname{cosec} x$ for $-360° \leqslant x \leqslant 360°$
  (c) $y = \cot x$ for $0° \leqslant x \leqslant 720°$
  (d) $y = \sec 2x$ for $0° \leqslant x \leqslant 360°$
  (e) $y = \operatorname{cosec} 3x$ for $0° \leqslant x \leqslant 360°$

5 Without using a calculator find the following.
  (a) $\sec 120°$
  (b) $\operatorname{cosec} 210°$
  (c) $\cot 210°$
  (d) $\sec 300°$
  (e) $\operatorname{cosec}(-150°)$

INVERSE TRIGONOMETRIC FUNCTIONS

When asked to solve $\sin\theta = 0.5$ you have used the inverse sine function to obtain the angle, that is

$$\sin\theta = 0.5 \implies \theta = \sin^{-1} 0.5 = 30°.$$

$\sin^{-1}\theta$ is the inverse sine of θ which is sometimes written as $\arcsin\theta$.

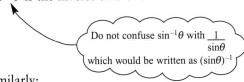

Do not confuse $\sin^{-1}\theta$ with $\dfrac{1}{\sin\theta}$ which would be written as $(\sin\theta)^{-1}$

Similarly:

- the inverse cosine of θ is written as $\cos^{-1}\theta$ or $\arccos\theta$
- the inverse tangent of θ is written as $\tan^{-1}\theta$ or $\arctan\theta$.

INVERSE TRIGONOMETRIC FUNCTIONS

Using functional notation, if

$$f: x \mapsto \sin x$$

then its inverse function f^{-1} is found by:

let $\quad y = \sin x$

and swapping x with y gives

$$x = \sin y$$
$$\therefore y = \sin^{-1} x$$

so the inverse function is

$$f^{-1}: x \mapsto \sin^{-1} x.$$

However for a function to have an inverse it must be a one-to-one mapping.

The functions sine, cosine and tangent are all many-to-one mappings, so their inverse mappings are one-to-many. Thus the problem 'find sin30°' has only one solution, 0.5, whilst 'find θ such that $\sin\theta = 0.5$' has infinitely many solutions. You can see this from the graph of $y = \sin\theta$ (figure 2.6).

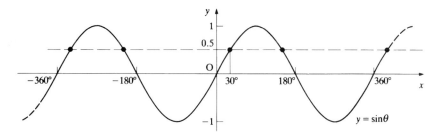

FIGURE 2.6

In order to define inverse functions for sine, cosine and tangent, a restriction has to be placed on the domain of each so that it becomes a one-to-one mapping.

The restriction of the domain determines the principal values for that trigonometrical function. The restricted domains are not all the same. They are listed below.

Function	Domain (degrees)	Domain (radians)
$y = \sin\theta$	$-90° \leq \theta \leq 90°$	$-\frac{\pi}{2} \leq \theta \leq \frac{\pi}{2}$
$y = \cos\theta$	$0° \leq \theta \leq 180°$	$0 \leq \theta \leq \pi$
$y = \tan\theta$	$-90° < \theta < 90°$	$-\frac{\pi}{2} < \theta < \frac{\pi}{2}$

Not only do these domains define parts of each curve to be one-to-one functions they are also the sets of values within which your calculator always gives an answer.

The graphs of the trigonometric functions over the defined domains are shown, with their inverse functions, in figure 2.7. The inverses are found by reflection in the line $y = x$.

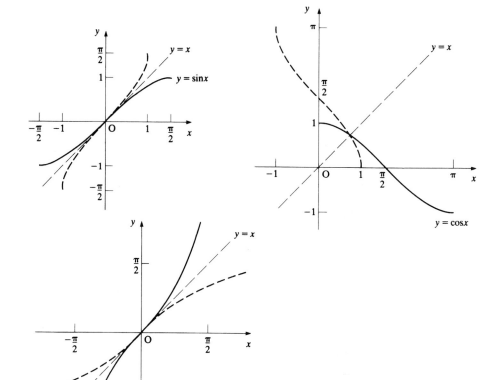

FIGURE 2.7

EXAMPLE 2.5

In a triangle OAB, OA = 5 cm, OB = 12 cm and angle AOB = 90°. Find $\cos^{-1} A$ and $\tan^{-1} B$.

Solution Using Pythagoras' theorem

$$AB^2 = 5^2 + 12^2$$

so   AB = 13.

Hence,

$$\cos^{-1} A = \frac{5}{13}$$

and

$$\tan^{-1} B = \frac{5}{12}.$$

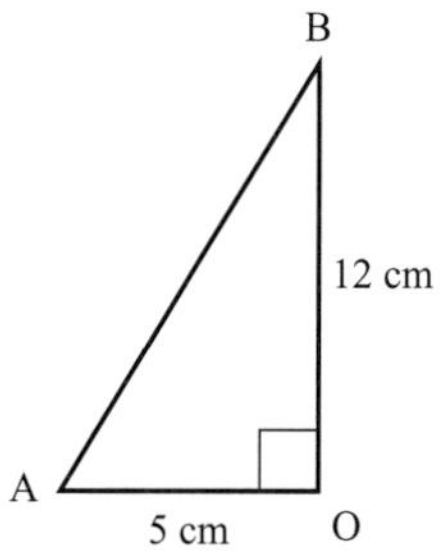

FIGURE 2.8

Fundamental trigonometric identity

Standard triangles

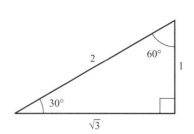

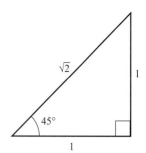

Figure 2.9

The two standard triangles are shown in figure 2.9.

Tabulating the information given in these triangles gives:

	30°	60°	45°	$\frac{\pi}{6}$	$\frac{\pi}{3}$	$\frac{\pi}{4}$
$\sin x$	$\frac{1}{2}$	$\frac{\sqrt{3}}{2}$	$\frac{1}{\sqrt{2}}$	$\frac{1}{2}$	$\frac{\sqrt{3}}{2}$	$\frac{1}{\sqrt{2}}$
$\cos x$	$\frac{\sqrt{3}}{2}$	$\frac{1}{2}$	$\frac{1}{\sqrt{2}}$	$\frac{\sqrt{3}}{2}$	$\frac{1}{2}$	$\frac{1}{\sqrt{2}}$
$\tan x$	$\frac{1}{\sqrt{3}}$	$\sqrt{3}$	1	$\frac{1}{\sqrt{3}}$	$\sqrt{3}$	1

The inverse results are:

$\sin^{-1}\frac{1}{2} = 30°$ or $\frac{\pi}{6}$ $\qquad$ $\sin^{-1}\frac{\sqrt{3}}{2} = 60°$ or $\frac{\pi}{3}$ $\qquad$ $\sin^{-1}\frac{1}{\sqrt{2}} = 45°$ or $\frac{\pi}{4}$

$\cos^{-1}\frac{1}{2} = 60°$ or $\frac{\pi}{3}$ $\qquad$ $\cos^{-1}\frac{\sqrt{3}}{2} = 30°$ or $\frac{\pi}{6}$ $\qquad$ $\cos^{-1}\frac{1}{\sqrt{2}} = 45°$ or $\frac{\pi}{4}$

$\tan^{-1}\frac{1}{\sqrt{3}} = 30°$ or $\frac{\pi}{6}$ $\qquad$ $\tan^{-1}\sqrt{3} = 60°$ or $\frac{\pi}{3}$ $\qquad$ $\tan^{-1}1 = 45°$ or $\frac{\pi}{4}$.

Fundamental trigonometric identity

From Chapter 9 of *Pure Mathematics: Core 1 and Core 2* we had the fundamental trigonometric identity

$$\sin^2\theta + \cos^2\theta \equiv 1.$$

If you take this equation and divide throughout by $\cos^2\theta$ then

$$\frac{\sin^2\theta}{\cos^2\theta} + \frac{\cos^2\theta}{\cos^2\theta} \equiv \frac{1}{\cos^2\theta}$$

giving $\qquad \tan^2\theta + 1 \equiv \sec^2\theta.$

Dividing both sides of the original equation by $\sin^2\theta$ gives

$$\frac{\sin^2\theta}{\sin^2\theta} + \frac{\cos^2\theta}{\sin^2\theta} \equiv \frac{1}{\sin^2\theta}.$$

giving $\quad \boxed{1 + \cot^2\theta \equiv \text{cosec}^2\theta.}$

EXAMPLE 2.5

Find the values of θ in the interval $0° \leqslant \theta \leqslant 360°$ for which $\sec^2\theta = 3 + \tan\theta$.

Solution First it is necessary to obtain an equation containing only one trigonometrical function, in this case $\tan\theta$.

$$\sec^2\theta = 3 + \tan\theta$$
$$\Rightarrow \quad \tan^2\theta + 1 = 3 + \tan\theta$$
$$\Rightarrow \quad \tan^2\theta - \tan\theta - 2 = 0 \quad \text{(This is a quadratic equation like } x^2 - x - 2 = 0\text{)}$$
$$\Rightarrow \quad (\tan\theta - 2)(\tan\theta + 1) = 0$$
$$\Rightarrow \quad \tan\theta = 2 \text{ or } \tan\theta = -1$$

$\tan\theta = 2 \Rightarrow \quad \theta = 63.4°$ (calculated)

$\quad\quad\quad\quad\quad$ or $\theta = 63.4° + 180°$ (from the graph)

$\quad\quad\quad\quad\quad\quad = 243.4°$.

$\tan\theta = -1 \Rightarrow \quad \theta = -45°$ (calculated).

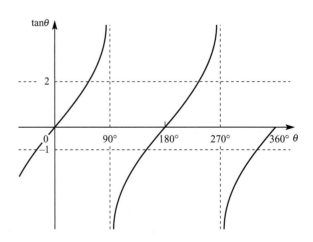

FIGURE 2.10

This is not in the range $0° \leqslant \theta \leqslant 360°$ so figure 2.10 is used to give

$\quad\quad \theta = -45° + 180° = 135°$

or $\quad \theta = -45° + 360° = 315°$.

The values of θ are $63.4°, 135°, 243.4°, 315°$.

EXAMPLE 2.7

Solve $3\sec^2 x - 5\tan x - 4 = 0$ for $0° \leq x \leq 360°$.

Solution Using $\sec^2 x \equiv \tan^2 x + 1$ gives

$$3(\tan^2 x + 1) - 5\tan x - 4 = 0$$
$$3\tan^2 x - 5\tan x - 1 = 0.$$

This does not factorise so use the formula $\dfrac{-b \pm \sqrt{b^2 - 4ac}}{2a}$ to give

$$\tan x = -0.1805 \text{ or } \tan x = 1.847$$

so $x = 61.6°, 169.8°, 241.6°, 349.8°$.

EXAMPLE 2.8

Prove that $(1 - \cos\theta)(1 + \sec\theta) \equiv \sin\theta\tan\theta$.

Solution Left-hand side $= (1 - \cos\theta)(1 + \sec\theta)$
$= 1 + \sec\theta - \cos\theta - \cos\theta\sec\theta$

$= \dfrac{1}{\cos\theta} - \cos\theta$

> But $\cos\theta\sec\theta = \cos\theta \times \dfrac{1}{\cos\theta} = 1$

$= \dfrac{1 - \cos^2\theta}{\cos\theta}$

$= \dfrac{\sin^2\theta}{\cos\theta}$

$= \sin\theta \times \dfrac{\sin\theta}{\cos\theta}$

$= \sin\theta\tan\theta$

$=$ right-hand side.

EXERCISE 2B

1 Solve the following.
  (a) $2\cos^2 x + \sin x - 1 = 0$ for $0° \leq x \leq 360°$
  (b) $2\text{cosec}^2\theta - 3\cot\theta - 1 = 0$ for $0° \leq \theta \leq 180°$
  (c) $2\tan^2 x - 7\sec x + 8 = 0$ for $0° \leq x \leq 360°$
  (d) $\text{cosec}^2\theta = 3\cot\theta - 1$ for $0° \leq \theta \leq 360°$
  (e) $2\sec^2 x + \tan x - 3 = 0$ for $0° \leq x \leq 360°$
  (f) $\tan^2 x - 5\sec x + 7 = 0$ for $0° \leq x \leq 180°$
  (g) $5\cot\theta = 1 + 2\text{cosec}^2\theta$ for $0° \leq \theta \leq 180°$
  (h) $\text{cosec}\, x + 1 = \cot^2 x$ for $0° \leq x \leq 2\pi$
  (i) $\tan^2 x + \sec x - 1 = 0$ for $0° \leq x \leq 2\pi$
  (j) $6\text{cosec}^2 x + \cot x = 8$ for $0° \leq x \leq 360°$

2 Prove the following identities.
   (a) $\sin\theta - \operatorname{cosec}\theta \equiv -\cos\theta\cot\theta$
   (b) $\cos^4\theta - \sin^4\theta \equiv 2\cos^2\theta - 1$
   (c) $\operatorname{cosec}^2\theta + \sec^2\theta \equiv \operatorname{cosec}^2\theta \sec^2\theta$
   (d) $\dfrac{\sec\theta}{\operatorname{cosec}^2\theta} \equiv \cos\theta\tan^2\theta$

3 In triangle ABC, angle $A = 90°$ and $\sec B = 2$.
   (a) Find the angles B and C.
   (b) Find $\tan B$.
   (c) Show that $1 + \tan^2 B = \sec^2 B$.

4 In triangle LMN, angle $M = 90°$ and $\cot N = 1$.
   (a) Find the angles L and N.
   (b) Find $\sec L$, $\operatorname{cosec} L$, and $\tan L$.
   (c) Show that $1 + \tan^2 L = \sec^2 L$.

5 Solve the quadratic equation
   $$8x^2 + 2x - 3 = 0.$$
   Hence find the values of θ between $0°$ and $360°$ which satisfy the equation
   $$8\sin^2\theta + 2\sin\theta - 3 = 0.$$

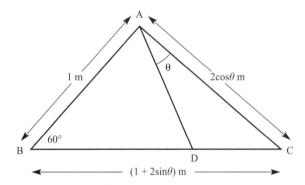

The diagram, which is not to scale, shows a metal framework in which rod AD is at an angle θ to rod AC. The length of AB is 1 m and the angle ABC is $60°$. For structural reasons, $AC = 2\cos\theta$ m and $BC = (1 + 2\sin\theta)$ m.

By applying the cosine rule to triangle ABC, show that

$$4\cos^2\theta = 4\sin^2\theta + 2\sin\theta + 1.$$

Write this equation as a quadratic equation in $\sin\theta$. Find the angle θ.

[MEI]

COMPOUND-ANGLE FORMULAE

You might think that $\sin(\theta + 60°)$ should equal $\sin\theta + \sin 60°$, but this is not so, as you can see by substituting a numerical value of θ. For example, putting $\theta = 30°$ gives $\sin(\theta + 60°) = 1$, but $\sin\theta + \sin 60° \approx 1.366$.

To find an expression for $\sin(\theta + 60°)$, you would use the *compound-angle formula*:

$$\sin(\theta + \phi) = \sin\theta\cos\phi + \cos\theta\sin\phi.$$

This is proved below in the case when θ and ϕ are acute angles. It is, however, true for all values of the angles. It is an *identity*.

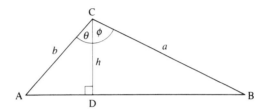

FIGURE 2.11

Using the trigonometric formula for the area of a triangle in figure 2.11

$$\text{area ABC} = \text{area ADC} + \text{area DBC}$$

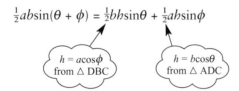

$$\tfrac{1}{2}ab\sin(\theta + \phi) = \tfrac{1}{2}bh\sin\theta + \tfrac{1}{2}ah\sin\phi$$

with $h = a\cos\phi$ from $\triangle$ DBC and $h = b\cos\theta$ from $\triangle$ ADC

$$ab\sin(\theta + \phi) = ab\sin\theta\cos\phi + ab\cos\theta\sin\phi$$

which gives

$$\sin(\theta + \phi) = \sin\theta\cos\phi + \cos\theta\sin\phi. \qquad ①$$

This is the first of the compound-angle formulae (or expansions), and it can be used to prove several more. These are true for all values of θ and ϕ.

Replacing ϕ by $-\phi$ in ① gives

$$\sin(\theta - \phi) = \sin\theta\cos(-\phi) + \cos\theta\sin(-\phi)$$

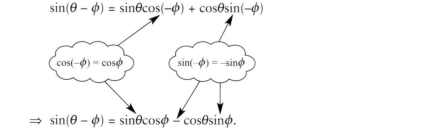

with $\cos(-\phi) = \cos\phi$ and $\sin(-\phi) = -\sin\phi$

$$\Rightarrow \sin(\theta - \phi) = \sin\theta\cos\phi - \cos\theta\sin\phi. \qquad ②$$

From the shape of the sine graph and the cosine graph you can show that

$$\sin(90° - \theta) = \cos\theta$$

and $\cos(90° - \theta) = \sin\theta$.

Hence replacing θ by $90° - \theta$ in $\sin(\theta + \phi) = \sin\theta\cos\phi + \cos\theta\sin\phi$ (equation ① above) gives

$$\sin(90° - \theta + \phi) = \sin(90° - \theta)\cos\phi + \cos(90° - \theta)\sin\phi$$
$$\sin(90° - (\theta - \phi)) = \cos\theta\cos\phi + \sin\theta\sin\phi$$

so $\cos(\theta - \phi) = \cos\theta\cos\phi + \sin\theta\sin\phi$.

Replacing ϕ by $-\phi$ gives

$$\cos(\theta + \phi) = \cos\theta\cos\phi - \sin\theta\sin\phi.$$

From these can be obtained the equivalent formulae for tangents. For example, put

$$\tan(\theta + \phi) = \frac{\sin(\theta + \phi)}{\cos(\theta + \phi)}$$

and expand and simplify as follows.

$$\tan(\theta + \phi) = \frac{\sin(\theta + \phi)}{\cos(\theta + \phi)}$$

$$= \frac{\sin\theta\cos\phi + \cos\theta\sin\phi}{\cos\theta\cos\phi - \sin\theta\sin\phi}$$

to simplify this expression you should divide throughout by $\cos\theta\cos\phi$:

$$= \frac{\dfrac{\sin\theta\cos\phi}{\cos\theta\cos\phi} + \dfrac{\cos\theta\sin\phi}{\cos\theta\cos\phi}}{\dfrac{\cos\theta\cos\phi}{\cos\theta\cos\phi} - \dfrac{\sin\theta\sin\phi}{\cos\theta\cos\phi}}$$

$$= \frac{\tan\theta + \tan\phi}{1 - \tan\theta\tan\phi}$$

The results are summarised as follows:

$$\sin(\theta + \phi) = \sin\theta\cos\phi + \cos\theta\sin\phi$$
$$\sin(\theta - \phi) = \sin\theta\cos\phi - \cos\theta\sin\phi$$
$$\cos(\theta + \phi) = \cos\theta\cos\phi - \sin\theta\sin\phi$$
$$\cos(\theta - \phi) = \cos\theta\cos\phi + \sin\theta\sin\phi$$
$$\tan(\theta + \phi) = \frac{\tan\theta + \tan\phi}{1 - \tan\theta\tan\phi}$$
$$\tan(\theta - \phi) = \frac{\tan\theta - \tan\phi}{1 + \tan\theta\tan\phi}.$$

EXAMPLE 2.9

By putting $75° = 45° + 30°$ find the exact value of $\cos 75°$.

Solution $\cos 75° = \cos(45° + 30°)$

$$= \cos 45° \cos 30° - \sin 45° \sin 30°$$

$$= \frac{1}{\sqrt{2}} \frac{\sqrt{3}}{2} - \frac{1}{\sqrt{2}} \frac{1}{2}$$

$$= \frac{\sqrt{3} - 1}{2\sqrt{2}}.$$

EXAMPLE 2.10

Find the acute angle for which $\sin(\theta + 60°) = \cos(\theta - 60°)$.

Solution To find an acute angle θ such that $\sin(\theta + 60°) = \cos(\theta - 60°)$, you expand each side using the compound-angle formulae:

$$\sin(\theta + 60°) = \sin\theta \cos 60° + \cos\theta \sin 60°$$

$$= \frac{1}{2}\sin\theta + \frac{\sqrt{3}}{2}\cos\theta \qquad \text{①}$$

$$\cos(\theta - 60°) = \cos\theta \cos 60° + \sin\theta \sin 60°$$

$$= \frac{1}{2}\cos\theta + \frac{\sqrt{3}}{2}\sin\theta \qquad \text{②}$$

From ① and ②

$$\frac{1}{2}\sin\theta + \frac{\sqrt{3}}{2}\cos\theta = \frac{1}{2}\cos\theta + \frac{\sqrt{3}}{2}\sin\theta$$

$$\sin\theta + \sqrt{3}\cos\theta = \cos\theta + \sqrt{3}\sin\theta$$

Collect like terms:

$$\Rightarrow \quad (\sqrt{3} - 1)\cos\theta = (\sqrt{3} - 1)\sin\theta$$

$$\cos\theta = \sin\theta.$$

Divide by $\cos\theta$:

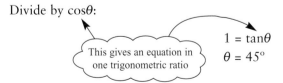

$$1 = \tan\theta$$
$$\theta = 45°$$

This gives an equation in one trigonometric ratio

Since an acute angle was required, this is the only root.

EXAMPLE 2.11

Simplify $\cos\theta\cos3\theta - \sin\theta\sin3\theta$.

Solution The formula which has the same pattern of cos cos − sin sin is

$$\cos(\theta + \phi) = \cos\theta\cos\phi - \sin\theta\sin\phi.$$

Using this, and replacing ϕ by 3θ, gives

$$\cos\theta\cos3\theta - \sin\theta\sin3\theta = \cos(\theta + 3\theta)$$
$$= \cos4\theta.$$

EXERCISE 2C

1 Use the compound-angle formulae to write the following as surds.
   (a) $\sin75° = \sin(45° + 30°)$
   (b) $\cos135° = \cos(90° + 45°)$
   (c) $\tan15° = \tan(45° - 30°)$
   (d) $\tan75° = \tan(45° + 30°)$

2 Expand each of the following expressions.
   (a) $\sin(\theta + 45°)$
   (b) $\cos(\theta - 30°)$
   (c) $\sin(60° - \theta)$
   (d) $\cos(2\theta + 45°)$
   (e) $\tan(\theta + 45°)$
   (f) $\tan(\theta - 45°)$

3 Simplify each of the following expressions.
   (a) $\sin2\theta\cos\theta - \cos2\theta\sin\theta$
   (b) $\cos\phi\cos3\phi - \sin\phi\sin3\phi$
   (c) $\sin120°\cos60° + \cos120°\sin60°$
   (d) $\cos\theta\cos\theta - \sin\theta\sin\theta$

4 Solve the following equations for values of θ in the range $0° \leq \theta \leq 180°$.
   (a) $\cos(60° + \theta) = \sin\theta$
   (b) $\sin(45° - \theta) = \cos\theta$
   (c) $\tan(45° + \theta) = \tan(45° - \theta)$
   (d) $2\sin\theta = 3\cos(\theta - 60°)$
   (e) $\sin\theta = \cos(\theta + 120°)$

5 Solve the following equations for values of θ in the range $0 \leq \theta \leq \pi$.
   (When the range is given in radians, the solutions should be in radians, using multiples of π where appropriate.)
   (a) $\sin\left(\theta + \frac{\pi}{4}\right) = \cos\theta$
   (b) $2\cos\left(\theta - \frac{\pi}{3}\right) = \cos\left(\theta + \frac{\pi}{2}\right)$

6 Prove the following identities.
   (a) $\dfrac{\sin(A + B)}{\cos A \cos B} \equiv \tan A + \tan B$
   (b) $\sin(A + B) + \cos(A - B) \equiv (\sin A + \cos A)(\sin B + \cos B)$
   (c) $\cos(45° + A) - \cos(45° - A) \equiv \sqrt{2}\sin A$
   (d) $\tan(A - B) = \dfrac{\tan A - \tan B}{1 + \tan A \tan B}$
   (e) $\sin2x\cos x - \cos2x\sin x \equiv \sin x$

DOUBLE-ANGLE FORMULAE

Substituting $\phi = \theta$ in the relevant compound-angle formulae leads immediately to expressions for $\sin 2\theta$, $\cos 2\theta$ and $\tan 2\theta$, as follows.

(a) $\quad \sin(\theta + \phi) = \sin\theta\cos\phi + \cos\theta\sin\phi$

When $\phi = \theta$, this becomes

$\sin(\theta + \theta) = \sin\theta\cos\theta + \cos\theta\sin\theta$

giving $\sin 2\theta = 2\sin\theta\cos\theta$.

(b) $\quad \cos(\theta + \phi) = \cos\theta\cos\phi - \sin\theta\sin\phi$

When $\phi = \theta$, this becomes

$\cos(\theta + \theta) = \cos\theta\cos\theta - \sin\theta\sin\theta$

giving $\cos 2\theta = \cos^2\theta - \sin^2\theta$.

Using the fundamental trigonometric identity $\cos^2\theta + \sin^2\theta = 1$, two other forms for $\cos 2\theta$ can be obtained:

$\cos 2\theta = (1 - \sin^2\theta) - \sin^2\theta \quad \Rightarrow \quad \cos 2\theta = 1 - 2\sin^2\theta$

$\cos 2\theta = \cos^2\theta - (1 - \cos^2\theta) \quad \Rightarrow \quad \cos 2\theta = 2\cos^2\theta - 1$.

These alternative forms are often more useful since thay contain only one trigonometric function.

(c) $\quad \tan(\theta + \phi) = \dfrac{\tan\theta + \tan\phi}{1 - \tan\theta\tan\phi}$

When $\phi = \theta$ this becomes

$\tan(\theta + \theta) = \dfrac{\tan\theta + \tan\theta}{1 - \tan\theta\tan\theta}$

giving $\tan 2\theta = \dfrac{2\tan\theta}{1 - \tan^2\theta}$.

In summary:

$$\sin 2\theta = 2\sin\theta\cos\theta$$
$$\cos 2\theta = \cos^2\theta - \sin^2\theta$$
$$= 2\cos^2\theta - 1$$
$$= 1 - 2\sin^2\theta$$
$$\tan 2\theta = \dfrac{2\tan\theta}{1 - \tan^2\theta}$$

EXAMPLE 2.12

Solve the equation $\sin 2\theta = \sin\theta$ for $0° \leq \theta \leq 360°$.

Solution

$$\sin 2\theta = \sin\theta$$
$$\Rightarrow 2\sin\theta\cos\theta = \sin\theta$$
$$\Rightarrow 2\sin\theta\cos\theta - \sin\theta = 0$$
$$\Rightarrow \sin\theta(2\cos\theta - 1) = 0$$
$$\Rightarrow \sin\theta = 0 \text{ or } \cos\theta = \tfrac{1}{2}$$

Be careful here: don't cancel by $\sin\theta$ or some roots will be lost

$\sin\theta = 0 \Rightarrow \theta = 0°$ (principal value) or $180°$ or $360°$ (see figure 2.12).

The principal value is the one which comes from your calculator

$\cos\theta = \tfrac{1}{2} \Rightarrow \theta = 60°$ (principal value) or $300°$ (see figure 2.13).

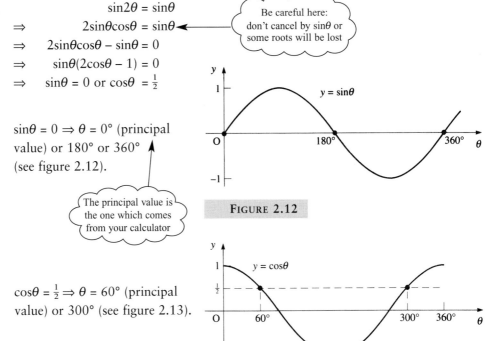

FIGURE 2.12

FIGURE 2.13

The full set of roots for $0 \leq \theta \leq 360°$ is $\theta = 0°, 60°, 180°, 300°, 360°$.

Note

When an equation contains $\cos 2\theta$, you will save time if you take care to choose the most suitable expansion.

EXAMPLE 2.13

Solve $2 + \cos 2\theta = \sin\theta$ for $0 \leq \theta \leq 2\pi$. (Notice that the request for $0 \leq \theta \leq 2\pi$, i.e. in radians, is an invitation to give the answer in radians.)

Solution Using $\cos 2\theta = 1 - 2\sin^2\theta$ gives

This is the most suitable expansion since the right-hand side contains $\sin\theta$

$$2 + (1 - 2\sin^2\theta) = \sin\theta$$
$$\Rightarrow 2\sin^2\theta + \sin\theta - 3 = 0$$
$$\Rightarrow (2\sin\theta + 3)(\sin\theta - 1) = 0$$
$$\Rightarrow \sin\theta = -\tfrac{3}{2} \text{ (not valid since } -1 \leq \sin\theta \leq 1)$$
or $\sin\theta = 1$.

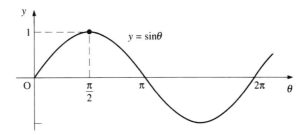

FIGURE 2.14

Figure 2.14 shows that the principal value $\theta = \frac{\pi}{2}$ is the only root for $0 \leq \theta \leq 2\pi$.

EXAMPLE 2.14

Prove that $\cos3\theta \equiv 4\cos^3\theta - 3\cos\theta$.

Solution
$$\cos3\theta \equiv \cos(\theta + 2\theta)$$
$$\equiv \cos\theta\cos2\theta - \sin\theta\sin2\theta$$
$$\equiv \cos\theta(2\cos^2\theta - 1) - \sin\theta \times 2\sin\theta\cos\theta$$
$$\equiv 2\cos^3\theta - \cos\theta - 2\cos\theta\sin^2\theta$$
$$\equiv 2\cos^3\theta - \cos\theta - 2\cos\theta(1 - \cos^2\theta)$$
$$\equiv 4\cos^3\theta - 3\cos\theta$$

HALF-ANGLE FORMULAE

The double-angle formulae can be written in terms of θ and its half-angle $\frac{1}{2}\theta$.

Since $\sin2\theta \equiv 2\sin\theta\cos\theta$

replacing θ by $\frac{1}{2}\theta$ gives

$$\sin\theta \equiv 2\sin\tfrac{1}{2}\theta\cos\tfrac{1}{2}\theta.$$

Similarly $\cos\theta \equiv \cos^2\tfrac{1}{2}\theta - \sin^2\tfrac{1}{2}\theta$
$$\equiv 2\cos^2\tfrac{1}{2}\theta - 1$$
$$\equiv 1 - 2\sin^2\tfrac{1}{2}\theta$$

and $\tan\theta \equiv \dfrac{2\tan\tfrac{1}{2}\theta}{1 - \tan^2\tfrac{1}{2}\theta}.$

EXAMPLE 2.15

Solve $\sin\theta = \cot\tfrac{1}{2}\theta$ for $0° \leq \theta \leq 360°$.

Solution

$$\sin\theta = \cot\tfrac{1}{2}\theta$$

$$2\sin\tfrac{1}{2}\theta\cos\tfrac{1}{2}\theta = \frac{\cos\tfrac{1}{2}\theta}{\sin\tfrac{1}{2}\theta}$$

$$2\sin^2\tfrac{1}{2}\theta\cos\tfrac{1}{2}\theta - \cos\tfrac{1}{2}\theta = 0$$

$$\cos\tfrac{1}{2}\theta(2\sin^2\tfrac{1}{2}\theta - 1) = 0$$

$$\cos\tfrac{1}{2}\theta = 0 \quad \text{or} \quad \sin\tfrac{1}{2}\theta = \pm\frac{1}{\sqrt{2}}$$

$$\tfrac{1}{2}\theta = 90° \quad \text{or} \quad \tfrac{1}{2}\theta = 45°, 135°$$

$$\theta = 90°, 180°, 270°$$

Note if $0° \leq \theta \leq 360°$, $0° \leq \tfrac{1}{2}\theta \leq 180°$

Note

θ can be replaced by any other multiple of θ. For example, replacing θ by 2θ would give $\sin 4\theta = 2\sin 2\theta \cos 2\theta$.

EXERCISE 2D

1. Solve the following equations for $0° \leq \theta \leq 360°$.
   (a) $2\sin 2\theta = \cos\theta$
   (b) $\tan 2\theta = 4\tan\theta$
   (c) $\cos 2\theta + \sin\theta = 0$
   (d) $\tan\theta \tan 2\theta = 1$
   (e) $2\cos 2\theta = 1 + \cos\theta$

2. Solve the following equations for $-\pi \leq \theta \leq \pi$.
   (a) $\sin 2\theta = 2\sin\theta$
   (b) $\tan 2\theta = 2\tan\theta$
   (c) $\cos 2\theta - \cos\theta = 0$
   (d) $1 + \cos 2\theta = 2\sin^2\theta$
   (e) $\sin 4\theta = \cos 2\theta$ (**Hint:** Express this as an equation in 2θ.)

3. By first writing $\sin 3\theta$ as $\sin(2\theta + \theta)$, express $\sin 3\theta$ in terms of $\sin\theta$. Hence solve the equation $\sin 3\theta = \sin\theta$ for $0 \leq \theta \leq 2\pi$.

4. Solve $\cos 3\theta = 1 - 3\cos\theta$ for $0° \leq \theta \leq 360°$.

5. Simplify $\dfrac{1 + \cos 2\theta}{\sin 2\theta}$.

6. Express $\tan 3\theta$ in terms of $\tan\theta$.

7. Show that $\dfrac{1 - \tan^2\theta}{1 + \tan^2\theta} = \cos 2\theta$.

8. (a) Show that $\tan(\tfrac{\pi}{4} + \theta)\tan(\tfrac{\pi}{4} - \theta) = 1$.
   (b) Given that $\tan 26.6° = 0.5$, solve $\tan\theta = 2$ without using your calculator. Give θ to 1 decimal place, where $0° < \theta < 90°$.

EXERCISE 2D

9 Use the half-angle formulae to solve the following equations for $0° \leq \theta \leq 360°$.
   (a) $\sin\theta + \cos\tfrac{1}{2}\theta = 0$
   (b) $\sin\theta = 2(\cos\theta + 1)$
   (c) $3\tan\theta = 8\tan\tfrac{1}{2}\theta$
   (d) $4\cos\theta = 1 - 2\sin\tfrac{1}{2}\theta$

10 Prove the following identities.
   (a) $\dfrac{\sin\theta}{1 + \cos\theta} \equiv \tan\tfrac{1}{2}\theta$
   (b) $\cot\theta + \csc\theta \equiv \cot\tfrac{1}{2}\theta$
   (c) $\sec\theta + \tan\theta \equiv \dfrac{\cos\tfrac{1}{2}\theta + \sin\tfrac{1}{2}\theta}{\cos\tfrac{1}{2}\theta - \sin\tfrac{1}{2}\theta}$

11 Prove the following identities where $t = \tan\tfrac{1}{2}\theta$.
   (a) $\sin\theta = \dfrac{2t}{1 + t^2}$
   (b) $\cos\theta = \dfrac{1 - t^2}{1 + t^2}$
   (c) $\tan\theta = \dfrac{2t}{1 - t^2}$

$a\cos\theta + b\sin\theta$ IN THE FORM $r\cos(\theta \pm \alpha)$ OR $r\sin(\theta \pm \alpha)$

$a\cos\theta + b\sin\theta$ can be combined to form a single function $r\cos(\theta \pm \alpha)$ or $r\sin(\theta \pm \alpha)$ using the compound-angle formulae. The following examples illustrate the process.

EXAMPLE 2.16

Express $4\cos\theta + 3\sin\theta$ in the form $r\sin(\theta + \alpha)$.

Solution Let $\quad 4\cos\theta + 3\sin\theta \equiv r\sin(\theta + \alpha)$
$\quad\quad\quad\quad\quad 4\cos\theta + 3\sin\theta \equiv r\sin\theta\cos\alpha + r\cos\theta\sin\alpha$

Comparing the coefficients of $\cos\theta$ and $\sin\theta$ gives

$\quad\cos\theta: \quad\quad\quad\quad\quad 4 = r\sin\alpha \quad\quad\quad ①$
$\quad\sin\theta: \quad\quad\quad\quad\quad 3 = r\cos\alpha \quad\quad\quad ②$

To find r $①^2 + ②^2$ $r^2\sin^2\alpha + r^2\cos^2\alpha = 4^2 + 3^2$
$\quad\quad\quad\quad\quad\quad\quad\quad r^2(\sin^2\alpha + \cos^2\alpha) = 25$
$\quad\quad\quad\quad\quad\quad\quad\quad r^2 = 25$
$\quad\quad\quad\quad\quad\quad\quad\quad r = 5.$ ← Since $\sin^2\alpha + \cos^2\alpha = 1$

To find α $① \div ②$ $\quad \dfrac{r\sin\alpha}{r\cos\alpha} = \tan\alpha = \dfrac{4}{3}$
$\quad\quad\quad\quad\quad\quad\quad\quad \alpha = 53.1°.$

So, the answer is $\quad 4\cos\theta + 3\sin\theta \equiv 5\sin(\theta + 53.1°).$

Note

From equations ① and ② we could have calculated r and α by drawing a right-angled triangle (figure 2.15):

$\quad 4 = r\sin\alpha$
$\quad 3 = r\cos\alpha$

from which r and α can be found.

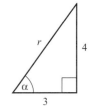

FIGURE 2.15

The value of r is always positive but $\cos\alpha$ and $\sin\alpha$ may be positive or negative, depending on the values of a and b. In all cases, it is possible to find an angle α for which $-180° \leq \alpha \leq 180°$.

If the question does not specify which expression of $r\cos(\theta \pm \alpha)$ or $r\sin(\theta \pm \alpha)$ to use then choose either the sine or the cosine function with the signs of their expansions matching the signs of $a\cos\theta + b\sin\theta$.

For example $3\cos\theta - 4\sin\theta$ matches up with $r\cos(\theta + \alpha)$ or $-r\sin(\theta - \alpha)$.

The result of replacing $a\cos\theta + b\sin\theta$ with $r\cos(\theta \pm \alpha)$ or $r\sin(\theta \pm \alpha)$ is the combination of two waves to form a single sine wave. In Example 2.16 the amplitude of the single wave is 5 and the phase shift is 53.1°. This is illustrated in figure 2.16 over the range 0° to 360°.

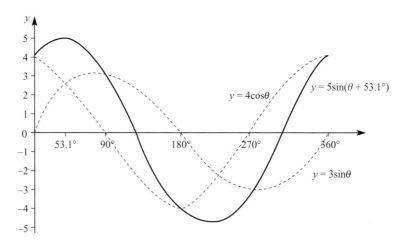

FIGURE 2.16

EXAMPLE 2.17

Express $12\cos\theta + 5\sin\theta$ in the form $r\cos(\theta - \alpha)$.
Use your answer to solve $12\cos\theta + 5\sin\theta = 4$ giving θ in the range $-180°$ to $180°$.

Solution Let $12\cos\theta + 5\sin\theta \equiv r\cos(\theta - \alpha)$
$12\cos\theta + 5\sin\theta \equiv r\cos\theta\cos\alpha + r\sin\theta\sin\alpha$

$\cos\theta$: $\qquad 12 = r\cos\alpha$
$\sin\theta$: $\qquad 5 = r\sin\alpha$

$\qquad r = \sqrt{12^2 + 5^2} = 13$

$\qquad \tan\alpha = \dfrac{5}{12}$

so $\qquad \alpha = 22.6°$
hence

$\qquad 12\cos\theta + 5\sin\theta \equiv 13\cos(\theta - 22.6°)$.

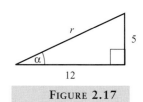

FIGURE 2.17

To solve $\quad 12\cos\theta + 5\sin\theta = 4$

put $\quad 13\cos(\theta - 22.6°) = 4$

$\therefore \cos(\theta - 22.6°) = \dfrac{4}{13}$

Note: if $-180° \leq \theta \leq 180°$ then $-202.6° \leq \theta - 22.6° \leq 157.4°$

$\therefore \theta - 22.6° = \pm 72.1°$

$\therefore \theta = -49.5°$ or $94.7°$.

EXAMPLE 2.18

(a) Express $\sqrt{3}\sin\theta - \cos\theta$ in the form $r\sin(\theta - \alpha)$, where $r > 0$ and $0 < \alpha < \dfrac{\pi}{2}$.

(b) State the maximum and minimum values of $\sqrt{3}\sin\theta - \cos\theta$.

(c) Sketch the graph of $y = \sqrt{3}\sin\theta - \cos\theta$ for $0 \leq \theta \leq 2\pi$.

(d) Solve the equation $\sqrt{3}\sin\theta - \cos\theta = 1$ for $0 \leq \theta \leq 2\pi$.

Solution

(a) Let

$\sqrt{3}\sin\theta - \cos\theta \equiv r\sin(\theta - \alpha)$

$\sqrt{3}\sin\theta - \cos\theta \equiv r\sin\theta\cos\alpha - r\cos\theta\sin\alpha$

Comparing coefficients

$\sin\theta: \quad \sqrt{3} = r\cos\alpha$

$\cos\theta: \quad 1 = r\cos\alpha$

From the triangle (figure 2.18)

$r = \sqrt{1 + 3} = 2$ and $\tan\alpha = \dfrac{1}{\sqrt{3}} \Rightarrow \alpha = \dfrac{\pi}{6}$

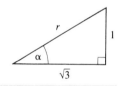

FIGURE 2.18

so $\quad \sqrt{3}\sin\theta - \cos\theta \equiv 2\sin\left(\theta - \dfrac{\pi}{6}\right)$.

(b) The sine function oscillates between 1 and -1, so $2\sin\left(\theta - \dfrac{\pi}{6}\right)$ oscillates between 2 and -2.

Maximum value = 2.
Minimum value = -2.

(c) The graph of $y = 2\sin\left(\theta - \dfrac{\pi}{6}\right)$ is obtained from the graph of $y = \sin\theta$ by a translation $\begin{pmatrix} \dfrac{\pi}{6} \\ 0 \end{pmatrix}$ and a stretch of factor 2 parallel to the y-axis.

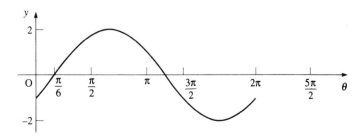

FIGURE 2.19

(d) The equation $\sqrt{3}\sin\theta - \cos\theta = 1$ is equivalent to
$$2\sin(\theta - \tfrac{\pi}{6}) = 1$$
$$\Rightarrow \quad \sin(\theta - \tfrac{\pi}{6}) = \tfrac{1}{2}$$

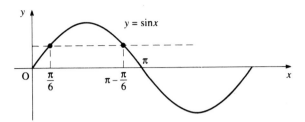

FIGURE 2.20

Solving $\sin(\theta - \tfrac{\pi}{6}) = \tfrac{1}{2}$ gives

$\theta - \tfrac{\pi}{6} = \tfrac{\pi}{6}$ (principal value) or $\theta - \tfrac{\pi}{6} = \pi - \tfrac{\pi}{6} = \tfrac{5\pi}{6}$ (from the graph)

giving $\theta = \tfrac{\pi}{6} + \tfrac{\pi}{6} = \tfrac{\pi}{3}$ or $\theta = \tfrac{5\pi}{6} + \tfrac{\pi}{6} = \pi$.

The roots in $0 \leq \theta \leq 2\pi$ are $\theta = \tfrac{\pi}{3}$ and π.

Note

Always check (for example, by reference to a sketch graph) that the number of roots you have found is consistent with the number you are expecting. When solving equations of the form $\sin(\theta - \alpha) = c$ by considering $\sin x = c$, it is sometimes necessary to go outside the range specified for θ since, for example, $0 \leq \theta \leq 2\pi$ is the same as $-\alpha \leq \theta - \alpha \leq 2\pi - \alpha$.

EXERCISE 2E

1 Express each of the following in the form $r\cos(\theta - \alpha)$, where $r > 0$ and $0 < \alpha < 90°$.
   (a) $\cos\theta + \sin\theta$
   (b) $3\cos\theta + 4\sin\theta$
   (c) $\cos\theta + \sqrt{3}\sin\theta$
   (d) $\sqrt{5}\cos\theta + 2\sin\theta$

2 Express each of the following in the form $r\cos(\theta + \alpha)$, where $r > 0$ and $0 < \alpha < \tfrac{\pi}{2}$.
   (a) $\cos\theta - \sin\theta$
   (b) $\sqrt{3}\cos\theta - \sin\theta$

3 Express each of the following in the form $r\sin(\theta + \alpha)$, where $r > 0$ and $0 < \alpha < 90°$.
   (a) $\sin\theta + 2\cos\theta$
   (b) $3\sin\theta + 4\cos\theta$

EXERCISE 2E

4 Express each of the following in the form $r\sin(\theta - \alpha)$, where $r > 0$ and $0 < \alpha < \frac{\pi}{2}$.
  (a) $\sin\theta - \cos\theta$
  (b) $\sqrt{3}\sin\theta - \cos\theta$

5 Express each of the following in the form $r\cos(\theta - \alpha)$, where $r > 0$ and $-180° < \alpha < 180°$.
  (a) $\cos\theta - \sqrt{3}\sin\theta$
  (b) $2\sqrt{2}\cos\theta - 2\sqrt{2}\sin\theta$
  (c) $\sin\theta + \sqrt{3}\cos\theta$
  (d) $5\sin\theta + 12\cos\theta$
  (e) $\sin\theta - \sqrt{3}\cos\theta$
  (f) $\sqrt{2}\sin\theta - \sqrt{2}\cos\theta$

6 (a) Express $5\cos\theta - 12\sin\theta$ in the form $r\cos(\theta + \alpha)$, where $r > 0$ and $0 < \alpha < 90°$.
  (b) State the maximum and minimum values of $5\cos\theta - 12\sin\theta$.
  (c) Sketch the graph of $y = 5\cos\theta - 12\sin\theta$ for $0 \leq \theta \leq 360°$.
  (d) Solve the equation $5\cos\theta - 12\sin\theta = 4$ for $0 \leq \theta \leq 360°$.

7 (a) Express $3\sin\theta - \sqrt{3}\cos\theta$ in the form $r\sin(\theta - \alpha)$, where $r > 0$ and $0 < \alpha < \frac{\pi}{2}$.
  (b) State the maximum and minimum values of $3\sin\theta - \sqrt{3}\cos\theta$ and the smallest positive values of θ for which they occur.
  (c) Sketch the graph of $y = 3\sin\theta - \sqrt{3}\cos\theta$ for $0 \leq \theta \leq 2\pi$.
  (d) Solve the equation $3\sin\theta - \sqrt{3}\cos\theta = \sqrt{3}$ for $0 \leq \theta \leq 2\pi$.

8 (a) Express $2\sin2\theta + 3\cos2\theta$ in the form $r\sin(2\theta + \alpha)$, where $r > 0$ and $0 < \alpha < 90°$.
  (b) State the maximum and minimum values of $2\sin2\theta + 3\cos2\theta$ and the smallest positive values of θ for which they occur.
  (c) Sketch the graph of $y = 2\sin2\theta + 3\cos2\theta$ for $0 \leq \theta \leq 360°$.
  (d) Solve the equation $2\sin2\theta + 3\cos2\theta = 1$ for $0 \leq \theta \leq 360°$.

9 (a) Express $\cos\theta + \sqrt{2}\sin\theta$ in the form $r\cos(\theta - \alpha)$, where $r > 0$ and $0 < \alpha < 90°$.
  (b) State the maximum and minimum values of $\cos\theta + \sqrt{2}\sin\theta$ and the smallest positive values of θ for which they occur.
  (c) Sketch the graph of $y = \cos\theta + \sqrt{2}\sin\theta$ for $0 \leq \theta \leq 360°$.
  (d) State the maximum and minimum values of
  $$\frac{1}{3 + \cos\theta + \sqrt{2}\sin\theta}$$
  and the smallest positive values for which they occur.

10 The diagram shows a table jammed in a corridor. The table is 120 cm long and 80 cm wide, and the width of the corridor is 130 cm.
   (a) Show that $12\sin\theta + 8\cos\theta = 13$.
   (b) Hence find the angle θ (there are two answers) by expressing $12\sin\theta + 8\cos\theta$ in the form $r\sin(\theta + \alpha)$.

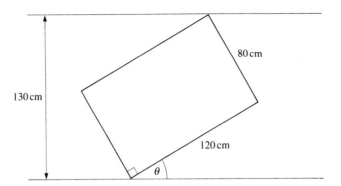

EXERCISE 2F

Examination-style questions

1 $7\cos x - 24\sin x \equiv R\cos(x + \alpha)$, where $R > 0$ and $0° \leq \alpha \leq 90°$.
   (a) Find the exact value of R and the value of α to 1 decimal place.
   (b) Solve $7\cos x - 24\sin x = 20$ giving your answers in the range $-180°$ to $180°$.
   (c) State the maximum value of $7\cos x - 24\sin x$ and the value of x for which this occurs.

2 (a) Using the expansion for $\cos(A + B)$ find an expression for $\cos 2x$ in terms of $\cos x$.
   (b) Solve $\cos 2x = \cos x$ for values of x in the interval 0 to 2π, giving your answers in terms of π.
   (c) Solve $\cos x = \cos\tfrac{1}{2}x$ for values of x in the interval 0 to 2π, giving your answers in terms of π.

3 Express $\cos A \cos B$ in terms of $\cos(A + B)$ and $\cos(A - B)$.
   (a) Find the exact value of $\cos 105° \cos 45°$.
   (b) Solve $\cos 5x + \cos 3x = 0$ for $0° \leq x \leq 360°$.

4 (a) On the same diagram, over the range $-360°$ to $360°$, sketch the graphs of $y = \cos x$ and $y = \sec x$.
   (b) Solve $3\sec^2 x = 10\tan x$ for $0° \leq x \leq 360°$.
   (c) Prove that $\dfrac{\cos\theta}{\sec\theta - \tan\theta} \equiv 1 + \sin\theta$.

5 $f(x) \equiv 6\sin x - 8\cos x$.
   Given that $f(x) \equiv R\sin(x - \alpha)$, where $R > 0$, $0 \leq \alpha \leq \tfrac{\pi}{2}$, and x and α are measured in radians,
   (a) find R and the value of α to 2 decimal places.

Hence:
(b) find the minimum value of f(x) and the value of x that gives this minimum.
(c) find the smallest angle x, in radians to 2 decimal places, for which

$$6\sin x - 8\cos x = 4.$$

6 Given that $y = \cos 2x + \sin x$, $0 < x < 2\pi$, and x is in radians, find, in terms of π, the values of x for which $y = 0$.

[Edexcel]

7 (a) Starting from the identity for $\cos(A + B)$ prove that

$$\cos 2x = 1 - 2\sin^2 x.$$

Find, in radians to 2 decimal places, the values of x in the interval $0 \leqslant x \leqslant 2\pi$ for which
(b) $2\cos 2x + 1 = \sin x$
(c) $2\cos x + 1 = \sin\frac{1}{2}x$.

[Edexcel]

8 (a) Solve the equation

$$2\cos^2 x + 5\sin x + 1 = 0, \quad 0° \leqslant x \leqslant 360°$$

giving your answer in degrees.

(b) Using the half-angle formulae, or otherwise,

(i) show that $\dfrac{1 - \cos\theta}{\sin\theta} = \tan\frac{1}{2}\theta$, $\quad 0 \leqslant \theta \leqslant \pi$,

(ii) solve $\dfrac{1 - \cos\theta}{\sin\theta} = \sqrt{3}\sin\theta$, $\quad 0 < \theta < \pi$,

giving your answer in radians to 3 significant figures.

[Edexcel]

9 Given that

$$7\cos\theta + 24\sin\theta \equiv R\cos(\theta - \alpha), \text{ where } R > 0, 0° \leqslant \alpha \leqslant 90°,$$

(a) find the values of the constants R and α.

Hence find:
(b) the solutions of the equation $7\cos\theta + 24\sin\theta = 15$ in the range $0° \leqslant \theta \leqslant 360°$
(c) the range of the function $f(\theta)$ where

$$f(\theta) = \dfrac{1}{5 + (7\cos\theta + 24\sin\theta)^2}, \quad 0° \leqslant \theta \leqslant 360°.$$

[Edexcel, adapted]

10 $f(\theta) \equiv 9\sin\theta + 12\cos\theta$.

Given that $f(\theta) \equiv R\sin(\theta + \alpha)$, where $R > 0$, $0° \leq \alpha \leq 90°$
(a) find the values of the constants R and α.
(b) Hence find the values of θ, $0° \leq \theta \leq 360°$, for which

$$9\sin\theta + 12\cos\theta = -7.5,$$

giving your answers to the nearest tenth of a degree.

(c) Find, in radians in terms of π, the solutions to the equation

$$\sqrt{3}\sin(\theta - \tfrac{1}{6}\pi) = \sin\theta$$

in the interval $0 \leq \theta \leq 2\pi$.

[Edexcel, adapted]

11 Solve, showing clear working and giving your answers in radians to 2 decimal places,

$$6\sec^2 2x + 5\tan 2x = 12, \; 0 \leq x \leq \pi.$$

[Edexcel]

12 Find, in terms of π, all solutions in the interval $0 \leq x < 2\pi$ of
(a) $\sin 2x = \sqrt{2}\cos x$,
(b) $2\sin\left(2x + \tfrac{\pi}{3}\right) = \cos\left(2x - \tfrac{\pi}{6}\right)$.

[Edexcel]

13 (a) Prove, by counter-example, that the statement

$$\sec(A + B) \equiv \sec A + \sec B, \text{ for all } A \text{ and } B$$

is false

(b) Prove that

$$\tan\theta + \cot\theta \equiv 2\operatorname{cosec}2\theta, \quad \theta \neq \frac{n\pi}{2}, n \in \mathbb{Z}$$

[Edexcel]

14 (a) Express $2\cos\theta + 5\sin\theta$ in the form $R\cos(\theta - \alpha)$, where $R > 0$ and $0 < \alpha < \tfrac{\pi}{2}$.
Give the values of R and α to 3 significant figures.
(b) Find the maximum value of $2\cos\theta + 5\sin\theta$ and the smallest positive value of θ for which this maximum occurs.

The temperature, $T°C$, of an unheated building is modelled using the equation

$$T = 15 + 2\cos\left(\frac{\pi t}{12}\right) + 5\sin\left(\frac{\pi t}{12}\right), \; 0 \leq t < 24,$$

where t hours is the number of hours after 1200.

EXERCISE 2F

(c) Calculate the maximum temperature predicted by this model and the value of t when this maximum occurs.

(d) Calculate, to the nearest half hour, the times when the temperature is predicted to be 12 °C.

[Edexcel]

15 (a) Given that $\cos(x + 30)° = 3\cos(x - 30)°$, prove that $\tan x° = -\dfrac{\sqrt{3}}{2}$.

(b) (i) Prove that $\dfrac{1 - \cos 2\theta}{\sin 2\theta} = \tan\theta$.

(ii) Verify that $\theta = 180°$ is a solution of the equation $\sin 2\theta = 2 - 2\cos 2\theta$.

(iii) Using the result in part (a), or otherwise, find the other two solutions, $0° < \theta < 360°$, of the equation $\sin 2\theta = 2 - 2\cos 2\theta$.

[Edexcel]

KEY POINTS

1 **Reciprocals**

$$\operatorname{cosec}\theta = \dfrac{1}{\sin\theta} \qquad \sec\theta = \dfrac{1}{\cos\theta} \qquad \cot\theta = \dfrac{1}{\tan\theta}$$

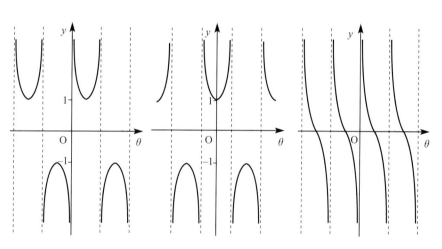

$$\tan\theta \equiv \dfrac{\sin\theta}{\cos\theta} \qquad \cot\theta \equiv \dfrac{\cos\theta}{\sin\theta}$$

2 Inverse functions

$\sin^{-1}\theta$ (arcsinθ) $\cos^{-1}\theta$ (arccosθ) $\tan^{-1}\theta$ (arctanθ)

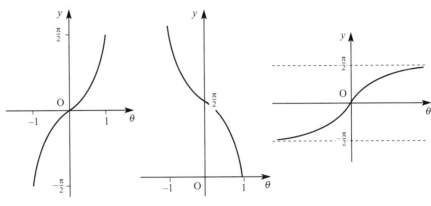

$-\frac{\pi}{2} \leq \theta \leq \frac{\pi}{2}$ $0 \leq \theta \leq \pi$ $-\frac{\pi}{2} \leq \theta \leq \frac{\pi}{2}$

3 Fundamental trigonometric identity

$\sin^2\theta + \cos^2\theta \equiv 1$ $\tan^2\theta + 1 \equiv \sec^2\theta$ $1 + \cot^2\theta \equiv \text{cosec}^2\theta$

4 Compound-angle formulae

$\sin(A \pm B) = \sin A \cos B \pm \cos A \sin B$

$\cos(A \pm B) = \cos A \cos B \mp \sin A \sin B$

$\tan(A \pm B) = \dfrac{\tan A \pm \tan B}{1 \mp \tan A \tan B}$

5 Double-angle formulae

$\sin 2\theta = 2\sin\theta\cos\theta$

$\cos 2\theta = \cos^2\theta - \sin^2\theta = 2\cos^2\theta - 1 = 1 - 2\sin^2\theta$

$\tan 2\theta = \dfrac{2\tan\theta}{1 - \tan^2\theta}$

6 $a\cos\theta + b\sin\theta$

Let $a\cos\theta + b\sin\theta = r\cos(\theta \pm \alpha)$ or $a\cos\theta + b\sin\theta = r\sin(\theta \pm \alpha)$.

Chapter three

EXPONENTIALS AND LOGARITHMS

We have short time to stay, as you,
We have as short a Spring,
As quick a growth to meet decay,
As you or anything.

Robert Herrick

THE FUNCTION e^x

You will recall from Chapter 10 of *Pure Mathematics: Core 1 and Core 2* that the exponential graph $y = a^x$ looks like that in figure 3.1 where $a > 0$.

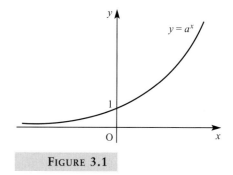

FIGURE 3.1

The graph cuts the y-axis at 1 and is asymptotic to the x-axis.

We said that there is a particular value of a for which the gradient of the curve at every point of the graph is equal to the value of y. This value of a is e. It has the special property that if $y = e^x$ then $\dfrac{dy}{dx} = e^x$.

The graph of $y = e^x$ looks like that in figure 3.2.

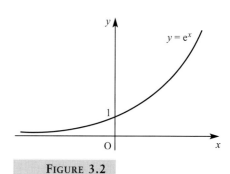

FIGURE 3.2

Note In some books e^x is written as $\exp(x)$.

If we reflect the graph of $y = e^x$ in the y-axis it looks like that shown in figure 3.3.

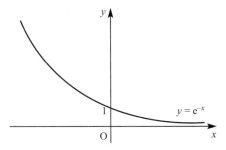

FIGURE 3.3

The equation of this graph is $y = e^{-x}$ and it describes *exponential decay*.

Historical Note

The symbol e is used after Leonhard Euler (1707–83). Euler is often thought of as a hero by mathematicians. He was highly productive throughout the eighteenth century. Although he was Swiss by birth he worked at St Peterburg's Academy in Russia. Despite becoming blind in 1766 he continued to work by dictation and he was noted for his phenomenal memory. Euler established the use of the symbols π, ∞, i (for $\sqrt{-1}$) and e. He combined some of these in the statement

$$e^{i\pi} + 1 = 0$$

which must be one of the most exciting equations imaginable.

e is rather like π in that it is an irrational number. It is therefore a never-ending decimal and you will see from your calculator that, to 8 decimal places, e^1 has the value

$$e = 2.182\,818\,28.$$

(On most calculators you will have to press the keys e^x then 1. Do not use the power key.)

You will notice that the graphs of $y = e^x$ and $y = e^{-x}$ are both asymptotic to the x-axis (that is they get closer and closer to the axis but do not quite reach it). Consequently if you translate the graph in the y-direction you should draw a dotted line to indicate the position of the asymptote.

EXAMPLE 3.1

Sketch the graph of $y = e^x - 1$.

Solution This is a translation of $y = e^x$ by $\begin{pmatrix} 0 \\ -1 \end{pmatrix}$. Since $y = e^x$ cuts the y-axis at 1 this graph will go through the origin when it is moved down one unit (see figure 3.4).

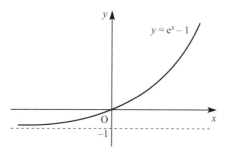

FIGURE 3.4

EXAMPLE 3.2

Sketch the graph of $y = 2 - e^{-x}$.

Solution From Chapter 1 you will recognise that this is a reflection of $y = e^{-x}$ in the x-axis followed by a translation $\begin{pmatrix} 0 \\ 2 \end{pmatrix}$. See figure 3.5.

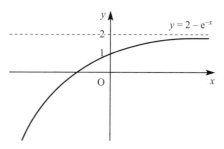

FIGURE 3.5

EXAMPLE 3.3

Sketch the graph of $y = e^{2x+4} - 3$.

Solution From Chapter 1 you will recognise this as a combination of several transformations.

Consider $y = e^x$. Replacing x by $2x + 4$ or $2(x + 2)$ gives a stretch scale factor $\frac{1}{2}$ parallel to the x-axis followed by a translation 2 to the left. The whole graph is then moved down 3 units. See figure 3.6.

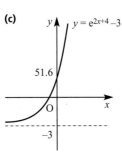

FIGURE 3.6

EXERCISE 3A

After you have sketched the graphs in this exercise you may wish to check them using a graphical calculator or a computer drawing package.

1 On separate diagrams sketch these curves.
   (a) $y = e^x$
   (b) $y = -e^x$
   (c) $y = e^{-x}$
   (d) $y = -e^{-x}$

2 On separate diagrams sketch the following curves, showing asymptotes by drawing dotted lines.
   (a) $y = e^x + 1$
   (b) $y = -e^x + 1$
   (c) $y = 2e^x - 1$
   (d) $y = e^{x+1}$

3 On separate diagrams sketch the following curves, showing asymptotes by drawing dotted lines.
   (a) $y = e^{-x} + 2$
   (b) $y = e^{-x} - 3$
   (c) $y = 2e^{-x} + 1$
   (d) $y = -e^{1-x}$

4 On separate diagrams sketch the following curves, showing asymptotes by drawing dotted lines.
   (a) $y = e^{x+1}$
   (b) $y = e^{x+1} - 2$
   (c) $y = e^{2x} + 1$
   (d) $y = e^{2x+1}$

5 On separate diagrams sketch the following curves, showing asymptotes by drawing dotted lines.
   (a) $y = e^{2x-4} + 3$
   (b) $y = e^{1-0.5x} + 1$
   (c) $y = e^{3x+6} - 2$
   (d) $y = 3 - e^{2-4x}$

The function ln$x$

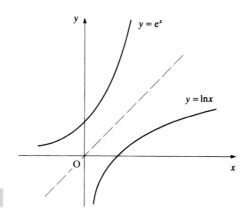

FIGURE 3.7

If the graph of $y = e^x$ is reflected in the line $y = x$ you obtain its inverse function as shown in figure 3.7.

The inverse function is the natural logarithm $y = \ln x$. Since the domain of $y = e^x$ was $x \in \mathbb{R}$ and the range was $y > 0$ then the domain of $y = \ln x$ is $x > 0$ with a range $y \in \mathbb{R}$.

Note On most calculators the functions e^x and ln$x$ are obtained using the same key.

ln is an abbreviation for the logarithm of x to the base e. This is written as

$$\ln x = \log_e x.$$

You will see the importance of natural logarithms when differentiating and integrating.

Look at the graph of $y = \ln x$ in figure 3.8.

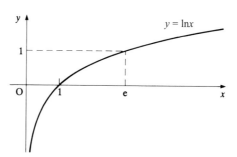

FIGURE 3.8

Note that:
- the curve crosses the x-axis at 1, where its gradient is 1
- it passes through the point (e, 1)
- it only exists for $x > 0$
- the y-axis is an asymptote
- there is no limit to the height of the curve for large values of x though the gradient progressively decreases.

The function $y = \ln x$ is also the inverse of $y = e^x$.

EXAMPLE 3.4

Sketch the graph of $y = \ln(x - 3)$.

Solution This is a translation of $y = \ln x$ by $\binom{3}{0}$. The graph is shown in figure 3.9.

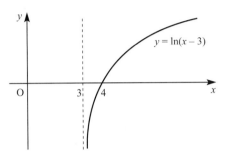

FIGURE 3.9

EXAMPLE 3.5

Sketch the graph of $y = \ln(2x + 4)$.

Solution Start with the graph of $y = \ln x$. Replacing x by $2x + 4$ or $2(x + 2)$ gives a stretch scale factor $\frac{1}{2}$ parallel to the x-axis followed by a translation 2 to the left. The result is shown in figure 3.10.

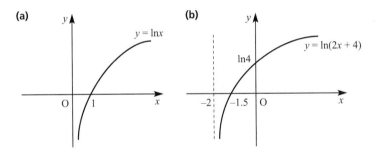

FIGURE 3.10

EXAMPLE 3.6

The function f(x) is defined by

$$f(x) \mapsto e^x + 2 \quad x \in \mathbb{R}$$

(a) Sketch the graph of $y = f(x)$.
(b) Sketch the graph of $y = f^{-1}(x)$.
(c) Find the inverse function $y = f^{-1}(x)$.
(d) State the domain of $f^{-1}(x)$.

EXERCISE 3B

Solution (a)
(b)

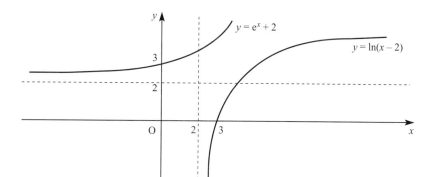

FIGURE 3.11

(c) The inverse function is $f^{-1}(x) = \ln(x - 2)$.
(d) The domain of $f^{-1}(x)$ is $x > 2$.

Historical note

Logarithms were discovered independently by John Napier (1550–1617), who lived at Merchiston Castle in Edinburgh, and Jolst Bürgi (1552–1632) from Switzerland. It is generally believed that Napier had the idea first, and so he is credited with their discovery. Natural logarithms are also called Naperian logarithms but there is no basis for this since Napier's logarithms were definitely not the same as natural logarithms. Napier was deeply involved in the political and religious events of his day and mathematics and science were little more than hobbies for him. He was a man of remarkable ingenuity and imagination and also drew plans for war chariots that look very like modern tanks, and for submarines.

EXERCISE 3B

After you have sketched the graphs in this exercise you may wish to check them using a graphical calculator or a computer drawing package.

1 On separate diagrams sketch these curves.
   (a) $y = \ln x$
   (b) $y = \ln(-x)$
   (c) $y = -\ln x$
   (d) $y = -\ln(-x)$

2 On separate diagrams sketch the following curves, showing asymptotes by drawing dotted lines.
   (a) $y = \ln(x - 2)$
   (b) $y = \ln(x + 3)$
   (c) $y = \ln(1 - x)$
   (d) $y = 2 + \ln x$

3 On separate diagrams sketch the following curves, showing asymptotes by drawing dotted lines.
   (a) $y = \ln(2x - 2)$
   (b) $y = \ln(2x + 3)$
   (c) $y = -\ln(1 - 2x)$
   (d) $y = 2 - \ln(0.5x - 1)$

4 The function f(x) is defined by

$$f(x) = e^{(x-1)} + 3, x \in \mathbb{R}.$$

(a) Sketch the graph of $y = f(x)$.
(b) Sketch the graph of $y = f^{-1}(x)$.
(c) Find the inverse function $f^{-1}(x)$.
(d) State the domain of $f^{-1}(x)$.

5 The function f(x) is defined by

$$f(x) = \ln(2x - 6), x > 3.$$

(a) Sketch the graph of $y = f(x)$.
(b) Sketch the graph of $y = f^{-1}(x)$.
(c) Find the inverse function $f^{-1}(x)$.
(d) State the domain and range of $f^{-1}(x)$.

EQUATIONS OF THE FORM $e^{ax+b} = p$ AND $\ln(ax+b) = q$

You have seen that the functions $y = e^x$ and $y = \ln x$ are inverses of each other. So you use this fact to solve equations of the form $e^x = p$ and $\ln x = q$.

The proof is that:

since $f(f^{-1}(x)) = x$ then $\ln(e^x) = x$ and $e^{\ln x} = x$

and so
 if $e^x = p$ then taking natural logarithms of both sides gives

$$\ln(e^x) = \ln p$$
$$\therefore x = \ln p$$

and if $\ln x = q$ taking the exponential of both sides gives

$$e^{\ln x} = e^q$$
$$\therefore x = e^q.$$

EXAMPLE 3.7

Solve $e^x = 5$.

Solution Taking natural logarithms of both sides of $e^x = 5$ gives

$$x = \ln 5$$

and the answer may be left in this exact form or, using a calculator, could be given to, say, 4 significant figures as

$$x = 1.609.$$

EXAMPLE 3.8

Solve $e^{2x+1} = 200$.

Solution Taking natural logarithms of both sides gives

$\ln(e^{2x+1}) = \ln 200$

$\therefore 2x + 1 = \ln 200$

$\therefore x = \frac{1}{2}(\ln 200 - 1) = 2.149$ (4 s.f.)

EXAMPLE 3.9

Solve $\ln x = 5$.

Solution Taking the exponential of both sides of $\ln x = 5$ gives

$x = e^5$

and this answer may be left in this form or approximated using your calculator, again to 4 significant figures as

$x = 148.4$.

Reminder: key in e^x followed by 5. Do not use the power key

EXAMPLE 3.10

Solve $\ln(2x+1) = 4$.

Solution Taking the exponential of both sides gives

$e^{\ln(2x+1)} = e^4$

$\therefore 2x + 1 = e^4$

$\therefore x = \frac{1}{2}(e^4 - 1) = 26.8$ (3 s.f.)

EXAMPLE 3.11

The number, N, of insects in a colony is given by $N = 2000e^{0.1t}$ where t is the number of days after observations have begun.
(a) Sketch the graph of N against t.
(b) What is the population of the colony after 20 days?
(c) How long does it take the colony to reach a population of 10 000?

Solution (a)

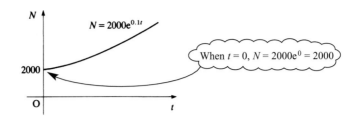

FIGURE 3.12

(b) When $t = 20$, $\qquad N = 2000e^{0.1 \times 20}$

$\qquad\qquad\qquad\qquad\quad = 14\,778.$

The population is $14\,778$ insects.

(c) When $N = 10\,000$, $\;10\,000 = 2000e^{0.1t}$

$\qquad\qquad\qquad\qquad 5 = e^{0.1t}.$

Taking natural logarithms of both sides,

$$\ln 5 = \ln(e^{0.1t})$$
$$= 0.1t$$

Remember $\ln(e^x) = x$

and so $\qquad\qquad t = 10\ln 5$

$\qquad\qquad\qquad t = 16.09\ldots$

It takes just over 16 days for the population to reach $10\,000$.

EXAMPLE 3.12

The radioactive mass, M g in a lump of material is given by $M = 25e^{-0.0012t}$ where t is the time in seconds since the first observation.

(a) Sketch the graph of M against t.
(b) What is the initial size of the mass?
(c) What is the mass after 1 hour?
(d) The half-life of a radioactive substance is the time it takes to decay to half of its mass. What is the half-life of this material?

Solution (a)

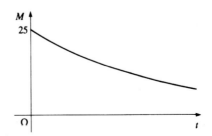

FIGURE 3.13

EXERCISE 3C

(b) When $t = 0$, $M = 25e^0 = 25$.

The initial mass is 25 g.

(c) After one hour, $t = 3600$ $M = 25e^{-0.0012 \times 3600}$.

The mass after one hour is 0.33 g (to 2 decimal places).

(d) The initial mass is 25 g, so after one half-life

$$M = \tfrac{1}{2} \times 25 = 12.5 \text{ g}.$$

At this point the value of t is given by

$$12.5 = 25e^{-0.0012t}.$$

Dividing both sides by 25 gives

$$0.5 = e^{-0.0012t}.$$

Taking logarithms of both sides gives

$$\ln 0.5 = \ln e^{-0.0012t} = -0.0012t$$

$$\Rightarrow \quad t = \frac{\ln 0.5}{-0.0012} = 557.6 \text{ (to 1 decimal place)}.$$

The half-life is 577.6 seconds. (This is just under 10 minutes, so the substance is highly radioactive.)

EXERCISE 3C

1 Solve the following equations giving your answers to 3 significant figures:
   (a) $e^x = 2$
   (b) $e^{(x-1)} = 8$
   (c) $e^{(x+3)} = 100$
   (d) $50e^x = 1$
   (e) $100e^x = 2 \times 10^6$
   (f) $0.5e^{-x} = 10$
   (g) $e^{(2-x)} = 0.5$
   (h) $2.5e^{(3-2x)} = 4$
   (i) $0.1e^{0.2x} = 7$
   (j) $e^{5x} = 2e^{2x}$

2 Solve the following equations giving your answers to 3 significant figures:
   (a) $\ln x = 2$
   (b) $\ln(x - 3) = 0.3$
   (c) $\ln(x + 5) = 5$
   (d) $\ln(2x - 1) = 1.5$
   (e) $\ln x = 0.04$
   (f) $\ln(1 - x) = 0.1$
   (g) $25 \ln x = 8$
   (h) $2 \ln x = 0.75$
   (i) $4 \ln(0.2x - 1) = 9$
   (j) $3 \ln(2 - 3x) = 0.5$

3 The number of widgets after time t, in hours, is given by N where

$$N = 100 + 10 \ln(5t + 1), \ t \geq 0$$

   (a) Initally how many widgets are there?
   (b) How many widgets are there after 24 hours?
   (c) How long does it take for the number of widgets to reach 150?

4 A function f is defined as $f(x) = \ln(ax + b)$.
It is given that $f(5) = 1.5$ and $f(7) = 1.7$.
   (a) Find a and b to 2 decimal places
   (b) Sketch the graph of $y = f(x)$

5 A colony of humans settles on a previously uninhabited planet. After t years, their population, P, is given by

$$P = 100e^{0.05t}.$$

   (a) Sketch the graph of P against t.
   (b) How many settlers land on the planet initially?
   (c) What is the population after 50 years?
   (d) How long does it take the population to reach 1 million?

6 Ela sits on a swing. Her father pulls it back and then releases it. The swing returns to its maximum backwards displacement once every 5 seconds, but the maximum displacement, $\theta°$, becomes progressively smaller because of friction. At time t seconds, θ is given by

$$\theta = 25e^{-0.03t} \;(t = 0, 5, 10, 15, \ldots).$$

   (a) Plot the values of θ for $0 \leq t \leq 30$ on graph paper.
   (b) To what angle did Ela's father pull the swing?
   (c) What is the value of θ after one minute?
   (d) After how many swings is the angle θ less than 1°?

7 Alexander lives 800 metres from school. One morning he sets out at 8.00 am and t minutes later the distance s m, which he has walked is given by

$$s = 800\,(1 - e^{-0.1t}).$$

   (a) Sketch the graph of s against t.
   (b) How far has Alexander walked by 8.15 am?
   (c) What time is it when Alexander is half-way to school?
   (d) When does Alexander get to school?

8 A parachutist jumps out of an aircraft and some time later opens the parachute. His speed at time t seconds from when the parachute opens is v m s^{-1}. It is given by

$$v = 8 + 22e^{-0.07t}.$$

   (a) Sketch the graph of v against t.
   (b) State the speed of the parachutist when the parachute opens, and the final speed that he would attain if he jumped from a very great height.
   (c) Find the value of v as the parachutist lands, 60 seconds later.
   (d) Find the value of t when the parachutist is travelling at 20 m s^{-1}.

9 A bacterium *Mathematicus estus funius* grows such that its population P at time t is given by

$$P = 250e^{0.2t}.$$

(a) What is the initial population (i.e. when $t = 0$)?
(b) Find the value of t when the population is four times the initial population.
(c) This is an example of exponential growth. Explain what is meant by exponential decay and sketch a graph to illustrate your answer.

10 The manufacturers of the cream Acno claim that the bacterium *Spoticus youthus* has its population halved within five days of its use. In his laboratory Professor Smiff finds that the number of bacteria N is connected by the number of days t by the model

$$N = 5000e^{-0.15t}.$$

Are the manufacturers accurate in their claim?

According to Professor Smiff's model
(a) how many *Spoticus youthus* bacteria will there be after 10 days?
(b) After how many days will there only be 100 *Spoticus youthus* bacteria left?

Exercise 3D   Examination-style questions

1 The population S of a sample of buglettes is given by the model

$$S = 6000e^{0.01t}$$

where t is the time in hours for which they have been growing.
(a) What is the initial population?
(b) Calculate the time to 3 significant figures when the population is three times its initial size.

A different strain of buglette is found to double its population P every 15 hours. If its initial population is P_0 find an equation connecting P, P_0 and t (the time in hours). Hence, or otherwise, find the number of hours it takes for its population to increase by a factor of 100.

2 (a) On the same diagram sketch the graphs of

$$y = \ln x \quad \text{and} \quad y = \ln(x - 3).$$

Describe the transformation that maps the first graph onto the second.

(b) $y = e^x$ is translated by moving it 2 units to the left. Write down the equation of the image.
State an alternative transformation which maps $y = e^x$ on to the image.

3 The number N of sproglets at time t (in weeks) is given by the equation

$$N = \frac{1300}{5 + 8e^{-0.2t}}.$$

(a) Explain why the graph of this function (shown below) indicates that the growth of sproglets is not exponential?
(b) How many sproglets are there after 5 weeks?
(c) After how many weeks are there 200 sproglets?
(d) What is the largest the population of sproglets can be?

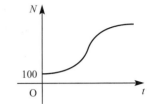

4 The function f is defined by f: $x \mapsto e^x + k$, $x \in \mathbb{R}$ and k is a positive constant.
(a) State the range of f.
(b) Find $f(\ln k)$, simplifying your answer.
(c) Find f^{-1}, the inverse function of f, in the form $f^{-1}: x \mapsto ...$, stating its domain.
(d) On the same axes, sketch the curves with equations $y = f(x)$, and $y = f^{-1}(x)$, giving the coordinates of all points where the graphs cut the axes.

[Edexcel]

5 The points P and Q lie on the curve with equation $y = e^{\frac{1}{2}x}$. The x coordinates of P and Q are ln4 and ln16 respectively.
(a) Find an equation of the line PQ.
(b) Show that this line passes through the origin O.
(c) Calculate the length, to 3 significant figures, of the line segment PQ.

[Edexcel]

6 A formula used to calculate the power gain of an amplifier has the form

$$G = h\ln\left(\frac{p_2}{p_1}\right)$$

Given that $G = 16$, $h = 4.3$ and $p_1 = 8$,
(a) calculate, to the nearest whole number, the value of p_2.

Given that the values of G and p_1 are exact but that the value of h has been given to 1 decimal place,
(b) find the range of possible values of p_2.

[Edexcel]

7 You are given that f: $x \to \ln(x + 1)$.
Sketch, on separate diagrams, the graphs of
(a) $y = f(x)$
(b) $y = f(x) + 1$
(c) $y = |f(x)|$
(d) $y = f(|x|)$.

Exercise 3D

8 The function f is defined as

$$f: x \to \tfrac{1}{4}\ln(3x + 6).$$

  (a) Sketch the graph of $y = f(x)$ stating its domain and range.
  (b) Is $f(x)$ a one-to-one function?
  (c) Find $f^{-1}(x)$.
  (d) Sketch the graph of $y = f^{-1}(x)$ stating its domain and range.

9 You are given that $f(x) = 5 + e^{2-x}$.

  (a) Find, to three significant figures,
   (i) the value of $f(0)$
   (ii) x if $f(x) = 5.5$.
  (b) Sketch the graph of $y = f(x)$ stating its domain and range.
  (c) Find $f^{-1}(x)$.
  (d) Sketch the graph of $y = f^{-1}(x)$ stating its domain and range.

10 Given that $f: x \to 3e^{2x}, x \in \mathbb{R}$, and $g: x \to x - 2, x \in \mathbb{R}$, sketch, on separate graphs

  (a) $y = f(x)$ (b) $y = f^{-1}(x)$
  (c) $y = gf(x)$ (d) $y = fg(x)$.

11 The functions f and g are defined as

$$f: x \to 5e^{0.5x-6}, x \in \mathbb{R} \text{ and } g: x \to x - 2, x \in \mathbb{R}.$$

Solve, to three significant figures,

  (a) $gf(x) = 7$
  (b) $fg(x) = 7$.

12 The temperature θ of a room in degrees Celsius is given by the equation

$$\theta = 25 + 5e^{-0.1t}$$

where t is the time in hours.
  (a) What is the initial temperature of the room?
  (b) What is the temperature of the room after a long period of time?
  (c) When is the temperature of the room $27\,°C$?
  (d) Sketch the graph of temperature against time.

13 You are given that

$$f(x) = \ln(ax + b), \text{ where } a \text{ and } b \text{ are integers.}$$

  (a) Find a and b if $f(3) = 2.3$ and $f(8) = 3.4$.
  (b) Solve $f(x) = 5$, giving your answer to 3 significant figures.
  (c) Sketch the graph of $y = f(x)$, stating its domain and range.
  (d) On the same diagram sketch the graph of $y = f^{-1}(x)$.
  (e) Find $f^{-1}(x)$.

14 The population P of a colony of ants is given by

$$P = 2000e^{0.1t-k}$$

where t is the time in days.
(a) If the initial population is 1000 ants show that k is approximately 0.693.
(b) What is the number of ants after 5 days?
(c) On which day does the population reach 4000 ants?
(d) Sketch a graph of population against time.
(e) Explain whether or not this is a realistic model.

15 Solve, to three significant figures, these simultaneous equations.

$$x = e^{2y-3}$$
$$y = \ln(4x - 1)$$

KEY POINTS

1 $y = e^x$

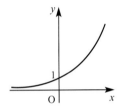

$y = e^{-x}$

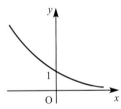

$y = \ln x$

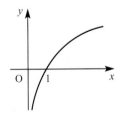

e^x and $\ln x$ are inverse functions
$e^{\ln x} = x$
$\ln(e^x) = x$

2 **Equations**

$e^{ax+b} = p \Rightarrow ax + b = \ln p$
$\ln(ax + b) = q \Rightarrow ax + b = e^q$

Chapter four

DIFFERENTIATION

Little by little does the trick.

Aesop

You will recall from *Pure Mathematics: Core 1* that the derivative of a function $y = f(x)$ is the gradient of the function and is written as

$$\frac{dy}{dx} \text{ or } f'(x).$$

In particular you learnt the rule that

$$\text{if } y = x^n \text{ then } \frac{dy}{dx} = nx^{n-1}.$$

You also learnt that the second derivative is found by differentiating the first derivative and is written as

$$\frac{d^2y}{dx^2} \text{ or } f''(x).$$

Since the derivative at a point is the gradient of the tangent at that point you could find the equation of the tangent to a curve using $y - y_1 = m(x - x_1)$ and you could also find the equation of the normal. The normal is the line at right angles to the tangent. If the tangent has gradient m then the normal has gradient $-\frac{1}{m}$.

In Chapter 11 of *Pure Mathematics: Core 1 and Core 2* you also calculated local maximum and minimum turning points. At a maximum turning point (figure 4.1), the gradient $\frac{dy}{dx} = 0$ and the gradient on the left of the turning point is positive and on the right negative. This means that the gradient must be decreasing so $\frac{d^2y}{dx^2} < 0$ and the turning point is a maximum one.

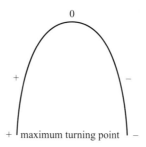

FIGURE 4.1

Similarly, if (as in figure 4.2), $\frac{dy}{dx} = 0$ and $\frac{d^2y}{dx^2} > 0$ the turning point is a minimum one.

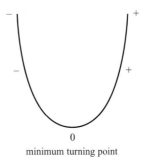

FIGURE 4.2

minimum turning point

A point of inflection (figure 4.3) is where the curve changes direction from being concave to convex, or vice versa. At such a point $\frac{d^2y}{dx^2} = 0$.

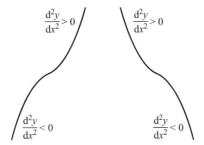

FIGURE 4.3

If, in addition, $\frac{dy}{dx} = 0$ (figure 4.4) it is a stationary point of inflection.

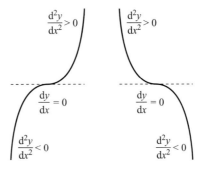

FIGURE 4.4

You will recall that if $\frac{dy}{dx} = 0$ and $\frac{d^2y}{dx^2} = 0$ it is not clear whether you have a turning point or a point of inflection. In such cases you need to check the values of $\frac{dy}{dx}$ on either side of the stationary point to determine its nature.

DIFFERENTIATION

EXAMPLE 4.1

A curve has equation $y = 4x^3 - 3x - 1$.

(a) Find the positions of the turning points.
(b) Sketch the curve.
(c) Find the equation of the normal at the point where $x = 1$.

Solution (a) $y = 4x^3 - 3x - 1$

Differentiate to give $\dfrac{dy}{dx} = 12x^2 - 3$.

For a turning point $\dfrac{dy}{dx} = 0$, therefore

$$12x^2 - 3 = 0$$
$$4x^2 = 1$$
$$x = \pm\tfrac{1}{2}.$$

When $x = \tfrac{1}{2}$, $y = -3$. Or $\dfrac{d^2y}{dx^2} = 24x > 0$ when $x = \tfrac{1}{2}$ so there is a minimum turning point.

x	0	$\tfrac{1}{2}$	1
$\dfrac{dy}{dx}$	−	0	+

So there is a minimum turning point at $(\tfrac{1}{2}, -3)$.

When $x = -\tfrac{1}{2}$, $y = 0$. Or $\dfrac{d^2y}{dx^2} = 24x < 0$ when $x = -\tfrac{1}{2}$ so there is a maximum turning point.

x	−1	$-\tfrac{1}{2}$	0
$\dfrac{dy}{dx}$	+	0	−

So there is a maximum turning point at $(-\tfrac{1}{2}, 0)$.

(b) When $x = 0$, $y = -1$.

When $x = -\tfrac{1}{2}$, $y = 0$, so $(2x + 1)$ is a factor of $4x^3 - 3x - 1$.

Factorising completely,

$$4x^3 - 3x - 1 = (2x + 1)(2x^2 - x - 1)$$
$$= (2x + 1)(2x + 1)(x - 1).$$

So when $y = 0$, $x = -\tfrac{1}{2}$ (twice) and $x = 1$.

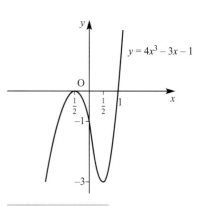

FIGURE 4.5

(c) Since $\frac{dy}{dx} = 12x^2 - 3$, at the point where $x = 1$, $\frac{dy}{dx} = 9$ and $y = 0$.

The gradient of the normal is $-\frac{1}{9}$.

Using $y - y_1 = m(x - x_1)$, the equation of the normal is

$$y - 0 = -\tfrac{1}{9}(x - 1)$$
$$9y = -x + 1$$
$$x + 9y = 1.$$

The next sections will deal with the derivatives of e^x and $\ln x$.

DERIVATIVE OF e^x

In Chapter 3 we defined the curve $y = e^x$ to be one for which the gradient at any point on the curve, say (x_1, y_1), is equal to the y value at that point. The graph of $y = e^x$ is shown in figure 4.6 with a tangent drawn at the point (x_1, y_1).

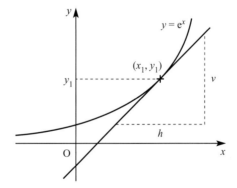

FIGURE 4.6

From figure 4.6 you can see that the gradient of the tangent is $\underline{v} = y_1 = e^{x_1}$. That is, the gradient of e^x at x_1 is e^{x_1}. So, in general, the curve $y = e^x$ has gradient or derivative function $\frac{dy}{dx} = e^x$, i.e.

$$y = e^x \Rightarrow \frac{dy}{dx} = e^x.$$

If the curve is stretched by a constant scale factor k parallel to the y-axis, with the x-axis invariant, then both the y-value and the gradient at each point will be multiplied by a factor of k, so that

$$y = ke^x \Rightarrow \frac{dy}{dx} = ke^x.$$

Note The derivative of $y = e^{kx}$ and the derivative of $y = e^{ax+b}$ will be studied later in this chapter.

EXAMPLE 4.2

Find the derivative of $y = 5e^x$.

Solution   Using $y = ke^x \Rightarrow \dfrac{dy}{dx} = ke^x$ means that

if $\quad y = 5e^x$

then $\dfrac{dy}{dx} = 5e^x$.

EXAMPLE 4.3

Show that the gradient of $e^x + \sqrt{x}$ at the point where $x = 1$ is $\tfrac{1}{2}(2e + 1)$.

Solution   Let $y = e^x + \sqrt{x}$

then $\quad\quad\quad\quad y = e^x + x^{\frac{1}{2}}$.

Differentiating $\quad \dfrac{dy}{dx} = e^x + \dfrac{1}{2}x^{-\frac{1}{2}} = e^x + \dfrac{1}{2}\dfrac{1}{\sqrt{x}}$.

At the point where $x = 1$

$$\dfrac{dy}{dx} = e^1 + \dfrac{1}{2}\dfrac{1}{\sqrt{1}} = e + \dfrac{1}{2} = \dfrac{2e}{2} + \dfrac{1}{2}$$

$$= \dfrac{1}{2}(2e + 1).$$

DERIVATIVE OF ln*x*

You will recall from Chapter 3 that ln*x* is the inverse of e^x. So if you take the graph of $y = e^x$ and reflect it in the line $y = x$ you will have the graph of $y = \ln x$ as shown on the left in figure 4.7. The tangent to the curve at the general point (x_1, y_1) is also shown on the graph. Its gradient is $\dfrac{v}{h}$.

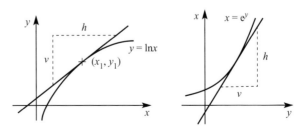

FIGURE 4.7

From figure 4.7 the gradient of the tangent to $y = \ln x$ is $\dfrac{v}{h}$. If you now reflect the whole of the graph in $y = x$ you obtain the graph shown on the right in figure 4.7. As the *x*- and *y*-axes have been reflected the equation of this graph is $x = e^y$. The gradient of the tangent to $x = e^y$ is $\frac{h}{v} = e^y = x$. Combining these results gives the gradient of $y = \ln x$ as $\frac{v}{h} = \frac{1}{x}$, leading to the general result

$$y = \ln x \Rightarrow \frac{dy}{dx} = \frac{1}{x}.$$

If $y = \ln(kx)$, where k is a constant, we use the laws of logarithms to write this expression as $y = \ln k + \ln x$. Since k is a constant the logarithm $\ln k$ is also a constant and when it is differentiated the result is 0. Hence

$$y = \ln(kx) \Rightarrow \frac{dy}{dx} = \frac{1}{x}.$$

You can also use the laws of logarithms to differentiate $y = \ln x^n$:

$$y = \ln x^n = n \ln x \Rightarrow \frac{dy}{dx} = n \times \frac{1}{x} = \frac{n}{x}.$$

Note

You will learn how to differentiate functions of the type $\ln(ax + b)$ later in this chapter.

EXAMPLE 4.4

Differentiate $y = \ln(5x)$ with respect to x.

Solution Using the laws of logarithms

$$y = \ln(5x) = \ln 5 + \ln x$$

then

$$\frac{dy}{dx} = \frac{1}{x}.$$

> ln5 is a constant so its deriative is 0

EXAMPLE 4.5

Find the gradient of $\ln x^2$ at the point where $x = \frac{1}{4}$.

Solution Using the laws of logarithms

$$y = \ln x^2 = 2\ln x$$

which gives

$$\frac{dy}{dx} = 2 \times \frac{1}{x} = \frac{2}{x}.$$

At the point where $x = \frac{1}{4}$ the gradient is $\frac{dy}{dx} = \frac{2}{\frac{1}{4}} = 8.$

Exercise 4A

1. Differentiate the following with respect to x.
   - (a) e^x
   - (b) $2e^x$
   - (c) $4 - 3e^x$
   - (d) $2x + 8e^x$
   - (e) $0.1e^x - 0.5x^2$
   - (f) $\sqrt{x} - \tfrac{1}{2}e^x$

2. Differentiate the following with respect to x.
   - (a) $\ln x$
   - (b) $2\ln x$
   - (c) $\ln(4x)$
   - (d) $\ln x^4$
   - (e) $0.8 \ln x^5$
   - (f) $\ln(\tfrac{1}{x})$
   - (g) $2\ln(\tfrac{3}{x})$
   - (h) $6\ln(\tfrac{3}{x})$
   - (i) $x - \ln x$
   - (j) $\tfrac{2}{x} - 4\ln(\tfrac{1}{x})$

3. Find the exact values of the gradients of the following at the given values of x.
   - (a) $e^x + \ln x$ $x = 1$
   - (b) $3\ln x - 4e^x$ $x = 2$
   - (c) $\ln x^3 + 2e^x + x$ $x = 1$
   - (d) $\dfrac{2}{x} - 4\ln(\sqrt{x}) + 2e^x$ $x = 4$
   - (e) $\ln\dfrac{3}{x} + \dfrac{4}{x^3} - 5e^x$ $x = 2$
   - (f) $\ln(3xe^x)$ $x = \tfrac{1}{2}$

4. Prove that if $y = \sqrt{x} + \ln(\sqrt{x})$ then $\dfrac{dy}{dx} = \dfrac{x + \sqrt{x}}{2x\sqrt{x}}$.

5. The population P of the agents of a virus is found to be proportional to e^t, where t is the time measured in years. The initial population is 1000.
   - (a) Find the population, to 3 significant figures, after 10 days.
   - (b) What was the initial rate of growth of the population?

6. Find the equation of the tangent to the curve
   $$y = 2 + 4\ln x$$
   at the point where $x = 1$.

7. Given that
   $$y = e^x - 4$$
   - (a) Find the value of x at the point where $y = 0$.
   - (b) Find the gradient at this point.
   - (c) Find the equation of the tangent at this point.

8 (a) On the same diagram sketch $y = \ln x$ and $y = \ln(x - 2)$ stating the geometric relationship between the two curves.

(b) Using your sketches explain why
$$\frac{d(\ln(x - 2))}{dx} = \frac{1}{x - 2}.$$

(c) The equation of a curve is $y = \ln \frac{x - 2}{x^2}$. Using the results given in part **(b)** and the laws of logarithms prove that
$$\frac{dy}{dx} = \frac{4 - x}{x(x - 2)}.$$

(d) Find the equation of the tangent at the point where $x = 3$.

9 Using the result that $a^{m+n} = a^m \times a^n$ prove that
$$\frac{d(e^{x+2})}{dx} = e^{x+2}.$$

Hence show that the equation of the normal to
$$y = e^{x+2} + x^2$$
at the point where $x = -1$ is
$$x + (e - 2)y = e^2 - e - 3.$$

10 Show that the tangent to the curve
$$y = 2 - \ln\left(\tfrac{x}{2}\right)$$
at the point where $x = 2$ cuts the x-axis at the point $(6, 0)$.

11 Show that the area of the triangle enclosed between the x-axis, the y-axis and the tangent to $y = e^x + 1$ at the point where $x = -1$ is
$$\tfrac{e}{2} + 2 + \tfrac{2}{e}.$$

12 Show that the tangent to $y = e^x - \ln(3x)$ at the point where $x = 2$ cuts the y-axis at $1 - e^2 - \ln 6$.

You have learnt that

$$\text{if } y = e^x \quad \text{then} \quad \frac{dy}{dx} = e^x$$

and if $y = \ln x$ then $\dfrac{dy}{dx} = \dfrac{1}{x}$.

However you have not learnt how to differentiate functions of the type

$\sqrt{x^2 + 1}$	function of a function
or xe^x	product
or $\dfrac{x + 1}{x^2 - 1}$	quotient

other than for simple cases where multiplying or dividing out led to a function that could be differentiated using the few rules you have already learnt.

The next sections will deal with derivatives of these types of functions.

THE CHAIN RULE

How would you differentiate an expression like

$$y = \sqrt{x^2 + 1}?$$

Your first thought may be to write it as $y = (x^2 + 1)^{\frac{1}{2}}$ and then get rid of the brackets, but that is not possible in this case because the power $\frac{1}{2}$ is not a positive integer. Instead you need to think of the expression as a composite function, a 'function of a function'.

In this case we use the substitution, $\quad u = x^2 + 1$
so that $\quad\quad\quad\quad\quad\quad\quad\quad\quad\quad\quad y = \sqrt{u} = u^{\frac{1}{2}}.$

This is now in a form which you can differentiate using the *chain rule*.

DIFFERENTIATING A COMPOSITE FUNCTION

To find $\frac{dy}{dx}$ for a function of a function, you consider the effect of a small change in x on the two variables, y and u, as follows. A small change δx in x leads to a small change δu in u and a corresponding small change δy in y, and by simple algebra,

$$\frac{\delta y}{\delta x} = \frac{\delta y}{\delta u} \times \frac{\delta u}{\delta x}.$$

In the limit, as $\delta x \to 0$, $\delta u \to 0$ and

$$\frac{\delta y}{\delta x} \to \frac{dy}{dx}, \quad \frac{\delta y}{\delta u} \to \frac{dy}{du} \quad \text{and} \quad \frac{\delta u}{\delta x} \to \frac{du}{dx}.$$

The relationship becomes

$$\boxed{\frac{dy}{dx} = \frac{dy}{du} \times \frac{du}{dx}.}$$

This is known as the *chain rule*.

EXAMPLE 4.6

Differentiate $y = (x^2 + 1)^{\frac{1}{2}}$.

Solution As you saw earlier, you can break down this expression as follows:

$$u = x^2 + 1 \Rightarrow y = u^{\frac{1}{2}}.$$

Differentiating these gives

$$\frac{dy}{du} = \frac{1}{2}u^{-\frac{1}{2}} = \frac{1}{2\sqrt{x^2+1}} \quad \text{and} \quad \frac{dy}{du} = 2x.$$

By the chain rule

$$\frac{dy}{dx} = \frac{dy}{du} \times \frac{du}{dx}$$

$$= \frac{1}{2\sqrt{x^2+1}} \times 2x$$

$$= \frac{x}{\sqrt{x^2+1}}.$$

Note The answer must be given in terms of the same variables as the question, in this case x and y. The variable u was our invention and so should not appear in the answer.

You can see that effectively you have made a substitution, in this case $u = x^2 + 1$. This transformed the problem into one that could easily be solved.

Note Notice that the substitution gave you two functions that you could differentiate. Some substitutions would not have worked. For example, the substitution $u = x^2$ would give you

$$y = (u+1)^{\frac{1}{2}}.$$

You would still not be able to differentiate y, so you would have gained nothing.

EXAMPLE 4.7

Use the chain rule to find $\frac{dy}{dx}$ when $y = (x^2 - 2)^6$.

Solution Let $u = x^2 - 2$, then $y = u^6$.

$$\frac{du}{dx} = 2x \quad \text{and} \quad \frac{dy}{du} = 6u^5$$
$$= 6(x^2 - 2)^5.$$

$$\frac{dy}{dx} = \frac{dy}{du} \times \frac{du}{dx}$$
$$= 6(x^2 - 2)^5 \times 2x$$
$$= 12x(x^2 - 2)^5.$$

You could verify this answer by expanding $(x^2 - 2)^6$, differentiating the expansion and simplifying your answer!

EXAMPLE 4.8

Use the chain rule to find $\frac{dy}{dx}$ when $y = \frac{5}{x^3 + 1}$.

Solution Let $u = x^3 + 1$, then $y = \frac{5}{u} = 5u^{-1}$.

Using the chain rule

$$\frac{dy}{dx} = \frac{dy}{du} \times \frac{du}{dx}$$
$$= -5u^{-2} \times 3x^2$$
$$= -15x^2(x^3 + 1)^{-2}$$
$$= \frac{-15x^2}{(x^3 + 1)^2}$$

With practice you may find that you can do some stages of questions like this in your head, and just write down the answer. If you have any doubt, however, you should write down the full method.

You will now see particular applications of the chain rule to help you learn how to write down the derivatives of some common functions.

DERIVATIVE OF $(ax + b)^n$

Let $y = (ax + b)^n$ and put $u = ax + b$ so that $y = u^n$. Then differentiating gives

$$\frac{dy}{du} = nu^{n-1} = n(ax + b)^{n-1} \quad \text{and} \quad \frac{du}{dx} = a.$$

By the chain rule

$$\frac{dy}{dx} = \frac{dy}{du} \times \frac{du}{dx} = n(ax + b)^{n-1} \times a.$$

That is

$$y = (ax + b)^n \implies \frac{dy}{dx} = an(ax + b)^{n-1}.$$

EXAMPLE 4.9

Differentiate $(2x + 3)^4$.

Solution Let $y = (2x + 3)^4$.

Using the chain rule

$$\frac{dy}{dx} = 2 \times 4(2x + 3)^3 = 8(2x + 3)^3.$$

EXAMPLE 4.10

Differentiate $\dfrac{3}{\sqrt{4x+1}}$.

Solution Put $y = \dfrac{3}{\sqrt{4x+1}}$ and rewrite it in the form

$$y = 3(4x+1)^{-\frac{1}{2}}.$$

Differentiating gives

$$\frac{dy}{dx} = 4 \times -\tfrac{1}{2} \times 3(4x+1)^{-\frac{3}{2}} = -6(4x+1)^{-\frac{3}{2}}.$$

DERIVATIVE OF e^{ax+b}

Given that $y = e^{ax+b}$, break down the expression as follows:

$$y = e^u, \quad u = ax+b.$$

Differentiating these gives

$$\frac{dy}{du} = e^u = e^{ax+b} \quad \text{and} \quad \frac{du}{dx} = a.$$

By the chain rule

$$\frac{dy}{dx} = \frac{dy}{du} \times \frac{du}{dx} = e^{ax+b} \times a.$$

That is

$$y = e^{ax+b} \implies \frac{dy}{dx} = ae^{ax+b}.$$

DERIVATIVE OF $e^{f(x)}$

Putting $y = e^{f(x)}$ and writing $y = e^u$ where $u = f(x)$, you have

$$\frac{dy}{du} = e^u = e^{f(x)} \quad \text{and} \quad \frac{du}{dx} = f'(x).$$

The chain rule then gives

$$y = e^{f(x)} \implies \frac{dy}{dx} = f'(x)e^{f(x)}.$$

Note

If you have a function of the type $e^{f(x)}$ you do not need to use the chain rule to find its derivative. Using the rule we have just found you can write down its derivative straight away.

EXAMPLE 4.11

Find the exact value of x at the point on the curve $y = e^{2-3x}$ for which the gradient is -6.

Solution

$y = e^{2-3x}$

Using $y = e^{f(x)} \Rightarrow \dfrac{dy}{dx} = f'(x)e^{f(x)}$ you can write down the derivative, that is

$\dfrac{dy}{dx} = -3e^{2-3x}.$

At the point where the gradient $\dfrac{dy}{dx} = -6$:

$-3e^{2-3x} = -6$
$e^{2-3x} = 2$
$2 - 3x = \ln 2$
$x = \tfrac{1}{3}(2 - \ln 2).$

> Remember that e^x and $\ln x$ are inverse functions

DERIVATIVE OF $\ln(f(x))$

Let $y = \ln(f(x))$ and put $y = \ln u$ where $u = f(x)$.

Differentiating gives

$\dfrac{dy}{du} = \dfrac{1}{u}$ and $\dfrac{du}{dx} = f'(x).$

Using the chain rule

$\dfrac{dy}{dx} = \dfrac{dy}{du} \times \dfrac{du}{dx} = \dfrac{1}{f(x)} \times f'(x) = \dfrac{f'(x)}{f(x)}.$

That is

$$y = \ln(f(x)) \quad \Rightarrow \quad \dfrac{dy}{dx} = \dfrac{f'(x)}{f(x)}.$$

In words, the derivative of the logarithm of a function is found by putting the function on the bottom of the fraction and its derivative on the top. Again, you may write down derivatives of this type without using the chain rule explicitly.

EXAMPLE 4.12

Find the equation of the tangent to $y = \ln(3x - 5)$ at the point where $x = 2$.

Solution

$y = \ln(3x - 5) \quad \Rightarrow \quad \dfrac{dy}{dx} = \dfrac{3}{3x - 5}.$

At the point where $x = 2$,

$$\frac{dy}{dx} = \frac{3}{3 \times 2 - 5} = 3 \quad \text{and} \quad y = \ln 1 = 0.$$

Using $y - y_1 = m(x - x_1)$, the equation of the tangent is

$$y - 0 = 3(x - 2)$$
$$y = 3x - 6.$$

DIFFERENTIATION WITH RESPECT TO DIFFERENT VARIABLES

The chain rule makes it possible to differentiate with respect to a variable which does not feature in the original expression. For example, the volume V of a sphere of radius r is given by $V = \frac{4}{3}\pi r^3$. Differentiating this with respect to r gives the rate of change of volume with radius, $\frac{dV}{dr} = 4\pi r^2$. However you might be more interested in finding $\frac{dV}{dt}$, the rate of change of volume with time, t.

To find this, you would use the chain rule:

$$\frac{dV}{dt} = \frac{dV}{dr} \times \frac{dr}{dt}$$

$$\frac{dV}{dt} = 4\pi r^2 \frac{dr}{dt}$$

Notice that the expression for $\frac{dV}{dt}$ includes $\frac{dr}{dt}$, the rate of increase of radius with time

You have now differentiated V with respect to t.

The use of the chain rule in this way widens the scope of differentiation and this means that you have to be careful how you describe the process.

Notes

1 'Differentiate $y = x^2$' could mean differentiation with respect to x, or t, or any other variable. In this book, and others in this series, we have adopted the convention that, unless otherwise stated, differentiation is with respect to the variable on the right-hand side of the expression. So when we write 'differentiate $y = x^2$' or simply 'differentiate x^2', it is to be understood that the differentiation is with respect to x.

2 The expression 'increasing at a rate of' is generally understood to imply differentiation with respect to time, t.

EXAMPLE 4.13

The radius r cm of a circular ripple made by dropping a stone into a pond is increasing at a rate of 8 cm s^{-1}. At what rate is the area A cm^2 enclosed by the ripple increasing when the radius is 25 cm?

EXERCISE 4B

Solution $A = \pi r^2$

$$\frac{dA}{dr} = 2\pi r$$

The question is asking for $\frac{dA}{dt}$, the rate of change of area (with respect to time).

Now $\frac{dA}{dt} = \frac{dA}{dr} \times \frac{dr}{dt}$

$\phantom{Now \frac{dA}{dt}} = 2\pi r \frac{dr}{dt}.$

When $r = 25$ and $\frac{dr}{dt} = 8$

$\frac{dA}{dt} = 2\pi \times 25 \times 8$

$\phantom{\frac{dA}{dt}} \approx 1260 \, \text{cm}^2 \, \text{s}^{-1}.$

EXERCISE 4B

In some of these questions you are asked to find the stationary points of a curve and then to use them as a guide for sketching it. You will find it helpful to use a graphical calculator to check your answers in these cases.

1 Use the chain rule to differentiate the following functions.

(a) $y = (x + 2)^3$
(b) $y = (2x + 3)^4$
(c) $y = (x^2 - 5)^3$
(d) $y = (x^3 + 4)^5$
(e) $y = (3x + 2)^{-1}$
(f) $y = \dfrac{1}{(x^2 - 3)^3}$
(g) $y = (x^2 - 1)^{\frac{3}{2}}$
(h) $y = \left(\dfrac{1}{x} + x\right)^3$
(i) $y = (\sqrt{x} - 1)^4$

2 Given that $y = (3x - 5)^3$:

(a) find $\frac{dy}{dx}$;
(b) find the equation of the tangent to the curve at $(2, 1)$;
(c) show that the equation of the normal to the curve at $(1, -8)$ can be written in the form

$$36y + x + 287 = 0.$$

3 Given that $y = (2x - 1)^4$:

(a) find $\frac{dy}{dx}$;
(b) find the coordinates of any stationary points and determine their nature;
(c) sketch the curve.

4 Given that $y = (x^2 - 4)^3$:

(a) find $\frac{dy}{dx}$;
(b) find the coordinates of any stationary points and determine their nature;
(c) sketch the curve.

5 Given that $y = (x^2 - x - 2)^4$:
  (a) find $\frac{dy}{dx}$;
  (b) find the coordinates of any stationary points and determine their nature;
  (c) sketch the curve.

6 The length of a side of a square is increasing at a rate of 0.2 cm s^{-1}. At what rate is the area increasing when the length of the side is 10 cm?

7 The force F newtons between two magnetic poles is given by the formula $F = \frac{1}{500r^2}$, where r m is their distance apart.

Find the rate of change of the force when the poles are 0.2 m apart and the distance between them is increasing at a rate of 0.03 m s^{-1}.

8 The radius of a circular fungus is increasing at a uniform rate of 5 cm per day. At what rate is the area increasing when the radius is 1 m?

9 The graph of $y = (x^3 - x^2 + 2)^3$, is shown in the diagram.
  (a) Find the gradient function $\frac{dy}{dx}$.
  (b) Verify, showing your working clearly, that when $x = -1$ the curve has a point of inflection and when $x = 0$ the curve has a maximum.
  (c) The curve has a minimum when $x = a$. Find a and verify that this corresponds to a minimum.
  (d) Find the gradient at $(1, 8)$ and the equation of the tangent to the curve at this point.

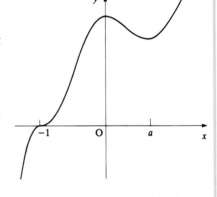

10 Some students on an expedition reach the corner of a very muddy field. They need to reach the opposite corner as quickly as possible as they are behind schedule. They estimate that they could walk along the edge of the field at 5 km h^{-1} and across the field at 3 km h^{-1}. They know from their map that the field is a square of side 0.5 km.

How far should they walk along the edge of the field before cutting across?

11 Write down the derivatives of the following functions.
  (a) $y = e^{5x}$
  (b) $y = 2e^{5+x}$
  (c) $y = 3e^{5-x}$
  (d) $y = \frac{1}{2}e^{3-5x}$
  (e) $y = \frac{1}{e^{3x}}$
  (f) $y = \frac{2}{e^{\frac{1}{2}x}}$
  (g) $y = e^{x^2}$
  (h) $y = e^{1+x+x^2}$
  (i) $y = e^{(1+x)^2}$
  (j) $y = e^{\sqrt{1+4x}}$

EXERCISE 4B

12 Write down the derivatives of the following functions.
   (a) $y = 3\ln x$
   (b) $y = \ln 2x$
   (c) $y = \ln(2 - x)$
   (d) $y = \ln(x^2)$
   (e) $y = (\ln x)^2$
   (f) $y = 4\ln(1 + x + x^2)$
   (g) $y = \ln\left(\dfrac{1}{x}\right)$
   (h) $y = \ln(\ln x)$
   (i) $y = 5\ln(\sqrt{x^2 + 1})$
   (j) $y = \ln(e^x)$

13 Find the derivative of $y = \ln\left(\dfrac{x^2}{2x - 1}\right)$, $x > \tfrac{1}{2}$. Show that your answer may be written as $-\dfrac{2}{x(x-1)}$.

   Determine the value of x at the stationary point and find whether it is a maximum or a minimum.

14 (a) Find $\dfrac{dy}{dx}$ if $y = \dfrac{1}{e^{3x}} - x^2$.
   (b) Show that the equation of the tangent to this curve at the point where $x = 0$ is $y = -3x + 1$.

15 Show that the tangent to the curve $y = x + e^{2x-2}$ at the point $(1, 2)$ cuts the x-axis at the point $(\tfrac{1}{3}, 0)$.

THE PRODUCT RULE

Figure 4.8 shows a sketch of the curve of $y = 20x(x - 1)^6$.

If you wanted to find the gradient function, $\dfrac{dy}{dx}$, for the curve, you could expand the right-hand side then differentiate it term by term – a long and cumbersome process!

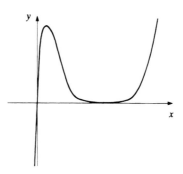

FIGURE 4.8

There are other functions like this, made up of the product of two or more simpler functions, which are not just time-consuming to expand – they are *impossible* to expand to give a finite number of real terms. One such function is

$$y = (x - 1)^{\tfrac{1}{2}}(x + 1)^6.$$

Clearly you need a technique for differentiating functions which are products of simpler ones, and a suitable notation with which to express it.

The most commonly used notation involves writing

$$y = uv,$$

where the variables u and v are both functions of x. Using this notation, $\frac{dy}{dx}$ is given by

$$\frac{dy}{dx} = u\frac{dv}{dx} + v\frac{du}{dx}.$$

This is called the *product rule* and it is derived from first principles in the next section.

THE PRODUCT RULE FROM FIRST PRINCIPLES

A small increase δx in x leads to corresponding small increases δu, δv and δy in u, v and y. And so

$$y + \delta y = (u + \delta u)(v + \delta v)$$
$$= uv + v\delta u + u\delta v + \delta u\delta v.$$

Since $y = uv$, the increase in y is given by

$$\delta y = v\delta u + u\delta v + \delta u\delta v.$$

Dividing both sides by δx, $\quad \dfrac{\delta y}{\delta x} = v\dfrac{\delta u}{\delta x} + u\dfrac{\delta v}{\delta x} + \delta u\dfrac{\delta v}{\delta x}.$

In the limit, as $\delta x \to 0$, so do δu, δv and δy, and

$$\frac{\delta u}{\delta x} \to \frac{du}{dx}, \quad \frac{\delta v}{\delta x} \to \frac{dv}{dx}, \quad \text{and} \quad \frac{\delta y}{\delta x} \to \frac{dy}{dx}.$$

The expression becomes $\quad \dfrac{dy}{dx} = v\dfrac{du}{dx} + u\dfrac{dv}{dx}.$

Notice that since $\delta u \to 0$ the last term on the right-hand side has disappeared.

EXAMPLE 4.14

Given that $y = (2x + 3)(x^2 - 5)$, find $\frac{dy}{dx}$ using the product rule.

Solution $\quad y = (2x + 3)(x^2 - 5)$

Let $u = 2x + 3$ and $v = x^2 - 5$.

Then $\dfrac{du}{dx} = 2$ and $\dfrac{dv}{dx} = 2x$.

The Product Rule

Using the product rule

$$\frac{dy}{dx} = v\frac{du}{dx} + u\frac{dv}{dx}$$
$$= (x^2 - 5) \times 2 + (2x + 3) \times 2x$$
$$= 2(x^2 - 5 + 2x^2 + 3x)$$
$$= 2(3x^2 + 3x - 5).$$

Note

In this case you could have multiplied out the expression for y:

$$y = 2x^3 + 3x^2 - 10x - 15$$
$$\frac{dy}{dx} = 6x^2 + 6x - 10$$
$$= 2(3x^2 + 3x - 5).$$

EXAMPLE 4.15

Differentiate $y = 20x(x - 1)^6$.

Solution

Let $u = 20x$ and $v = (x - 1)^6$.

Then $\frac{du}{dx} = 20$, and $\frac{dv}{dx} = 6(x - 1)^5$ (using the chain rule).

Using the product rule

$$\frac{dy}{dx} = v\frac{du}{dx} + u\frac{dv}{dx}$$
$$= (x - 1)^6 \times 20 + 20x \times 6(x - 1)^5$$
$$= 20(x - 1)^5 \times (x - 1) + 20(x - 1)^5 \times 6x$$
$$= 20(x - 1)^5[(x - 1) + 6x]$$
$$= 20(x - 1)^5(7x - 1).$$

$20(x - 1)^5$ is a common factor

The factorised result is the most useful form for the solution, as it allows you to find stationary points easily. You should always try to factorise your answer as much as possible. Once you have used the product rule, look for factors straight away and do not be tempted to multiply out.

The Quotient Rule

In the last section, you met a technique for differentiating the product of two functions. In this section you will see how to differentiate a function which is the quotient of two simpler functions.

As before, you start by identifying the simpler functions. For example, the function:

$$y = \frac{3x + 1}{x - 2}$$

can be written as $y = \frac{u}{v}$ where $u = 3x + 1$ and $v = x - 2$. Using this notation, $\frac{dy}{dx}$ is given by

$$\frac{dy}{dx} = \frac{v\frac{du}{dx} - u\frac{dv}{dx}}{v^2}.$$

This is called the *quotient rule*, and it is derived from first principles below.

The Quotient Rule from First Principles

A small increase δx in x results in corresponding small increases δu, δv and δy in u, v and y. The new value of y is given by

$$y + \delta y = \frac{u + \delta u}{v + \delta v}$$

and since $y = \frac{u}{v}$, you can rearrange this to obtain an expression for δy in terms of u and v:

$$\delta y = \frac{u + \delta u}{v + \delta v} - \frac{u}{v}$$

$$= \frac{v(u + \delta u) - u(v + \delta v)}{v(v + \delta v)}$$

$$= \frac{uv + v\delta u - uv - u\delta v}{v(v + \delta v)}$$

$$= \frac{v\delta u - u\delta v}{v(v + \delta v)}.$$

Dividing both sides by δx gives

$$\frac{\delta y}{\delta x} = \frac{v\frac{\delta u}{\delta x} - u\frac{\delta v}{\delta x}}{v(v + \delta v)}.$$

> To divide the right-hand side by δx you only divide the numerator by δx

In the limit as $\delta x \to 0$, this is written in the form you met above:

$$\frac{dy}{dx} = \frac{v\frac{du}{dx} - u\frac{dv}{dx}}{v^2}.$$

EXAMPLE 4.16

Given that $y = \dfrac{3x + 1}{x - 2}$, find $\dfrac{dy}{dx}$ using the quotient rule.

Solution Letting $u = 3x + 1$ and $v = x - 2$ gives

$$\frac{du}{dx} = 3 \quad \text{and} \quad \frac{dv}{dx} = 1.$$

Using the quotient rule

$$\frac{dy}{dx} = \frac{v\dfrac{du}{dx} - u\dfrac{dv}{dx}}{v^2}$$

$$= \frac{(x - 2)3 - (3x + 1)1}{(x - 2)^2}$$

$$= \frac{3x - 6 - 3x - 1}{(x - 2)^2}$$

$$= \frac{-7}{(x - 2)^2}.$$

EXAMPLE 4.17

Given that $y = \dfrac{x^2 + 1}{3x - 1}$, find $\dfrac{dy}{dx}$ using the quotient rule.

Solution Letting $u = x^2 + 1$ and $v = 3x - 1$ gives

$$\frac{du}{dx} = 2x \quad \text{and} \quad \frac{dv}{dx} = 3.$$

Using the quotient rule

$$\frac{dy}{dx} = \frac{v\dfrac{du}{dx} - u\dfrac{dv}{dx}}{v^2}$$

$$= \frac{(3x - 1)2x - (x^2 + 1)3}{(3x - 1)^2}$$

$$= \frac{6x^2 - 2x - 3x^2 - 3}{(3x - 1)^2}$$

$$= \frac{3x^2 - 2x - 3}{(3x - 1)^2}.$$

Exercise 4C

1 Differentiate the following functions using the product rule or the quotient rule.

(a) $y = (x^2 - 1)(x^3 + 3)$

(b) $y = xe^x$

(c) $y = x^2(2x + 1)^4$

(d) $y = \dfrac{2x}{3x - 1}$

(e) $y = \dfrac{x^3}{x^2 + 1}$

(f) $y = (2x + 1)^2(3x^2 - 4)$

(g) $y = \dfrac{2x - 3}{2x^2 + 1}$

(h) $y = \dfrac{x - 2}{(x + 3)^2}$

(i) $y = (x + 1)\sqrt{x - 1}$

(j) $y = (x^2 + 1)\ln x$

2 The graph of $y = \dfrac{x}{x - 1}$ is shown below.

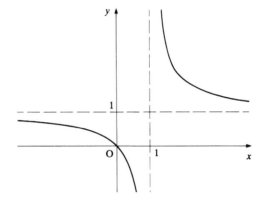

(a) Find $\dfrac{dy}{dx}$.

(b) Find the gradient of the curve at $(0, 0)$, and the equation of the tangent at $(0, 0)$.

(c) Find the gradient of the curve at $(2, 2)$, and the equation of the tangent at $(2, 2)$.

(d) What can you deduce about the two tangents?

3 Given that $y = (x + 1)(x - 2)^2$:

(a) find $\dfrac{dy}{dx}$;

(b) find any stationary points and determine their nature;

(c) sketch the curve.

4 Given that $y = (2x - 1)^3(x + 1)^3$:

(a) find $\dfrac{dy}{dx}$ and factorise the expression you obtain;

(b) find the values of x for which $\dfrac{dy}{dx} = 0$, and determine the nature of the corresponding stationary points.

The graph of $y = (2x - 1)^3(x + 1)^3$ is shown below.

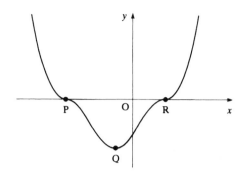

(c) Write down the coordinates of P, Q and R.

5 The graph of $y = \dfrac{2x}{\sqrt{x - 1}}$, which is undefined for $x < 0$ and $x = 1$, is shown below. P is a minimum point.

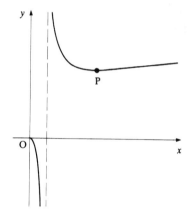

(a) Find $\dfrac{dy}{dx}$.
(b) Find the gradient of the curve at $(9, 9)$, and show that the equation of the normal at $(9, 9)$ is $y = -4x + 45$.
(c) Find the coordinates of P and verify that it is a minimum point.
(d) Write down the equation of the tangent and the normal to the curve at P.
(e) Write down the point of intersection of:
  (i) the normal found in (b) and the tangent found in (d), call it Q;
  (ii) the normal found in (b) and the normal found in (d), call it R.
(f) Show that the area of the triangle PQR is $\dfrac{441}{8}$.

6 Given that $y = \dfrac{x-3}{x-4}$:

(a) find $\dfrac{dy}{dx}$;

(b) find the equation of the tangent to the curve at the point $(6, 1.5)$;

(c) find the equation of the normal to the curve at the point $(5, 2)$;

(d) use your answer from (a) to deduce that the curve has no turning points, and sketch the graph.

7 Given that $y = \dfrac{x^2 - 2x - 5}{2x + 3}$:

(a) find $\dfrac{dy}{dx}$;

(b) use your answer from (a) to find any stationary points of the curve;

(c) classify each of the stationary points.

8 A curve has equation $y = \dfrac{x^2}{2x + 1}$.

(a) Find $\dfrac{dy}{dx}$.

Hence find the coordinates of the stationary points on the curve.

(b) You are given that $\dfrac{d^2y}{dx^2} = \dfrac{2}{(2x+1)^3}$.

Use this information to determine the nature of the stationary points in (a).

[MEI]

9 Given that $y = \sqrt{\dfrac{x+4}{x-1}}$:

(a) for what values of x does y have real values?

(b) Find $\dfrac{dy}{dx}$.

(c) Prove that the graph of this equation has no stationary points.

(d) Show that the equation of the normal to the curve at the point where $x = 5$ is $10y = 96x - 465$.

[MEI]

10 You are given that $f(x) = \dfrac{4x}{x^2 + 1}$.

(a) Find $f(0), f(1), f(2)$.

(b) Show that $f'(x) = \dfrac{4(1 - x^2)}{(x^2 + 1)^2}$.

Hence show that there is only one stationary point for $x \geq 0$ and state its coordinates.

(c) State what happens to $f(x)$ as $x \to \infty$.

(d) Using the information gained so far, sketch the graph of $y = f(x)$ for $x \geq 0$.

(e) Show that $f(x)$ is an odd function and hence complete the sketch graph for all values of x.

(f) Given that $g(x) = \dfrac{1}{x}$ $(x \neq 0)$, prove that $fg(x) = f(x)$.

[MEI]

11 Given that $y = xe^{2x}$:
   (a) find $\frac{dy}{dx}$;
   (b) show that there is one stationary point and find its coordinates;
   (c) find $\frac{d^2y}{dx^2}$;
   (d) determine whether the stationary point is a maximum or a minimum.

12 A curve has equation $y = x^2 \ln x$, $x > 0$.
   (a) Find $\frac{dy}{dx}$.
   (b) Show that the gradient at the point where $x = e$ is $3e$.
   (c) Find the equation of the tangent at the point where $x = e$.
   (d) Find where the tangent cuts the x-axis.

13 Given that $y = x^3 e^{2x}$:
   (a) find $\frac{dy}{dx}$ and factorise your answer;
   (b) show that the gradient of the curve at $x = -1$ is $\frac{1}{e^2}$;
   (c) find the equation of the normal at $x = -1$ in the form $ax + by + c = 0$.

14 You are given that $f(x) = xe^x \ln x$, $x > 0$.
   (a) Find $f'(x)$.
   (b) Show that the gradient of $f(x)$ at $x = 1$ is $\frac{1}{2}e(1 + \ln 16)$.
   (c) Prove that the graph of $y = f(x)$ has no turning points for $x > 0$.

15 (a) Show that the derivative of $(x+1)e^x$ is $(x+2)e^x$.

   A curve has equation $y = \frac{(x+1)e^x}{x^2}$, $x \neq 0$.

   (b) Find $\frac{dy}{dx}$.
   (c) Show that the curve has two turning points and determine their nature.

16 The graph of $y = xe^x$ is shown below.

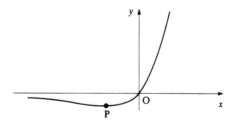

   (a) Find $\frac{dy}{dx}$ and $\frac{d^2y}{dx^2}$.
   (b) Find the coordinates of the minimum point P.

17 The graph of $f(x) = x\ln(x^2)$ is shown below.

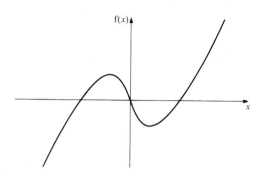

  (a) Describe, giving a reason, any symmetries of the graph.
  (b) Find $f'(x)$ and $f''(x)$.
  (c) Find the coordinates of any stationary points.

18 Given that $y = \dfrac{e^x}{x}$, $x \neq 0$:

  (a) find $\dfrac{dy}{dx}$;
  (b) find the coordinates of any stationary points on the curve of the function;
  (c) sketch the curve.

19 The function $f(x)$ is defined by

$$f(x) = \dfrac{x+1}{x^2 + 2x - 3}.$$

  (a) Use the quotient rule to find $f'(x)$

  (b) When expressed in partial fractions

$$f(x) \equiv \dfrac{1}{2(x+3)} + \dfrac{1}{2(x-1)}.$$

  Differentiate this form of $f(x)$ and show that the result is the same as in part **(a)**.

20 A curve C has equation

$$y = \dfrac{10x}{(x^2 + 1)(x - 3)}.$$

  (a) Show that $\dfrac{1-3x}{(x^2+1)} + \dfrac{3}{x-3} \equiv \dfrac{10x}{(x^2+1)(x-3)}$.
  (b) Find $\dfrac{dy}{dx}$.
  (c) Show that the gradient of C at the point P where $x = 2$ is $-\dfrac{14}{5}$.
  (d) Find the equation of the tangent to the curve at the point P.

THE DERIVATIVE $\frac{dx}{dy}$

What is the relationship between $\frac{dy}{dx}$ and $\frac{dx}{dy}$? Follow the steps below to help you to answer this question.

1. Differentiate $y = x^3$.
2. Rearrange $y = x^3$ in the form $x = f(y)$, and hence find $\frac{dy}{dx}$ as a function of y.
3. Write $\frac{dx}{dy}$ as a function of x.
4. Write down a relationship between $\frac{dy}{dx}$ and $\frac{dx}{dy}$.
5. Repeat steps 1–4 for other functions such as $y = 2x$, $y = x^2$ and $y = x^4$.
6. Use your results to propose a general rule relating $\frac{dy}{dx}$ and $\frac{dx}{dy}$.

You may have proposed the general result

$$\frac{dy}{dx} = \frac{1}{\frac{dx}{dy}}.$$

If so, well done! The result looks algebraically obvious, but remember that $\frac{dy}{dx}$ and $\frac{dx}{dy}$ are not fractions. The function $\frac{dy}{dx}$ is the rate of change of y with x, and $\frac{dx}{dy}$ is the rate of change of x with y.

The geometrical interpretation of this result can be seen in figure 4.9 where, as an example, the line $y = \tfrac{1}{2}x$ is drawn first with the axes the normal way round, and then with them interchanged.

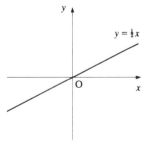
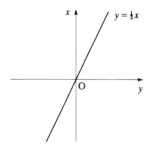

FIGURE 4.9

This demonstrates (but does not prove) that $\frac{dy}{dx} = \frac{1}{\frac{dx}{dy}}$. A proof is given on page 126.

EXAMPLE 4.18

Given that $x = y^{\frac{1}{3}}$, find $\frac{dy}{dx}$:

(a) by first finding $\frac{dx}{dy}$;
(b) by first making y the subject.

Solution (a) $x = y^{\frac{1}{5}} \implies \frac{dx}{dy} = \frac{1}{5} \times y^{\frac{1}{5}-1}$

$$= \frac{1}{5} y^{-\frac{4}{5}}$$

$$= \frac{1}{5(y^{\frac{1}{5}})^4}$$

$$= \frac{1}{5x^4}.$$

Since $\frac{dy}{dx} = \frac{1}{\frac{dx}{dy}}$ it follows that $\frac{dy}{dx} = \frac{1}{\frac{1}{5x^4}}$

$$= 5x^4.$$

(b) $x = y^{\frac{1}{5}} \implies y = x^5 \implies \frac{dy}{dx} = 5x^4.$

Note

The result $\frac{dy}{dx} = \frac{1}{\frac{dx}{dy}}$ has applications in two areas which you have met earlier: inverse functions and their gradients, and differentiation with respect to different variables.

EXAMPLE 4.19

(a) Sketch the graphs of $y = x^2 + 1$ for $x > 0$ and its inverse function $y = \sqrt{x - 1}$.

(b) Find the gradient of $y = \sqrt{x - 1}$ at the point (5, 2) by
  (i) direct differentiation;
  (ii) relating it to the gradient of $y = x^2 + 1$ at the point (2, 5).

Solution (a)

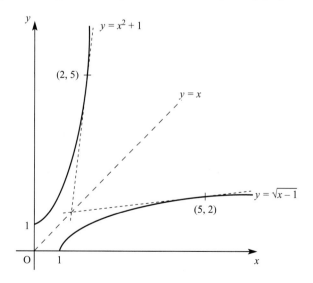

FIGURE 4.10

(b) (i) $y = \sqrt{x-1}$

$= (x-1)^{\frac{1}{2}}$

$\frac{dy}{dx} = \frac{1}{2}(x-1)^{-\frac{1}{2}} \times 1$ (chain rule)

$= \frac{1}{2\sqrt{x-1}}.$

At $(5, 2)$ the gradient is $\frac{1}{2\sqrt{5-1}} = \frac{1}{4}.$

(ii) Since any function and its inverse are reflections of each other in the line $y = x$, the two tangents shown in the sketch will meet on the line $y = x$.

So the gradient of $y = \sqrt{x-1}$ at $(5, 2) = \frac{1}{\text{gradient of } y = x^2 + 1 \text{ at } (2, 5)}$

$= \frac{1}{2x \text{ when } x = 2}$

$= \frac{1}{4}.$

Note

This is an illustration of the general result:

$$\text{gradient of } f^{-1}(x) \text{ at } (a, b) = \frac{1}{\text{gradient of } f(x) \text{ at } (b, a)}$$

and this is particularly useful when the equation of the inverse function cannot be found easily.

EXAMPLE 4.20

The area of a circular patch of mould is increasing at a rate of $0.4 \, \text{cm}^2 \, \text{h}^{-1}$. Calculate the rate at which the radius is increasing when the radius is 5 cm.

Solution You are required to find $\frac{dr}{dt} = \frac{dr}{dA} \times \frac{dA}{dt}.$

But $A = \pi r^2$, so $\frac{dA}{dr} = 2\pi r.$

$\frac{dr}{dA} = \frac{1}{\frac{dA}{dr}} = \frac{1}{2\pi r},$

so $\frac{dr}{dt} = \frac{1}{2\pi r} \times \frac{dA}{dt}.$

When $r = 5$ and $\dfrac{dA}{dt} = 0.4$

$$\dfrac{dr}{dt} = \dfrac{1}{2\pi \times 5} \times 0.4$$

$$= \dfrac{1}{25\pi}$$

$$= 0.0127 \text{ cm h}^{-1}.$$

PROOF OF THE RELATIONSHIP BETWEEN $\dfrac{dy}{dx}$ AND $\dfrac{dx}{dy}$

You will recall that we defined a derivative as the limiting value as $\delta x \to 0$ of the gradient of the chord. That is

$$\dfrac{dy}{dx} = \lim_{\delta x \to 0} \dfrac{\delta y}{\delta x}.$$

But $\dfrac{\delta y}{\delta x}$ is a fraction, so

$$\dfrac{dy}{dx} = \lim_{\delta x \to 0} \dfrac{1}{\frac{\delta x}{\delta y}}.$$

As $\delta x \to 0$ then $\delta y \to 0$, giving

$$\dfrac{dy}{dx} = \lim_{\delta x \to 0} \dfrac{1}{\frac{\delta x}{\delta y}} = \dfrac{1}{\lim_{\delta y \to 0} \frac{\delta x}{\delta y}} = \dfrac{1}{\frac{dx}{dy}}.$$

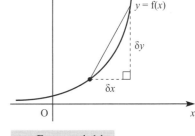

FIGURE 4.11

This proves the result.

EXERCISE 4D

1 The area of a circle is increasing at the uniform rate of 8 cm² min⁻¹. Calculate the rate at which the radius is increasing when the circumference is 50 cm.

2 (a) Sketch $f(x) = x(x + 1)(x + 2)$ for $x \in \mathbb{R}$.
   (b) The function $g(x) = x(x + 1)(x + 2)$ for $x \in \mathbb{R}^+$. Sketch $g(x)$ and $g^{-1}(x)$ on the same axes.
   (c) Find the gradient of $g(x)$ at the point $(2, 24)$ and hence find the gradient of $g^{-1}(x)$ at $(24, 2)$.

3 Sand is poured on to a horizontal floor at a rate of 4 cm³ s⁻¹ and forms a pile in the shape of a right circular cone, of which the height is three-quarters of the radius. Calculate the rate of change of the radius when the radius is 4 cm, leaving your answer in terms of π.

(The formula for the volume of a cone is $V = \tfrac{1}{3}\pi r^2 h$.)

4 A filter funnel is in the shape of a cone (vertex downwards) of vertical angle 90° with a small tube leaving at the vertex as shown in the diagram.

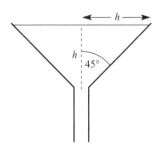

(a) When the depth of liquid in the funnel is 4 cm, the level is falling at 0.2 cm s^{-1}. At what rate is the volume decreasing?
(b) If the rate found in (a) is steady, how fast is the level falling when the depth is 2 cm?

5 You are given a curve with the equation $y = x^2(4 - x)$.
(a) Find the values of x for which $\frac{dy}{dx} = 0$.
(b) Denoting the values of x which you have just calculated by a and b, where $a < b$, show that $\frac{dy}{dx}$ is positive when $a < x < b$.
(c) Sketch the graph of y in the interval $a < x < b$, using the same scale on each axis.
(d) A function $f(x) = x^2(4 - x)$ is defined over the domain $a \leqslant x \leqslant b$ and has inverse function $f^{-1}(x)$.

Find the gradient of $f^{-1}(x)$ at the point $(\frac{7}{8}, \frac{1}{2})$.
[MEI]

6 The sketch below shows the graph of $y = f(x)$.
(a) Sketch the graphs of
(i) $y = 2f(x)$,
(ii) $y = f(-x)$,
(iii) $y = f(x + 2)$,
in each case superimposing them on a copy of the graph of $y = f(x)$.

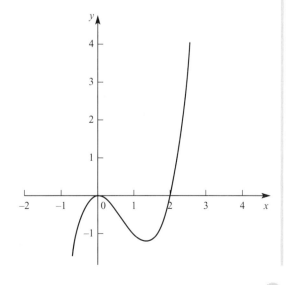

(b) Explain why the function f(x) does not have an inverse function.
(c) The function f(x) restricted to the domain $x > 2$ is called g(x). The inverse function of g(x) is $g^{-1}(x)$. Sketch the graphs of $y = g^{-1}(x)$ and $y = g(x)$ on the same axes.
(d) Given that $g(x) = x^2(x - 2)$, for $x > 2$, calculate the gradient of the graph of $y = g(x)$ at the point (3, 9). Deduce the gradient of the graph $y = g^{-1}(x)$ at the point (9, 3).

[MEI]

7 You are given that $y^2 = x + 1$.

Show that $\frac{dy}{dx} = \frac{1}{2\sqrt{x + 1}}$ by

(a) using $\frac{dy}{dx} = \frac{1}{\frac{dx}{dy}}$;

(b) making y the subject of the formula and differentiating using the chain rule.

8 (a) Make x the subject of $y = \ln\left(\frac{1}{x + 3}\right)$, $x > -3$.

(b) Find $\frac{dx}{dy}$ in terms of y.

(c) Find $\frac{dy}{dx}$ in terms of x.

9 By differentiating with respect to y, or otherwise, find the gradient of

$$x = (y^2 - 3)^4$$

at the point where $y = 2$.

Show that the equation of the tangent at this point is

$$16y = x + 31.$$

10 A curve has equation $y = \frac{x - 1}{x + 1}$.

(a) Sketch the curve.
(b) Prove that the gradient of the curve is always positive.
(c) The part of the curve for which $x > -1$ is reflected in the line $y = x$. Calculate the gradient of this reflected curve at the point where $y = 2$.

Differentiating trigonometric functions

Differentiating $\sin x$ and $\cos x$

You will remember the graph of $y = \sin x$, with x measured in radians. It is shown in figure 4.12.

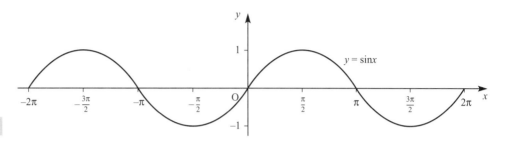

FIGURE 4.12

Let us see what happens if you measure the gradient at each point on the curve.

At $x = \frac{-3\pi}{2}, \frac{-\pi}{2}, \frac{\pi}{2}$ and $\frac{3\pi}{2}$ the gradient of the graph of $y = \sin x$ is, clearly, zero. At $x = -2\pi, 0$ and 2π the gradient looks to be about 1, whereas at $x = -\pi$ and π the gradient looks to be about −1. If you were to do this carefully for many points, noting where the gradient is positive or where it is negative, and you plotted your results, then you should obtain the graph shown in figure 4.13.

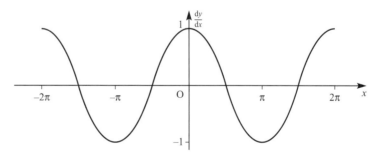

FIGURE 4.13

This graph looks like the cosine graph so it appears that the gradient function of $y = \sin x$ is $\cos x$.

If you repeat the same exercise for $y = \cos x$ then the result should be as shown in figure 4.14.

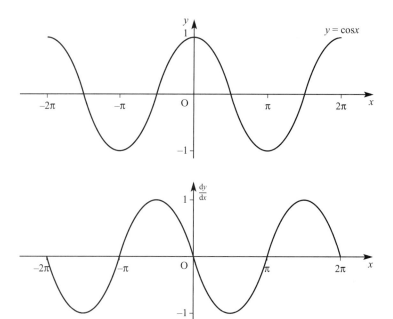

FIGURE 4.14

This time it looks as though the gradient function of $y = \cos x$ is $-\sin x$.

In other words,

if $y = \sin x$ then $\dfrac{dy}{dx} = \cos x$

if $y = \cos x$ then $\dfrac{dy}{dx} = -\sin x$.

The proof of the first of these results follows.

DIFFERENTIATING $y = \sin x$ FROM FIRST PRINCIPLES

The activity that you have just done has led you to form an idea of what the derivative of $\sin x$ might be. You can prove this by using differentiation from first principles.

Figure 4.15 shows part of the graph of $y = \sin x$. The point P is a general point $(x, \sin x)$ on the graph. The point Q is a very small distance further on, so it has x-coordinate $x + \delta x$, where δx is very small, and y-coordinate $\sin(x + \delta x)$.

You can find the gradient at the point P by finding the limit of the gradient of the chord PQ as δx approaches 0:

$$\dfrac{dy}{dx} = \lim_{\delta x \to 0} \dfrac{\sin(x + \delta x) - \sin x}{\delta x}.$$

$\sin(x + \delta x)$ may be simplified by using the compound-angle formula:

$$\sin(x + \delta x) = \sin x \cos \delta x + \cos x \sin \delta x.$$

DIFFERENTIATING TRIGONOMETRIC FUNCTIONS

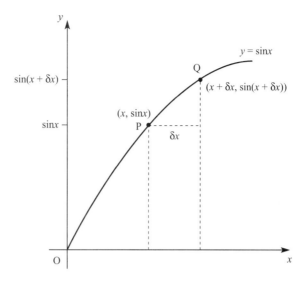

FIGURE 4.15

As δx is small, you can replace $\cos\delta x$ and $\sin\delta x$ by their small-angle approximation (the proof of which is a Further Pure 3 topic):

$$\cos\delta x \approx 1 - \tfrac{1}{2}(\delta x)^2 \text{ and } \sin\delta x \approx \delta x.$$

This leads to

$$\sin(x + \delta x) \approx (\sin x)[1 - \tfrac{1}{2}(\delta x)^2] + (\cos x)\delta x$$
$$= \sin x - \tfrac{1}{2}(\sin x)(\delta x)^2 + (\cos x)\delta x.$$

Substituting this in the expression $\dfrac{\sin(x + \delta x) - \sin x}{\delta x}$ gives

$$\frac{\sin x - \tfrac{1}{2}(\sin x)(\delta x)^2 + (\cos x)\delta x - \sin x}{\delta x} = \frac{-\tfrac{1}{2}(\sin x)(\delta x)^2 + (\cos x)\delta x}{\delta x}$$
$$= -\tfrac{1}{2}(\sin x)\delta x + \cos x.$$

In the limit as $\delta x \to 0$, this becomes simply $\cos x$. So

$$\frac{dy}{dx} = \cos x.$$

You have now proved the result which you found from the gradient graph sketching activity.

DERIVATIVE OF $\tan x$

By expressing $\tan x$ as $\dfrac{\sin x}{\cos x}$ the quotient rule can be used to find the derivative.

$$y = \tan x = \frac{\sin x}{\cos x}$$

$$\Rightarrow \quad \frac{dy}{dx} = \frac{\cos x \cos x - \sin x(-\sin x)}{\cos^2 x}$$

$$= \frac{\cos^2 x + \sin^2 x}{\cos^2 x}$$

Using $\frac{d}{dx}\left(\frac{u}{v}\right) = \frac{v\frac{du}{dx} - u\frac{dv}{dx}}{v^2}$

$$= \frac{1}{\cos^2 x}$$

But $\cos^2 x + \sin^2 x = 1$

$$= \sec^2 x$$

That is

$$\frac{d}{dx}(\tan x) = \sec^2 x.$$

DERIVATIVES OF THE RECIPROCAL TRIGONOMETRIC FUNCTIONS

Using the chain rule and the quotient rule the following results can be obtained:

$$\frac{d}{dx}(\sec x) = \sec x \tan x$$

$$\frac{d}{dx}(\operatorname{cosec} x) = -\operatorname{cosec} x \cot x$$

$$\frac{d}{dx}(\cot x) = -\operatorname{cosec}^2 x$$

Notes

1 When differentiating any trigonometric function the angle must be measured in radians.

2 To help you remember which derivatives are positive and which are negative, if the first letter of the function begins with 'c' you change the sign; for example in

$\frac{d}{dx}(\cot x) = -\operatorname{cosec}^2 x$, cot begins with 'c' and its derivative is negative.

EXAMPLE 4.21

Differentiate $y = \cos 2x$.

Solution As $\cos 2x$ is a function of a function, you may use the chain rule.

Let $u = 2x \quad \Rightarrow \quad \frac{du}{dx} = 2$

$y = \cos u \quad \Rightarrow \quad \frac{dy}{du} = -\sin u.$

Using the chain rule

$$\frac{dy}{dx} = \frac{dy}{du} \times \frac{du}{dx}$$
$$= -\sin u \times 2$$
$$= -2\sin 2x.$$

With practice it should be possible to do this in your head, without needing to write down the substitution.

Note

This result may be generalised:

$$y = \cos ax \implies \frac{dy}{dx} = -a\sin ax.$$

Similarly $y = \sin ax \implies \frac{dy}{dx} = a\cos ax.$

EXAMPLE 4.22

Differentiate $y = x^2 \sin x$.

Solution $x^2 \sin x$ is of the form uv, so the product rule can be used with $u = x^2$ and $v = \sin x$.

$$\frac{du}{dx} = 2x \qquad \frac{dv}{dx} = \cos x$$

Using the product rule

$$\frac{dy}{dx} = v\frac{du}{dx} + u\frac{dv}{dx}$$

$$\implies \frac{dy}{dx} = 2x\sin x + x^2\cos x.$$

EXAMPLE 4.23

Differentiate $y = e^{\tan x}$.

Solution $e^{\tan x}$ is a function of a function, so the chain rule may be used.

Let $u = \tan x \implies \frac{du}{dx} = \sec^2 x$

$y = e^u \implies \frac{dy}{du} = e^u.$

Using the chain rule

$$\frac{dy}{dx} = \frac{dy}{du} \times \frac{du}{dx}$$
$$= e^u \sec^2 x$$
$$= \sec^2 x \, e^{\tan x}.$$

EXAMPLE 4.24

Differentiate $y = \dfrac{1 + \sin x}{\cos x}$.

Solution $\dfrac{1 + \sin x}{\cos x}$ is of the form $\dfrac{u}{v}$ so the quotient rule can be used with

$$u = 1 + \sin x \quad \text{and} \quad v = \cos x$$

$$\Rightarrow \quad \dfrac{du}{dx} = \cos x \quad\quad\quad \dfrac{dv}{dx} = -\sin x.$$

The quotient rule is

$$\dfrac{dy}{dx} = \dfrac{v \dfrac{du}{dx} - u \dfrac{dv}{dx}}{v^2}.$$

Substituting for u and v and their derivatives gives

$$\dfrac{dy}{dx} = \dfrac{(\cos x)(\cos x) - (1 + \sin x)(-\sin x)}{(\cos x)^2}$$

$$= \dfrac{\cos^2 x + \sin x + \sin^2 x}{\cos^2 x}$$

$$= \dfrac{1 + \sin x}{\cos^2 x} \quad (\text{using } \sin^2 x + \cos^2 x = 1).$$

EXERCISE 4E

1 Prove from first principles that

$$\dfrac{d}{dx}(\cos x) = -\sin x.$$

2 Using the chain rule and the quotient rule prove that
   (a) $\dfrac{d}{dx}(\sec x) = \sec x \tan x;$
   (b) $\dfrac{d}{dx}(\cosec x) = -\cosec x \cot x;$
   (c) $\dfrac{d}{dx}(\cot x) = -\cosec^2 x.$

3 Differentiate each of the following functions.
   (a) $2\cos x + \sin x$ (b) $\tan x + 5$
   (c) $\sin x - \cos x$ (d) $\cos 3x$

4 Use the product rule to differentiate each of the following functions.
   (a) $x \tan x$ (b) $\sin x \cos x$
   (c) $e^x \sin x$ (d) $x^2 \sin 2x$

EXERCISE 4E

5 Use the quotient rule to differentiate each of the following functions.
   (a) $\dfrac{\sin x}{x}$
   (b) $\dfrac{e^x}{\cos x}$
   (c) $\dfrac{x + \cos x}{\sin x}$
   (d) $\cot x$

6 Use the chain rule to differentiate each of the following functions.
   (a) $\tan(x^2 + 1)$
   (b) $\cos^2 x$
   (c) $\ln(\sin x)$
   (d) $\sin(x^2 + 1)$

7 Use an appropriate method to differentiate each of the following functions.
   (a) $\sqrt{\cos x}$
   (b) $e^x \tan x$
   (c) $\sin 4x^2$
   (d) $e^{\cos 2x}$
   (e) $\dfrac{\sin x}{1 + \cos x}$
   (f) $\ln(\tan x)$

8 Differentiate the following with respect to x.
   (a) $\sec 3x$
   (b) $5 \operatorname{cosec} 4x$
   (c) $\cot^3 x$
   (d) $\sec x \tan x$

9 (a) Differentiate $y = x \cos x$.
   (b) Find the gradient of the curve $y = x \cos x$ at the point where $x = \pi$.
   (c) Find the equation of the tangent to the curve $y = x \cos x$ at the point where $x = \pi$.
   (d) Find the equation of the normal to the curve $y = x \cos x$ at the point where $x = \pi$.

10 The function $y = \sin^3 x$ has five stationary points in $-\pi \leqslant x \leqslant \pi$.
   (a) Find $\dfrac{dy}{dx}$ for this function.
   (b) Find the coordinates of the five stationary points.
   (c) Determine whether each of the five points is a maximum, minimum or point of inflection.
   (d) Use this information to sketch the graph of $y = \sin^3 x$ for values of x in $-\pi \leqslant x \leqslant \pi$.

11 For the curve $y = x + \sin 2x$:
   (a) find $\dfrac{dy}{dx}$;
   (b) find the coordinates of the stationary points in $0 \leqslant x \leqslant 2\pi$, and determine their nature;
   (c) sketch the curve for $0 \leqslant x \leqslant 2\pi$.

12 If $y = e^x \cos 3x$, find $\dfrac{dy}{dx}$ and $\dfrac{d^2y}{dx^2}$ and hence show that

$$\dfrac{d^2y}{dx^2} - 2\dfrac{dy}{dx} + 10y = 0.$$

[MEI]

13 Consider the function $y = e^{-x}\sin x$, where $-\pi \leq x \leq \pi$.
   (a) Find $\frac{dy}{dx}$.
   (b) Show that, at stationary points, $\tan x = 1$.
   (c) Determine the coordinates of the stationary points, correct to 2 significant figures.
   (d) Explain how you could determine whether your stationary points are maxima or minima. You are not required to do any calculations.
   [MEI]

14 (a) (i) Show that $(\cos x + \sin x)^2 = 1 + \sin 2x$, for all x.
       (ii) Hence, or otherwise, find the derivative of $(\cos x + \sin x)^2$.
   (b) (i) By expanding $(\cos^2 x + \sin^2 x)^2$, find and simplify an expression for $\cos^4 x + \sin^4 x$ involving $\sin 2x$.
       (ii) Hence, or otherwise, show that the derivative of $\cos^4 x + \sin^4 x$ is $-\sin 4x$.
   [MEI]

15 You are given that $y = e^{-x}\sin 2x$.
   (a) Find $\frac{dy}{dx}$.
   (b) Show that, at stationary points, $\tan 2x = 2$.
   (c) Find the solutions in radians of the equation $\tan 2x = 2$ which lie in the range $0 \leq x \leq \pi$ correct to 2 decimal places.
   (d) Show that $\frac{dy}{dx}$ can be written as $re^{-x}\cos(2x + \alpha)$ where r and α are to be determined.
   [MEI]

16 You are given that $y = e^{-2x}\tan x$. Find $\frac{dy}{dx}$, and show that at all stationary points $\sin 2x = 1$.
   [MEI, part]

EXERCISE 4F   Examination-style questions

1 The curve with equation $y = e^x - 1$ meets the line $y = 3$ at the point $(h, 3)$.
   (a) Find h, giving your answer in terms of natural logarithms.
   (b) Find the equation of the tangent at $(h, 3)$ leaving natural logarithms in your answer.

2 It is given that $y = 1 - \frac{x^2}{2} + \ln\frac{x}{4}$.
   (a) Show that $\frac{dy}{dx} = -\frac{15}{4}$ when $x = 4$.
   (b) Obtain the equation of the normal to the curve at the point where $x = 4$.

3 Given that a curve has equation $y = 4x^2 - \ln(2x) + 3$
  (a) Show that the gradient of the tangent at $x = \frac{1}{2}$ is 2.
  (b) Find the equation of the tangent at the point where $x = \frac{1}{2}$.
  (c) Show that the area of the triangle enclosed between the tangent and the axes is $\frac{9}{4}$.

4 Find the equation of the tangent and the normal to the curve
$$y = e^x + 2\ln x + 3$$
at the point where $x = 1$. Leave your answer in terms of e.

5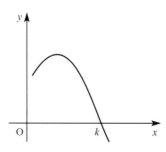

This diagram shows a sketch of the curve with equation $y = f(x)$ where
$$f(x) = 10 + \ln(3x) - \tfrac{1}{2}e^x, \quad 0.1 \leqslant x \leqslant 3.3$$

Given that $f(k) = 0$,

  (a) show, by calculation, that $3.1 < k < 3.2$.
  (b) find $f'(x)$.

The tangent to the graph at $x = 1$ intersects the y–axis at the point P.

  (c) (i) Find an equation of this tangent.
      (ii) Find the exact y-coordinate of P, giving you answer in the form $a + \ln b$.
  [Edexcel]

6 The curve with equation $y = \tfrac{1}{2}e^x$ meets the y-axis at the point A.
  (a) Prove that the tangent at A to the curve has equation $2y = x + 1$.

The point B has x-coordinate $\ln 4$ and lies on the curve. The normal at B to the curve meets the tangent at A to the curve at the point C.
  (b) Prove that the x-coordinate of C is $\tfrac{3}{2} + \ln 2$ and find the y-coordinate of C.
  [Edexcel]

7 (a) Differentiate the following with respect to x.
   (i) $x^2 \sin 2x$
   (ii) $\sqrt{\dfrac{x+1}{x-1}}$
   (iii) $e^{x \ln x}$
  (b) Given that $y = 5^x$ find the value of $\dfrac{dy}{dx}$ when $x = 2$.

8 Differentiate, with respect to x,

(a) $\dfrac{\sin x}{x}$, $x > 0$

(b) $\ln\left(\dfrac{1}{x^2 + 9}\right)$.

(c) Given that $y = x^x$, $x > 0$, $y > 0$, by taking logarithms show that
$$\dfrac{dy}{dx} = x^x(1 + \ln x).$$

[Edexcel]

9 A curve is defined by the equation $x = (3y - 7)^5$.

(a) By differentiating with respect to y, or otherwise, find the gradient of the curve at the point where $y = 2$.

(b) Show that the equation of the tangent to the curve at the point where $y = 2$ is $15y = x + 31$.

10 Given that $y = x^2 e^{3x}$:

(a) find $\dfrac{dy}{dx}$ and factorise your answer;

(b) show that the gradient of the curve at the point where $x = \tfrac{1}{3}$ is e;

(c) find the x-values of the stationary points and determine whether they are maxima or minima.

11 A curve has equation
$$y = \sin 2x + 2\cos x, \qquad 0 \leq x \leq 2\pi.$$

(a) Find $\dfrac{dy}{dx}$ and $\dfrac{d^2y}{dx^2}$.

(b) Find, for $0 \leq x \leq 2\pi$, the values of x for which $\dfrac{dy}{dx} = 0$.

(c) For the values of x found in part (b), determine the nature of the stationary points.

(d) Sketch the curve.

12 $f(x) \equiv e^{2x}\sin 2x$, $0 \leq x \leq \pi$

(a) Find the values of x for which $f(x) = 0$, giving your answers in terms of π.

(b) Use calculus to find the coordinates of the turning points on the graph of $y = f(x)$.

(c) Show that $f''(x) = 8e^{2x}\cos 2x$.

(d) Hence, or otherwise, determine which point is a maximum and which is a minimum.

[Edexcel]

13 $f(x) = \dfrac{x}{x^2 + 2}$, $x \in \mathbb{R}$

Find the set of values of x for which $f'(x) < 0$.

[Edexcel]

14 Given that $x = y^2 \ln y$, $y > 0$,

(a) find $\dfrac{dy}{dx}$.

(b) Use your answer to part (a) to find, in terms of e, the value of $\dfrac{dy}{dx}$ at $y = e$.

[Edexcel]

EXERCISE 4F

15 Prove the following.

(a) $\dfrac{d(\tan x)}{dx} = \sec^2 x$

(b) $\dfrac{d(\cot x)}{dx} = -\operatorname{cosec}^2 x$

(c) If $x = \tan y$ then $\dfrac{dy}{dx} = \dfrac{1}{1+x^2}$

(d) $\dfrac{d}{dx}(\sin^2 x) + \dfrac{d}{dx}(\cos^2 x) = 0$

KEY POINTS

1 Exponentials

$y = e^x \Rightarrow \dfrac{dy}{dx} = e^x$

$y = e^{f(x)} \Rightarrow \dfrac{dy}{dx} = f'(x)e^{f(x)}$

2 Logarithms

$y = \ln x \Rightarrow \dfrac{dy}{dx} = \dfrac{1}{x}$

$y = \ln x^a \Rightarrow \dfrac{dy}{dx} = \dfrac{a}{x}$

$y = \ln(f(x)) \Rightarrow \dfrac{dy}{dx} = \dfrac{f'(x)}{f(x)}$

3 Chain rule

$\dfrac{dy}{dx} = \dfrac{dy}{du} \times \dfrac{du}{dx}$

4 Product rule

$y = uv \Rightarrow \dfrac{dy}{dx} = u\dfrac{dv}{dx} + v\dfrac{du}{dx}$

5 Quotient rule

$y = \dfrac{u}{v} \Rightarrow \dfrac{dy}{dx} = \dfrac{v\dfrac{du}{dx} - u\dfrac{dv}{dx}}{v^2}$

6 Inverses

$\dfrac{dy}{dx} = \dfrac{1}{\dfrac{dx}{dy}}$

7 Trigonometric functions

$y = \sin x \Rightarrow \dfrac{dy}{dx} = \cos x \qquad y = \operatorname{cosec} x \Rightarrow \dfrac{dy}{dx} = -\operatorname{cosec} x \cot x$

$y = \cos x \Rightarrow \dfrac{dy}{dx} = -\sin x \qquad y = \sec x \Rightarrow \dfrac{dy}{dx} = \sec x \tan x$

$y = \tan x \Rightarrow \dfrac{dy}{dx} = \sec^2 x \qquad y = \cot x \Rightarrow \dfrac{dy}{dx} = -\operatorname{cosec}^2 x$

Chapter five

NUMERICAL METHODS

It is the true nature of mankind to learn from his mistakes.

Fred Hoyle

NUMERICAL SOLUTIONS OF EQUATIONS

Which of the following equations can be solved algebraically, and which cannot? For each equation find a solution, accurate or approximate.

- $x^2 - 4x + 3 = 0$
- $x^2 + 10x + 8 = 0$
- $x^5 - 5x + 3 = 0$
- $x^3 - x = 0$
- $e^x = 4x$

You probably realised that the equations $x^5 - 5x + 3 = 0$ and $e^x = 4x$ cannot be solved algebraically. You may have decided to draw their graphs, either manually or using a graphical calculator or computer package (see figure 5.1).

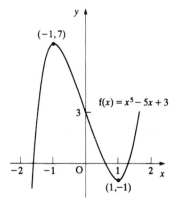 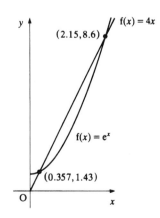

FIGURE 5.1

The graphs show you that:

- $x^5 - 5x + 3 = 0$ has three roots, lying in the intervals $[-2, -1]$, $[0, 1]$ and $[1, 2]$.
- $e^x = 4x$ has two roots, lying in the intervals $[0, 1]$ and $[2, 3]$.

The problem now is how to find the roots to any required degree of accuracy, and as efficiently as possible.

In many real problems, equations are obtained for which solutions using algebraic or analytical methods are not possible, but for which you nonetheless want to know the answers. In this chapter you will be introduced to numerical methods for solving such equations. In applying these methods, keep the following points in mind.

- Only use numerical methods when algebraic or trigonometric ones are not available. If you can solve an equation algebraically (or trigonometrically), that is the right method to use.
- Before starting to use a calculator or computer program, always start by drawing a sketch graph of the function or functions involved. This will show you how many roots the equation has and their approximate positions. It will also warn you of possible difficulties with particular methods.
- Always give a statement about the accuracy of an answer (e.g. to 5 decimal places, or ±0.000 005). An answer obtained by a numerical method is worthless without this; the fact that at some point in the procedure your calculator display reads, say, 1.676 470 588 2 does not mean that all these figures are valid.
- Your statement about the accuracy must be obtained from within the numerical method itself. Usually you find a sequence of estimates of ever-increasing accuracy.
- Remember that the most suitable method for one equation may not be the best for another.

Note

An interval written as [a, b] means the interval between a and b, including a and b. This notation is used in this chapter. If a and b are not included, the interval is written (a, b). You may also elsewhere meet the notation]a, b[, indicating that a and b are *not* included.

CHANGE OF SIGN METHOD

Assume that you are looking for the roots of the equation $f(x) = 0$. This means that you want the values of x for which the graph of $y = f(x)$ crosses the x-axis. As the curve crosses the x-axis, $f(x)$ changes sign, so provided that $f(x)$ is a continuous function (its graph has no asymptotes or other breaks in it), once you have located an interval in which $f(x)$ changes sign, you know that that interval must contain a root. In both of the graphs in figure 5.2, there is a root lying between a and b.

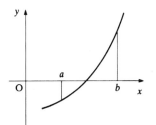
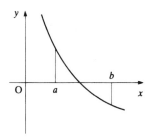

FIGURE 5.2

You have seen that $x^5 - 5x + 3 = 0$ has roots in the intervals [−2, −1], [0, 1] and [1, 2]. A decimal search will be used as an example to find the root in the interval [0, 1].

DECIMAL SEARCH

In this method you first take increments in x of size 0.1 within the interval [0, 1], working out the value of the function f(x), where f(x) = $x^5 - 5x + 3$ for each one. You do this until you find a change of sign.

x	0	0.1	0.2	0.3	0.4	0.5	0.6	0.7
f(x)	3.00	2.50	2.00	1.50	1.01	0.53	0.08	−0.33

There is a sign change, and therefore a root, in the interval [0.6, 0.7] since the function is continuous. Having narrowed down the interval, you can now continue with increments of 0.01 within the interval [0.6, 0.7].

x	0.60	0.61	0.62
f(x)	0.08	0.03	−0.01

This shows that the root lies in the interval [0.61, 0.62].

Alternative ways of expressing this information are:

(a) the root can be taken as 0.615 with a maximum error of ±0.005, or
(b) the root is 0.6 (to 1 decimal place).

This process can be continued by considering x = 0.611, x = 0.612, ... to obtain the root to any required number of decimal places.

When you use this procedure on a computer or calculator you should be aware that the machine is working in base 2, and that the conversion of many simple numbers from base 10 to base 2 introduces small rounding errors. This can lead to simple roots such as 2.7 being missed and only being found as 2.699 999.

EXAMPLE 5.1

Show that $2 + x^2 - x^3 = 0$ has a root between 1 and 2. By considering changes of sign find this root to 1 decimal place.

Solution Let $f(x) = 2 + x^2 - x^3$.

Using a graphical calculator the graph of $y = f(x)$ looks like that in figure 5.3.

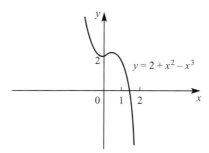

FIGURE 5.3

The graph is continuous over the interval [1, 2] and

$$f(1) = 2$$
$$f(2) = -2$$

so the change of sign implies that the root lies between $x = 1$ and $x = 2$.
Improving this root to 1 decimal place:

$$f(1.5) = 0.875$$
$$f(1.6) = 0.464$$
$$f(1.7) = -0.023$$

so the change of sign implies that the root lies in [1.6, 1.7].

Putting $x = 1.65$ confirms whether the root is nearer 1.6 or 1.7 (even when, as in this case, it appears obvious you should use this type of check).

$$f(1.65) = 0.23$$

So the root is close to 1.7.

EXAMPLE 5.2

By using a change of sign method show that $\sin\theta + \cos\theta = 1$ has a root between 1 and 2 and find θ to 1 decimal place.

Solution First rewrite in the form $f(\theta) = 0$.
Let $f(\theta) = \sin\theta + \cos\theta - 1$.
(Note that when angles are used they should be measured in radians.)

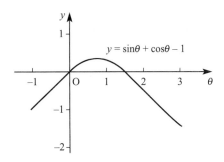

FIGURE 5.4

A calculator gives a sketch of $y = f(\theta)$ to be as in figure 5.4.
Using a decimal search

$$f(1) = 0.3818$$
$$f(2) = -0.5068$$

so the change in sign implies that the root lies between $\theta = 1$ and $\theta = 2$.

$$f(1.5) = 0.0682$$
$$f(1.6) = -0.0296$$
$$f(1.55) = 0.0206$$

so, to 1 decimal place, $\theta = 1.6$.

Check your answer by using the $a\sin\theta + b\cos\theta = R\sin(\theta + \alpha)$ method.

PROBLEMS WITH CHANGE OF SIGN METHODS

There are a number of situations which can cause problems for change of sign methods if they are applied blindly, for example by entering the equation into a computer program without prior thought. In all cases you can avoid problems by first drawing a sketch graph, provided that you know what dangers to look out for.

The curve touches the x-axis

In this case there is no change of sign, so change of sign methods are doomed to failure (figure 5.5).

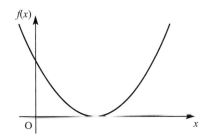

FIGURE 5.5

There are several roots close together

Where there are several roots close together, it is easy to miss a pair of them. The equation:

$$f(x) = x^3 - 1.9x^2 + 1.11x - 0.189 = 0$$

has roots at 0.3, 0.7 and 0.9. A sketch of the curve of f(x) is shown in figure 5.6.

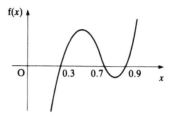

FIGURE 5.6

In this case $f(0) < 0$ and $f(1) > 0$, so you know there is a root between 0 and 1.

A decimal search would show that $f(0.3) = 0$, so that 0.3 is a root. You would be unlikely to search further in this interval.

There is a discontinuity in f(x)

The curve $y = \dfrac{1}{x - 2.7}$ has a discontinuity at $x = 2.7$, as shown in figure 5.7.

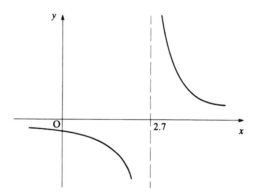

FIGURE 5.7

The equation $\dfrac{1}{x - 2.7} = 0$ has no root, but all change of sign methods will converge on a false root at $x = 2.7$.

None of these problems will arise if you start by drawing a sketch graph.

Note: Use of technology It is important that you understand how each method works and are able, if necessary, to perform the calculations using only a scientific calculator. However, these repeated operations lend themselves to the use of a spreadsheet or a programmable calculator and if possible you will benefit from using a variety of approaches when working through the following exercises.

Exercise 5A

1. Use a decimal search to find the roots of $x^5 - 5x + 3 = 0$ in the intervals $[-2, -1]$ and $[1, 2]$, correct to 2 decimal places.

2. (a) Use a systematic search for a change of sign, starting with $x = -2$, to locate intervals of unit length containing each of the three roots of $x^3 - 4x^2 - 3x + 8 = 0$.
   (b) Sketch the graph of $f(x) = x^3 - 4x^2 - 3x + 8$.
   (c) Find each of the roots correct to 2 decimal places.
   (d) Use your last intervals to give each of the roots in the form $x = a$ with a maximum error of $(0.5)n$, stating your values of a and n.

3. The diagram shows a sketch of the graph of $f(x) = e^x - x^3$ without scales.

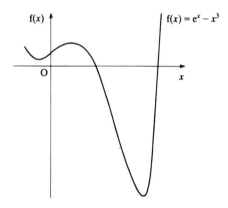

   (a) Use a systematic search for a change of sign to locate intervals of unit length containing each of the roots.
   (b) Find each of the roots correct to 3 decimal places.

4. (a) Show that the equation $x^3 + 3x - 5 = 0$ has no turning points.
   (b) Show with the aid of a sketch that the equation can have only one root, and that this root must be positive.
   (c) Find the root, correct to 3 decimal places.

5. (a) How many roots has the equation
       $e^x - 3x = 0$?
   (b) Find an interval of unit length containing each of the roots.
   (c) Find each root correct to 2 decimal places.

EXERCISE 5A

6. (a) Sketch $y = 2^x$ and $y = x + 2$ on the same axes.
   (b) Use your sketch to deduce the number of roots of the equation $2^x = x + 2$.
   (c) Find each root, correct to 3 decimal places if appropriate.

7. Find all the roots of $x^3 - 3x + 1 = 0$, giving your answers correct to 2 decimal places.

8. For each of the equations below:
   (i) sketch the curve;
   (ii) write down any roots;
   (iii) investigate what happens when you use a change of sign method with a starting interval of $[-0.3, 0.7]$.

   (a) $y = \dfrac{1}{x}$
   (b) $y = \dfrac{x}{x^2 + 1}$
   (c) $y = \dfrac{x^2}{x^2 + 1}$

9. Given that $f(x) = x^3 + ax^2 + b$ and $f(1) = 2$ with $f(2) = 21$, find the value of a and show that $b = -3$.

   $f(x) = 0$ has one positive root. By evaluating $f(x)$ for appropriate values of x, find this root correct to 1 decimal place.

10. On a single diagram sketch, over $-\pi \leq x \leq \pi$, the graphs of

    $y = \sin x$ and $y = \tfrac{1}{5}x + \tfrac{2}{5}$.

    From your sketch state the approximate values of x for which

    $\sin x = \tfrac{1}{5}x + \tfrac{2}{5}$ over $-\pi \leq x \leq \pi$.

    This equation has a root α in the range 0 to $\dfrac{\pi}{2}$ radians. By rewriting the equation as $5\sin x - x - 2 = 0$ show that α lies between 0.5 and 0.6 and find the value of α correct to 2 decimal places.

ITERATIVE METHODS TO SOLVE $f(x) = 0$

The method requires the equation $f(x) = 0$ to be rewritten in the form $x = g(x)$ and using an *iterative process* with this equation to generate a sequence of numbers by continued repetition of the same procedure. If the numbers obtained in this manner approach some limiting value, then they are said to *converge* to this value.

Figure 5.8 illustrates the fact that $f(x) = 0$ and its rearrangement $x = g(x)$ have the same root, using the equations $f(x) = x^2 - x - 2$ and $x = g(x) = x^2 - 2$.

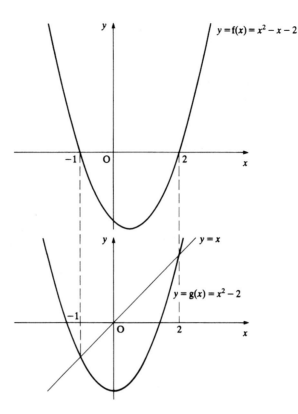

FIGURE 5.8

The equation $x^5 - 5x + 3 = 0$, which you met earlier, can be rewritten in a number of ways. One of these is $5x = x^5 + 3$, giving

$$x = g(x) = \frac{x^5 + 3}{5}.$$

Figure 5.9 shows the graphs of $y = x$ and $y = g(x)$ in this case.

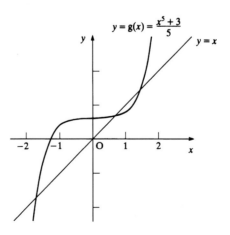

FIGURE 5.9

The rearrangement provides the basis for the iterative formula

$$x_{n+1} = \frac{x_n^5 + 3}{5}.$$

(You may wish to revise recurrence relations from Chapter 3 of *Pure Mathematics: Core 1 and Core 2*.)

It is known that there is a root in the interval [0, 1] so take $x_1 = 1$ as a starting point and use the iteration to find successive approximations:

$x_1 = 1$
$x_2 = 0.8$
$x_3 = 0.6655$
$x_4 = 0.6261$
$x_5 = 0.6192$
$x_6 = 0.6182$
$x_7 = 0.6181$
$x_8 = 0.6180$
$x_9 = 0.6180$

> On your calculator if you have an ANS key, key in 1 then = followed by (ANS⁵ + 3) ÷ 5, then repeatedly press the = key

In this case the iteration has converged quite rapidly to the root for which you were looking.

EXAMPLE 5.3

Show that $x^5 - 5x + 3 = 0$ can be rearranged to $x = \sqrt[5]{5x - 3}$. Use this to provide an iterative formula to find the root between 1 and 2 to 4 decimal places.

Solution Rearranging:

$$x^5 - 5x + 3 = 0$$
$$x^5 = 5x - 3$$
$$x = \sqrt[5]{5x - 3}$$

This becomes the iterative formula

$$x_{n+1} = \sqrt[5]{5x_n - 3}.$$

Starting with $x_1 = 1$:

$x_2 = 1.148\ 69$
$x_3 = 1.223\ 65$
$x_4 = 1.255\ 40$
$\vdots$
$x_{12} = 1.275\ 67$
$x_{13} = 1.275\ 68$
$x_{14} = 1.275\ 68$

So the root between 1 and 2 is 1.2757 correct to 4 decimal places.

Note

To give an answer to a particular number of decimal places it is often sufficient to carry out the iteration until the next decimal place does not change, and then round off. Care is needed, particularly if convergence is slow, since the iteration may continue to change the rounded off value. The more iterations you can do the better!

The iteration process is easiest to understand if you consider the graph. Rewriting the equation $f(x) = 0$ in the form $x = g(x)$ means that instead of looking for points where the graph of $y = f(x)$ crosses the x-axis, you are now finding the points of intersection of the curve $y = g(x)$ and the line $y = x$.

What you do	**What it looks like on the graph**
• Choose a value, x_1, of x.	Take a starting point on the x-axis.
• Find the corresponding value of $g(x_1)$.	Move vertically to the curve $y = g(x)$.
• Take this value, $g(x_1)$, as the new value of x, i.e. $x_2 = g(x_1)$.	Move horizontally to the line $y = x$.
• Find the value of $g(x_2)$ and so on.	Move vertically to the curve.

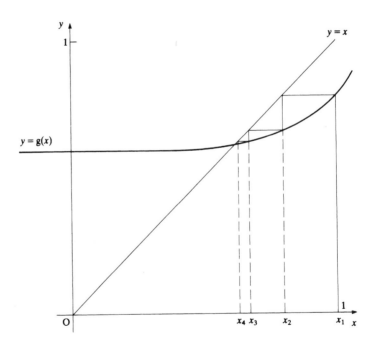

FIGURE 5.10

The effect of several repeats of this procedure is shown in figure 5.10. The successive steps look like a staircase approaching the root: this type of diagram is called a *staircase diagram*. In other examples, a *cobweb diagram* may be produced, as shown in figure 5.11.

Iterative methods to solve f(x) = 0

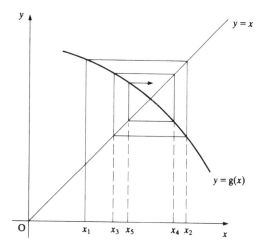

FIGURE 5.11

Successive approximations to the root are found by using the formula

$$x_{n+1} = g(x_n).$$

This is an example of an *iterative formula*. If the resulting values of x_n approach some limit, a, then $a = g(a)$, and so a is a *fixed point* of the iteration. It is also a root of the original equation, $f(x) = 0$.

Note

In the staircase diagram, the values of x_n approach the root from one side, but in a cobweb diagram they oscillate about the root. From figures 5.10 and 5.11 it is clear that the error (the difference between a and x_n) is decreasing in both diagrams.

Using different arrangements of the equation

So far we have used two possible arrangements of the equation $x^5 - 5x + 3 = 0$.

The first, $x = \dfrac{x^5 + 3}{5}$, converged to 0.618 and the second, $x = \sqrt[5]{5x - 3}$, converged to 1.2757 when a starting point $x_1 = 1$ was used.

The processes have clearly converged. If instead you had taken $x_1 = 0$ as your starting point and applied the second formula, you would have obtained a sequence converging to the value −1.6180, the root in the interval [−2, −1]. The second formula does not appear to converge to the root in the interval [0, 1].

The choice of g(x)

A particular rearrangement of the equation $f(x) = 0$ into the form $x = g(x)$ will allow convergence to a root a of the equation, provided that $-1 < g'(a) < 1$ for values of x close to the root.

Look again at the two rearrangements of $x^5 - 5x + 3 = 0$ which were suggested.

When you look at the graph of:

$$y = g(x) = \sqrt[5]{5x - 3},$$

you can see that its gradient near A, the root you were seeking, is greater than 1 (figure 5.12). This makes

$$x_{n+1} = \sqrt[5]{5x_n - 3},$$

an unsuitable iterative formula for finding the root in the interval [0, 1], as you saw earlier.

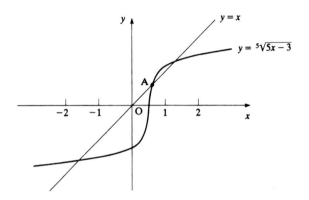

FIGURE 5.12

When an equation has two or more roots, a single rearrangement will not usually find all of them. This is demonstrated in figure 5.13.

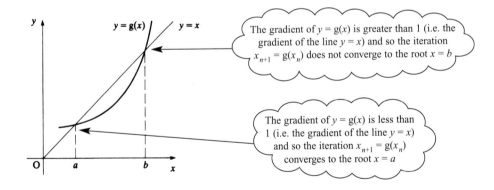

FIGURE 5.13

EXAMPLE 5.4

Show that $x_{n+1} = \sqrt{5x_n - 3}$, is an iterative formula which may be used to solve $x^2 - 5x + 3 = 0$.

Use the formula with $x_1 = 4$ to find one root of the equation correct to 3 decimal places.

Illustrate the convergence to the root on an appropriate diagram.

Iterative methods to solve f(x) = 0

Solution Rearranging the equation
$$x^2 - 5x + 3 = 0$$
gives $\quad x^2 = 5x - 3$
then $\quad x = \sqrt{5x - 3}$

which, as an iterative formula, is
$$x_{n+1} = \sqrt{5x_n - 3}.$$

The iteration gives
$$x_1 = 4$$
$$x_2 = 4.123...$$
$$x_3 = 4.197...$$
$$x_4 = 4.240...$$

which eventually converges to 4.303 to 3 decimal places.

To illustrate the convergence sketch $y = x$ and $y = \sqrt{5x - 3}$, on the same diagram (see figure 5.14).

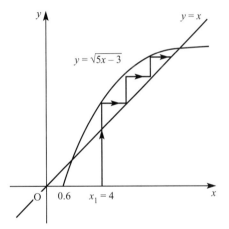

FIGURE 5.14

WHEN DOES THIS METHOD FAIL?

It is always possible to rearrange an equation $f(x) = 0$ into the form $x = g(x)$, but this only leads to a successful iteration if:

(a) successive iterations converge;

(b) they converge to the root for which you are looking.

Exercise 5B

1. (a) Show that the equation $x^3 - x - 2 = 0$ has a root between 1 and 2.
   (b) The equation is rearranged into the form $x = g(x)$, where $g(x) = \sqrt[3]{x + 2}$. Sketch $y = g(x)$ and show that starting values of both 1 and 2 will converge to the root in the interval $[1, 2]$.
   (c) Use the iterative formula suggested by this rearrangement to find the value of the root to 3 decimal places.

2. (a) Show that the equation $e^{-x} - x + 2 = 0$ has a root in the interval $[2, 3]$.
   (b) The equation is rearranged into the form $x = g(x)$ where $g(x) = e^{-x} + 2$. Use the iterative formula suggested by this rearrangement to find the value of the root to 3 decimal places.

3. (a) By considering $f'(x)$, where $f(x) = x^3 + x - 3$, show that there is exactly one real root of the equation $x^3 + x - 3 = 0$.
   (b) Show that the root lies in the interval $[1, 2]$.
   (c) Rearrange the equation to give the iterative formula $x_{n+1} = \sqrt[3]{3 - x_n}$.
   (d) Hence find the root correct to 4 decimal places.

4. (a) Show that the equation $e^x + x - 6 = 0$ has a root in the interval $[1, 2]$.
   (b) Show that this equation may be written in the form $x = \ln(6 - x)$.
   (c) Hence find the root correct to 3 decimal places.

5. (a) Sketch the curves $y = e^x$ and $y = x^2 + 2$ on the same graph.
   (b) Use your sketch to explain why the equation $e^x - x^2 - 2 = 0$ has only one root.
   (c) Rearrange this equation to give the iteration $x_{n+1} = \ln(x_n^2 + 2)$.
   (d) Find the root correct to 3 decimal places.

6. (a) On the same diagram sketch $y = x$ and $y = 3\ln(x + 1)$.
   (b) State one solution of $x = 3\ln(x + 1)$ and find the second, to 3 decimal places, by an iterative process.
   (c) Show on your diagram how the iterative process has converged to the root.

7. (a) Sketch the graphs of $y = x$ and $y = \cos x$ on the same axes, for $0 \leq x \leq \frac{\pi}{2}$.
   (b) Find the solution of the equation $x = \cos x$ to 5 decimal places.

8. Given that $x_{n+1} = \dfrac{(x_n^2 + 1)}{3}$.
   (a) Find the equation that this iteration solves.
   (b) Use the iteration with a starting value of 1 to find a solution of the equation correct to 3 significant figures.
   (c) By drawing $y = x$ and $y = \dfrac{(x^2 + 1)}{3}$ explain the convergence of your solutions.
   (d) Find an alternative rearrangement that gives a convergence to the other root, and find this root to 3 significant figures.

EXERCISE 5C

9. (a) On the same diagram sketch $y = \sin x$, $0 \leq x \leq \pi$, and $y = \frac{1}{2}x$.
   (b) Show that these graphs intersect at $x = 0$ and at some value of x between $x = 1.8$ and $x = 2.0$.
   (c) Using the iteraton $x_{n+1} = 2\sin x_n$ find the second value of x to 4 decimal places.

10. (a) For the domain $0 \leq x \leq \frac{\pi}{2}$ sketch, on the same diagram, the graphs of $y = e^x$ and $y = \cos x + 1$.
    (b) By considering a sign change in f(x), where f(x) = $e^x - \cos x - 1$, show that f(x) = 0 in the interval [0.5, 1].
    (c) By a suitable iterative process find the solution of $e^x - \cos x - 1 = 0$ to 3 decimal places.

Examination-style questions

1. A function f(x) is given by

   $$f(x) = e^x - 2x - 6.$$

   (a) Show that f(x) = 0 has a root in the interval [2, 3].
   (b) Find the value of the integer N, such that the root found in part (a) lies in the interval $\left[\frac{N}{10}, \frac{N+1}{10}\right]$.
   (c) Find between which two integers the negative root of f(x) = 0 lies.

2. A function f is given by

   $$f(x) = x^3 - 2x^2 - 8x + 8.$$

   (a) By a change of sign method show that f(x) = 0 has a root in the interval [0, 1].
   (b) Find the value of the negative integer, N, for which the negative root of f(x) = 0 lies in the interval [N, N + 1]. Show sufficient working to justify your answer.
   (c) How many other roots does f(x) = 0 have?

3. A circle of radius r has a sector OAB making an angle of θ radians at the centre O.

   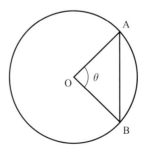

   (a) If the area of the sector is twice the area of the triangle OAB show that

   $$\sin\theta = \tfrac{1}{2}\theta.$$

(b) On the same diagram sketch $y = \sin\theta$ and $y = \frac{1}{2}\theta$ for $0 \leq \theta \leq \pi$.

(c) From your diagram give an approximate value for the non-zero solution to $\sin\theta = \frac{1}{2}\theta$ and, by a change of sign method, find the range of numbers within which this value lies correct to 2 decimal places.

4 A curve has equation $x^2 - 2x - 5 = 0$.

(a) Show that the equation can be rearranged to give the iteration

$$x_{n+1} = \sqrt{ax_n + b}$$

stating the values of a and b. Use the iteration with $x_1 = 3$ to find the positive solution to the quadratic equation correct to 3 decimal places, showing the result of each iteration.

(b) Prove that an alternative rearrangement gives the iteration

$$x_{n+1} = \frac{5}{x_n - 2}$$

and use this iteration to find the negative root to 3 decimal places.

5 Show that the equation $x^5 - 5x - 6 = 0$ has a root in the interval $[1, 2]$. Stating the values of the constants p, q and r, use an iteration of the form

$$x_{n+1} = (px_n + q)^{\frac{1}{r}}$$

the appropriate number of times to calculate this root of the equation $x^5 - 5x - 6 = 0$ correct to 3 decimal places. Show sufficient working to justify your final answer.

[Edexcel]

6 (a) By sketching the curves with equations $y = 4 - x^2$ and $y = e^x$ show that the equation $x^2 + e^x - 4 = 0$ has one negative root and one positive root.

(b) Use the iteration formula

$$x_{n+1} = -(4 - e^{x_n})^{\frac{1}{2}}$$

with $x_0 = -2$ to find in turn x_1, x_2, x_3 and x_4 and hence write down an approximation to the negative root of the equation, giving your answer to 4 decimal places.

An attempt to evaluate the positive root of the equation is made using the iteration formula

$$x_{n+1} = (4 - e^{x_n})^{\frac{1}{2}}$$

with $x_0 = 1.3$.

(c) Describe the result of such an attempt.

[Edexcel]

EXERCISE 5C

7 (a) Rearrange the cubic equation $x^3 - 6x - 2 = 0$ into the form
$$x = \pm\sqrt{a + \frac{b}{x}}.$$
State the values of a and b.

(b) Use the iterative formula
$$x_{n+1} = \sqrt{a + \frac{b}{x_n}}.$$
with $x_0 = 2$ and your values of a and b to find the approximate solution x_4 of the equation, to an appropriate degree of accuracy. Show all your intermediate answers.

[Edexcel]

8 The equation $x^x = 2$ has a solution near $x = 1.5$.

(a) Use the iteration formula
$$x_{n+1} = 2^{\frac{1}{x_n}}$$
with $x_0 = 1.5$ to find the approximate solution x_5 of the equation. Show the intermediate interations and give your final answer to 4 decimal places.

(b) Use the iteration formula
$$x_{n+1} = 2x_n^{(1-x_n)}$$
with $x_0 = 1.5$ to find x_1, x_2, x_3, x_4. Comment briefly on this sequence.

[Edexcel]

9 (a) Show that the equation $2^{1-x} = 4x + 1$ can be arranged in the form $x = \tfrac{1}{2}(2^{-x}) + q$, stating the value of the constant q.

(b) Using the iteration formula
$$x_{n+1} = \tfrac{1}{2}(2^{-x_n}) + q, \quad x_0 = 0.2,$$
with the value of q found in part **(a)**, find x_1, x_2, x_3 and x_4. Give the value of x_4 to 4 decimal places.

[Edexcel]

10 $f(x) = e^{0.8x} - \dfrac{1}{3 - 2x}, \quad x \neq \dfrac{3}{2}$

(a) Show that the equation $f(x) = 0$ can be written as
$$x = 1.5 - 0.5e^{-0.8x}.$$

(b) Use the iteration formula
$$x_{n+1} = 1.5 - 0.5e^{-0.8x_n},$$
with $x_0 = 1.3$, to obtain $x_1, x_2,$ and x_3. Give the value of x_3, an approximation to a root of $f(x) = 0$, to 3 decimal places.

(c) Show that the equation $f(x) = 0$ can be written in the form $x = p\ln(3 - 2x)$, stating the value of p.

(d) Use the iteration formula

$$x_{n+1} = p\ln(3 - 2x_n),$$

with $x_0 = -2.6$ and the value of p found in part (c), to obtain x_1, x_2 and x_3. Give the value of x_3, an approximation to the second root of $f(x) = 0$, to 3 decimal places.

[Edexcel]

11

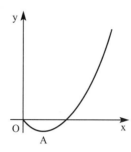

The diagram shows part of the curve with equation $y = f(x)$ where

$$f(x) = x(1 + x)\ln x, \quad x > 0.$$

The point A is the minimum point of the curve.

(a) Find $f'(x)$.

(b) Hence show that the x-coordinate of A is the solution of the equation $x = g(x)$, where

$$g(x) = e^{-\frac{1+x}{1+2x}}.$$

(c) Use the iteration $x_{n+1} = g(x_n)$, with $x_0 = 0.45$, to find x_1, x_2 and x_3. Give the value of x_3 to 4 decimal places.

(d) Using your value of x_3 find, to 2 decimal places, the coordinates of A.

[Edexcel]

12 The root of the equation $f(x) = 0$, where

$$f(x) = x + \ln 2x - 4,$$

is to be estimated using the iterative formula $x_{n+1} = 4 - \ln 2x_n$, with $x_0 = 2.4$.

(a) Showing your values of $x_1, x_2, x_3, \ldots$, obtain the value, to 3 decimal places, of the root.

(b) By considering the change of sign of $f(x)$ in a suitable interval, justify the accuracy of your answer to part (a).

[Edexcel]

13

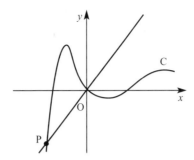

The diagram shows part of the curve C with equation $y = f(x)$, where

$$f(x) = (x^3 - 2x)e^{-x}.$$

(a) Find $f'(x)$.

The normal to C at the origin O intersects C at a point P, as shown in the diagram.

(b) Show that the x-coordinate of P is the solution of the equation

$$2x^2 = e^x + 4.$$

(c) Use the iteration $x_{n+1} = -\sqrt{0.5e^{x_n} + 2}$ with $x_0 = -1.5$ to obtain x_1, x_2, x_3 and x_4 and give, to 3 decimal places, the x-coordinate of P.

[Edexcel]

14

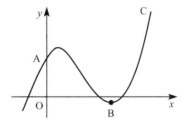

The diagram shows part of the curve C with equation $y = f(x)$, where

$$f(x) = 0.5e^x - x^2.$$

The curve C cuts the y-axis at A and there is a minimum at the point B.

(a) Find an equation of the tangent to C at A.

The x-coordinate of B is approximately 2.15. A more exact estimate is to be made of this coordinate using iterations $x_{n+1} = \ln g(x_n)$.

(b) Show that a possible form for $g(x)$ is $g(x) = 4x$.

(c) Using $x_{n+1} = \ln 4x_n$, with $x_0 = 2.15$, calculate x_1, x_2 and x_3. Give the value of x_3 to 4 decimal places.

[Edexcel]

15 (a) Sketch, on the same set of axes, the graphs of

$$y = 2 - e^{-x} \text{ and } y = \sqrt{x}.$$

[It is not necessary to find the coordinates of any points of intersection with the axes.]

Given that $f(x) = e^{-x} + \sqrt{x} - 2$, $x \geq 0$,

(b) explain how your graphs show that the equation $f(x) = 0$ has only one solution,

(c) show that the solution of $f(x) = 0$ lies between $x = 3$ and $x = 4$.

The iterative formula $x_{n+1} = (2 - e^{-x_n})^2$ is used to solve the equation $f(x) = 0$.

(d) Taking $x_0 = 4$, write down the values of x_1, x_2, x_3 and x_4, and hence find an approximation to the solution of $f(x) = 0$, giving your answer to 3 decimal places.

KEY POINTS

1 Location of roots by change of sign

If $y = f(x)$ is continuous over $[a, b]$ then $f(x) = 0$ has a root α in $[a, b]$ if $f(a)$ and $f(b)$ are of opposite sign. For example:

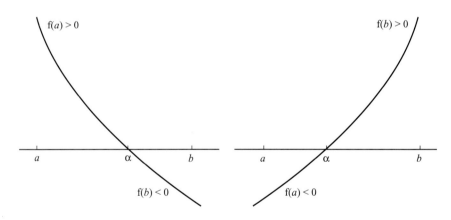

2 Iterative methods

To solve $f(x) = 0$ rewrite the equation as $x = g(x)$ to give the iterative formula

$$x_{n+1} = g(x_n).$$

The iteration will converge if $|g'(x)| < 1$ close to the root.

Core 4

C4

Chapter six

ALGEBRA

We will grieve not, rather find
Strength in what remains behind.

William Wordsworth

RATIONAL FUNCTIONS

You will recall from Chapter 1 that a rational function is an algebraic fraction of the type $\frac{f(x)}{g(x)}$, where f(x) and g(x) are polynomials in the variable x. The following examples will remind you of the type of problems you studied.

EXAMPLE 6.1

Simplify $\frac{2x+1}{x-2} \times \frac{x^2-4}{2x^2-x-1}$.

Solution

$$\frac{2x+1}{x-2} \times \frac{x^2-4}{2x^2-x-1} = \frac{\cancel{2x+1}}{\cancel{x-2}} \times \frac{(x+2)\cancel{(x-2)}}{\cancel{(2x+1)}(x-1)}$$

Factorise then cancel

$$= \frac{x+2}{x-1}$$

EXAMPLE 6.2

Express $\frac{1}{1+x} + \frac{2}{2-x}$ as a single fraction.

Solution

$$\frac{1}{1+x} + \frac{2}{2-x} = \frac{(2-x) + 2(1+x)}{(1+x)(2-x)}$$

Find a common denominator

$$= \frac{4+x}{(1+x)(2-x)}$$

EXAMPLE 6.3

Prove that $\dfrac{2x+3}{x^2+1} - \dfrac{1}{x-1} \equiv \dfrac{x^2+x-4}{(x^2+1)(x-1)}$.

(You will recall that the symbol '≡' means 'is equivalent to'.)

Solution

$$\dfrac{2x+3}{x^2+1} - \dfrac{1}{x-1} \equiv \dfrac{(2x+3)(x-1) - (x^2+1)}{(x^2+1)(x-1)}$$

$$\equiv \dfrac{x^2+x-4}{(x^2+1)(x-1)}$$

EXERCISE 6A

1. Simplify $\dfrac{x^2-9}{x^2-4} \times \dfrac{x+2}{x+3}$.

2. Express $\dfrac{1}{x+2} - \dfrac{1}{x+3}$ as a single fraction.

3. Prove that $\dfrac{1}{x-1} + \dfrac{1-x}{x^2+4} \equiv \dfrac{2x+3}{(x-1)(x^2+4)}$.

4. Simplify $\dfrac{x-1}{x-5} \times \dfrac{2x^2+4x}{x^2+x-2}$.

5. Express $\dfrac{2}{x-1} - \dfrac{2}{x-2} + \dfrac{3}{(x-2)^2}$ as a single fraction.

6. Prove that $\dfrac{3}{x} - \dfrac{3}{x^2} + \dfrac{2}{x+1} \equiv \dfrac{5x^2-3}{x^2(x+1)}$.

PARTIAL FRACTIONS

Until this point, any instruction to simplify an algebraic fractional expression was asking you to give the expression as a single fraction. Sometimes you will need to do the reverse process, that is, to start with a single fraction, for example $\dfrac{4+x}{(1+x)(2-x)}$, and write it as two (or more) simple fractions. In this case they would be $\dfrac{1}{1+x} + \dfrac{2}{2-x}$ (this comes from Example 6.2).

This process of separating a single algebraic fraction into two or more simple fractions is called expressing the algebraic fraction in *partial fractions*.

Binomial expansions sometimes require algebraic expressions to be written in partial fractions. You will see how to do this in Chapter 8. Some derivatives are best found after the functions are expressed in partial fractions. Expressing in partial fractions also allows integration of some algebraic expressions to be done and you will study this in Chapter 10.

An algebraic fraction is said to be *proper* if the order (the highest power of x) of the numerator is strictly less than that of the denominator; for example

$$\frac{2}{1+x}, \quad \frac{5x-1}{x^2-3} \quad \text{and} \quad \frac{7x}{(x+1)(x-2)} \quad \text{are all } \textit{proper} \text{ fractions.}$$

If the order of the numerator is equal to or greater than that of the denominator then the fraction is said to be *improper*; for example

$$\frac{x^2}{x-2} \quad \text{and} \quad \frac{x^2-1}{(x+2)(x+3)} \quad \text{are } \textit{improper} \text{ fractions.}$$

It can be shown that, when a proper algebraic fraction is decomposed into its partial fractions, each of the partial fractions will be a proper fraction.

The methods required to write an algebraic expression in terms of its partial fractions depend upon the denominator of the fraction. These methods are described in the following pages as different types of problems.

TYPE 1: DENOMINATORS OF THE FORM $(ax + b)(cx + d)$

EXAMPLE 6.4

Express $\dfrac{4+x}{(1+x)(2-x)}$ as a sum of partial fractions.

Solution Assume

$$\frac{4+x}{(1+x)(2-x)} \equiv \frac{A}{1+x} + \frac{B}{2-x}$$

> Remember: a linear denominator $\Rightarrow$ a constant numerator if the fraction is to be a proper fraction

Multiplying both sides by $(1+x)(2-x)$ gives

$$4 + x \equiv A(2-x) + B(1+x). \quad \text{①}$$

This is an identity; it is true for all values of x.

There are two possible ways in which you can find the constants A and B.

You can either

- substitute *any two* values of x in ① (two values are needed to give two equations to solve for the two unknowns A and B); or
- equate the constant terms to give one equation (this is the same as putting $x = 0$) and the coefficients of x to give another.

Sometimes one method is easier than the other, and in practice you will often want to use a combination of the two.

Substitution

Although you can substitute any two values of x, the easiest to use are $x = 2$ and $x = -1$, since each makes the value of one bracket zero in the identity.

$$4 + x \equiv A(2 - x) + B(1 + x)$$
$$x = 2 \quad \Rightarrow \quad 4 + 2 = A(2 - 2) + B(1 + 2)$$
$$6 = 3B \quad \Rightarrow \quad B = 2$$
$$x = -1 \quad \Rightarrow \quad 4 - 1 = A(2 + 1) + B(1 - 1)$$
$$3 = 3A \quad \Rightarrow \quad A = 1$$

Substituting these values for A and B gives

$$\frac{4 + x}{(1 + x)(2 - x)} \equiv \frac{1}{1 + x} + \frac{2}{2 - x}.$$

Equating coefficients

In this method, you write the right-hand side of

$$4 + x \equiv A(2 - x) + B(1 + x)$$

as a polynomial in x, and then compare the coefficients of the various terms:

$$4 + x \equiv 2A - Ax + B + Bx$$
$$4 + 1x \equiv (2A + B) + (-A + B)x.$$

Equating the constant terms: $4 = 2A + B$.

Equating the coefficients of x: $1 = -A + B$.

These are simultaneous equations in A and B

Solving these simultaneous equations gives $A = 1$ and $B = 2$ as before.

Note In some cases it is necessary to factorise the denominator before finding the partial fractions.

EXAMPLE 6.5

Express $\dfrac{2}{4 - x^2}$ as a sum of partial fractions.

Solution
$$\frac{2}{4 - x^2} \equiv \frac{2}{(2 + x)(2 - x)}$$
$$\equiv \frac{A}{2 + x} + \frac{B}{2 - x}$$

Multiplying both sides by $(2 + x)(2 - x)$ gives

$$2 \equiv A(2 - x) + B(2 + x)$$

$$x = -2 \quad \Rightarrow \quad 2 = 4A \quad \Rightarrow \quad A = \tfrac{1}{2}$$
$$x = 2 \quad \Rightarrow \quad 2 = 4B \quad \Rightarrow \quad B = \tfrac{1}{2}$$

Using these values

$$\frac{2}{(2+x)(2-x)} \equiv \frac{\frac{1}{2}}{2+x} + \frac{\frac{1}{2}}{2-x}$$

$$\therefore \frac{2}{4-x^2} \equiv \frac{1}{2(2+x)} + \frac{1}{2(2-x)}.$$

Type 2: Denominators of the form $(ax+b)(cx+d)(ex+f)$

EXAMPLE 6.6

Express $\dfrac{1}{(x-2)(x-1)(x+1)}$ as a sum of partial fractions.

Solution

$$\frac{1}{(x-2)(x-1)(x+1)} \equiv \frac{A}{x-2} + \frac{B}{x-1} + \frac{C}{x+1}$$

Multiplying both sides by $(x-2)(x-1)(x+1)$ gives

$$1 \equiv A(x-1)(x+1) + B(x-2)(x+1) + C(x-2)(x-1)$$

$x = 2 \Rightarrow 1 = 3A \Rightarrow A = \frac{1}{3}$

$x = 1 \Rightarrow 1 = -2B \Rightarrow B = -\frac{1}{2}$

$x = -1 \Rightarrow 1 = 6C \Rightarrow C = \frac{1}{6}$

Substituting the values for A, B and C gives

$$\frac{1}{(x-2)(x-1)(x+1)} \equiv \frac{1}{3(x-2)} - \frac{1}{2(x-1)} + \frac{1}{6(x+1)}.$$

EXERCISE 6B

Express each of the following fractions as a sum of partial fractions.

1. $\dfrac{5}{(x-2)(x+3)}$

2. $\dfrac{1}{x(x+1)}$

3. $\dfrac{6}{(x-1)(x-4)}$

4. $\dfrac{x+5}{(x-1)(x+2)}$

5. $\dfrac{3x}{(2x-1)(x+1)}$

6. $\dfrac{4}{x^2-2x}$

7. $\dfrac{2}{(x-1)(3x-1)}$

8. $\dfrac{x-1}{x^2-3x-4}$

9. $\dfrac{x+2}{2x^2-x}$

10. $\dfrac{7}{2x^2+x-6}$

11. $\dfrac{2x-1}{2x^2+3x-20}$

12. $\dfrac{2x+5}{18x^2-8}$

13. $\dfrac{6}{(x-3)(x-1)(x+2)}$

14. $\dfrac{2x+1}{(x-1)(x-2)(x-4)}$

15. $\dfrac{1}{(2x+1)(2x-1)(3x-1)}$

PARTIAL FRACTIONS

TYPE 3: DENOMINATORS OF THE FORM $(ax + b)(cx + d)^2$

The factor $(cx + d)^2$ is of order 2, so it would have an order 1 numerator in the partial fractions. However, in the case of a repeated factor there is a simpler form.

Consider $\dfrac{4x + 5}{(2x + 1)^2}$.

This can be written as $\dfrac{2(2x + 1) + 3}{(2x + 1)^2} \equiv \dfrac{2(2x + 1)}{(2x + 1)^2} + \dfrac{3}{(2x + 1)^2}$

$$\equiv \dfrac{2}{(2x + 1)} + \dfrac{3}{(2x + 1)^2}.$$

Notice how, in this form, both the numerators are constant.

In a similar way, any fraction of the form $\dfrac{px + q}{(cx + d)^2}$ can be written as

$$\dfrac{A}{(cx + d)} + \dfrac{B}{(cx + d)^2}.$$

When expressing an algebraic fraction in partial fractions, you are aiming to find the simplest partial fractions possible, so here you would want the form where the numerators are constant.

EXAMPLE 6.7

Express $\dfrac{x + 1}{(x - 1)(x - 2)^2}$ as a sum of partial fractions.

Solution   Let $\dfrac{x + 1}{(x - 1)(x - 2)^2} \equiv \dfrac{A}{x - 1} + \dfrac{B}{x - 2} + \dfrac{C}{(x - 2)^2}.$

> Notice that you only need $(x - 2)^2$ here and not $(x - 2)^3$

Multiplying both sides by $(x - 1)(x - 2)^2$ gives

$$x + 1 \equiv A(x - 2)^2 + B(x - 1)(x - 2) + C(x - 1).$$

$x = 1$ (so that $x - 1 = 0$) $\Rightarrow$ $2 = A(-1)^2$ $\Rightarrow$ $A = 2$
$x = 2$ (so that $x - 2 = 0$) $\Rightarrow$ $3 = C$

Equating coefficients of x^2: $0 = A + B$ $\Rightarrow$ $B = -2$.

This gives

$$\dfrac{x + 1}{(x - 1)(x - 2)^2} \equiv \dfrac{2}{x - 1} - \dfrac{2}{x - 2} + \dfrac{3}{(x - 2)^2}.$$

EXAMPLE 6.8

Express $\dfrac{5x^2 - 3}{x^2(x + 1)}$ as a sum of partial fractions.

Solution Let $\dfrac{5x^2 - 3}{x^2(x + 1)} \equiv \dfrac{A}{x} + \dfrac{B}{x^2} + \dfrac{C}{x + 1}$.

Multiplying both sides by $x^2(x + 1)$ gives

$$5x^2 - 3 \equiv Ax(x + 1) + B(x + 1) + Cx^2$$

$$x = 0 \quad \Rightarrow \quad -3 = B$$

$$x = -1 \quad \Rightarrow \quad 2 = C.$$

Equating coefficients of x^2: $5 = A + C \quad \Rightarrow \quad A = 3$.

This gives $\dfrac{5x^2 - 3}{x^2(x + 1)} \equiv \dfrac{3}{x} - \dfrac{3}{x^2} + \dfrac{2}{x + 1}$.

EXERCISE 6C

Express each of the following fractions as a sum of partial fractions.

1. $\dfrac{4}{(1 - 3x)(1 - x)^2}$
2. $\dfrac{5 + 2x}{(2x - 1)(x + 1)^2}$
3. $\dfrac{5 - 2x}{(x - 1)^2(x + 2)}$
4. $\dfrac{2x + 5}{(x - 2)(x + 4)^2}$
5. $\dfrac{23 - 13x}{(2x - 3)(x + 2)}$
6. $\dfrac{x^2 - 1}{x^2(2x + 1)}$
7. $\dfrac{x + 1}{x(3x - 1)^2}$
8. $\dfrac{2x^2 + x + 2}{(2x + 1)^2(x + 1)}$
9. $\dfrac{4x^2 - 3}{x(2x - 1)^2}$
10. $\dfrac{4}{(1 - x)(1 + x)^2}$
11. $\dfrac{2x + 3}{(1 + x)^2(2x - 1)}$
12. $\dfrac{x + 3}{(4 - x)(2 + x)^2}$
13. $\dfrac{4x - 1}{(2 - x)(1 + x)(2 + x)^2}$
14. $\dfrac{x + 2}{(x - 2)(x - 3)^2}$
15. $\dfrac{x + 1}{(x + 1)^2(x - 2)^2}$

TYPE 4: IMPROPER FRACTIONS

If the order of the numerator is greater than or equal to that of the denominator then the algebraic fraction is improper. Before the partial fractions are found you must first divide the denominator into the numerator.

EXAMPLE 6.9

Express $\dfrac{x^2 + 4x - 7}{(x - 3)(x + 4)}$ as a sum of partial fractions.

Solution $\dfrac{x^2 + 4x - 7}{(x - 3)(x + 4)}$ *Both numerator and denominator are quadratic so you have to divide*

The denominator is $(x-3)(x+4) = x^2 + x - 12$ so dividing gives

$$\begin{array}{r} 1 \\ x^2 + x - 12 \overline{\smash{)}x^2 + 4x - 7} \\ \underline{x^2 + x - 12} \\ 3x + 5 \end{array}$$

The algebraic fraction can be written as

$$\frac{x^2 + 4x - 7}{(x-3)(x+4)} \equiv 1 + \frac{3x+5}{(x-3)(x+4)} \qquad \text{①}$$

Now express this part in partial fractions

$$\frac{3x+5}{(x-3)(x+4)} \equiv \frac{A}{x-3} + \frac{B}{x+4}$$

$$3x + 5 \equiv A(x+4) + B(x-3)$$

$x = 3 \quad \Rightarrow \quad 14 = 7A \quad \Rightarrow \quad A = 2$
$x = -4 \quad \Rightarrow \quad -7 = -7B \quad \Rightarrow \quad B = 1.$

Substituting into ① gives the partial fractions

$$\frac{x^2 + 4x - 7}{(x-3)(x+4)} \equiv 1 + \frac{2}{x-3} + \frac{1}{x+4}.$$

EXAMPLE 6.10

Express $\dfrac{x^3 + 1}{(x+2)(x-1)}$ in partial fractions.

Solution

$$\frac{x^3 + 1}{(x+2)(x-1)} \equiv \frac{x^3 + 1}{x^2 + x - 2}$$

This is an improper fraction so you have to divide the denominator into the numerator.

$$\begin{array}{r} x - 1 \\ x^2 + x - 2 \overline{\smash{)}x^3 + 0x^2 + 0x + 1} \\ \underline{x^3 + x^2 - 2x} \\ -x^2 + 2x + 1 \\ \underline{-x^2 - x + 2} \\ 3x - 1 \end{array}$$

The algebraic fraction becomes

$$\frac{x^3 + 1}{(x+2)(x-1)} \equiv x - 1 + \frac{3x - 1}{(x+2)(x-1)}.$$

Expressing the final part in partial fractions gives:

$$\frac{3x-1}{(x+2)(x-1)} \equiv \frac{A}{x+2} + \frac{B}{x-1}$$

$$3x - 1 \equiv A(x-1) + B(x+2)$$

$x = -2 \implies -7 = -3A \implies A = \frac{7}{3}$

$x = 1 \implies 2 = 3B \implies B = \frac{2}{3}$.

The final answer is

$$\frac{x^3+1}{(x+2)(x-1)} \equiv x - 1 + \frac{7}{3(x+2)} + \frac{2}{3(x-1)}.$$

EXERCISE 6D

Express each of the following in partial fractions.

1. $\dfrac{x^2 + 3x - 5}{(x-1)(x-3)}$

2. $\dfrac{x^3 + x^2}{(x-2)(x+3)}$

3. $\dfrac{(x+1)(x-1)}{(x+2)(x-2)}$

4. $\dfrac{x^2 + 3x + 5}{x^2 + 5x + 6}$

5. $\dfrac{x^3 + x}{(x-2)(x-5)}$

6. $\dfrac{x^3 - 2}{(x+1)(x^2+1)}$

7. $\dfrac{x^4 + 2x^3 - 3x^2}{(x-2)(x+2)^2}$

8. $\dfrac{x^3 - 15x - 15}{(x+1)(x+3)(x-5)}$

9. $\dfrac{4x^2}{(2x+1)(x-3)}$

10. $\dfrac{x^4 + 1}{(x^2+1)(x-1)^2}$

PARTIAL FRACTIONS AND DIFFERENTIATION

If rational functions can be decomposed into partial fractions then it is usually best to do so before differentiating.

EXAMPLE 6.11

You are given that $f(x) = \dfrac{4x^2 - 18x + 8}{(x-2)(x-3)(x-5)}$.

(a) Express $f(x)$ in partial fractions.
(b) Show that $f'(4) = -7$.

Solutions (a) Let $\dfrac{4x^2 - 18x + 8}{(x-2)(x-3)(x-5)} \equiv \dfrac{A}{x-2} + \dfrac{B}{x-3} + \dfrac{C}{x-5}$

$\therefore 4x^2 - 18x + 8 \equiv A(x-3)(x-5) + B(x-2)(x-5) + C(x-2)(x-3)$

$x = 2 \quad -12 = 3A \implies A = -4$

$x = 3 \quad -10 = -2B \implies B = 5$

$x = 5 \quad 18 = 6C \implies C = 3$

$$\therefore f(x) = -\frac{4}{x-2} + \frac{5}{x-3} + \frac{3}{x-5}$$

(b) $f'(x) = +\frac{4}{(x-2)^2} - \frac{5}{(x-3)^2} - \frac{3}{(x-5)^2}$

$\left(\text{since if } y = -\frac{4}{(x-2)} = -4(x-2)^{-1}, \frac{dy}{dx} = -4 \times -1 \times (x-2)^{-2} = \frac{4}{(x-2)^2}\cdots\right)$

$\therefore f'(4) = \frac{4}{4} - \frac{5}{1} - \frac{3}{1} = -7$

EXAMPLE 6.12

$$f(x) = \frac{2x^2 + 17x + 24}{(x+2)^2(x+4)}$$

Express $f(x)$ in partial fractions. Hence find $f'(-3)$.

Solution Let $\dfrac{2x^2 + 17x + 24}{(x+2)^2(x+4)} = \dfrac{A}{x+2} + \dfrac{B}{(x+2)^2} \equiv \dfrac{C}{x+4}$

$\therefore 2x^2 + 17x + 24 \equiv A(x+2)(x+4) + B(x+4) + C(x+2)^2$

$x = -2 \qquad -2 = 2B \qquad \Rightarrow \quad B = -1$
$x = -4 \qquad -12 = 4C \qquad \Rightarrow \quad C = -3$
$x = 0 \qquad 24 = 8A - 4 - 12 \Rightarrow \quad A = 5$

$\therefore f(x) = \dfrac{5}{x+2} - \dfrac{1}{(x+2)^2} - \dfrac{3}{(x+4)}$

$f'(x) = -\dfrac{5}{(x+2)^2} + \dfrac{2}{(x+2)^3} + \dfrac{3}{(x+4)^2}$

$\Rightarrow f'(-3) = -5 - 2 + 3 = -4$

We shall use partial fractions with binomial expansions and with integration later in this book.

EXERCISE 6E

1 You are given that $f(x) = \dfrac{5x - 180}{(x+4)(x-6)}$

  (a) Express $f(x)$ in partial fractions.
  (b) Find $f'(x)$ and show that $f'(1) = -\frac{1}{5}$.

2 The curve C has equation $y = \dfrac{5x - 5}{(3x+1)(2x-1)}$

  (a) Express $\dfrac{5x-5}{(3x+1)(2x-1)}$ in the form $\dfrac{A}{3x+1} + \dfrac{B}{2x-1}$
  where A and B are constants to be found.
  (b) Show that the gradient of C at the point where $x = 1$ is $\frac{5}{4}$.

3 $f(x) \equiv \dfrac{x - 8}{(x + 2)(2x - 1)}$

   (a) Express $f(x)$ in partial fractions.
   (b) Find $f'(0)$.

4 You are given that
   $$f(x) = \dfrac{7x + 13}{x^2 + 2x - 3}.$$

   (a) Use the quotient rule to find $f'(x)$.
   (b) Express $f(x)$ in partial fractions. Differentiate your answer to obtain another expression for $f'(x)$.
   (c) Show that your answers to (a) and (b) are equivalent to each other.

5 $f(x) \equiv \dfrac{5x^2 - 15x + 4}{(1 - x)(2 - x)(4 - x)}$

   (a) Express $f(x)$ in partial fractions.
   (b) Show that $f'(3) = \tfrac{13}{2}$.

6 You are given that $f(x) = \dfrac{4x^2 + 9x - 1}{(x + 2)(x^2 - 1)}$.

   (a) Express $f(x)$ in the form $\dfrac{A}{x + 2} + \dfrac{B}{x + 1} + \dfrac{C}{x - 1}$ where A, B and C are constants to be found.
   (b) Find $f'(x)$.

7 $f(x) \equiv \dfrac{8x^2 - x + 1}{(4 - x)(x + 1)^2}$

   (a) Express $f(x)$ in partial fractions.
   (b) Show that $f'(x) = \dfrac{5}{(4 - x)^2} + \dfrac{3}{(x + 1)^2} - \dfrac{4}{(x + 1)^3}$.
   (c) Find $f'(3)$.

8 $f(x) = \dfrac{14x^2 + 7x - 32}{(1 - 2x)(2 + x)^2}$

   (a) Write $f(x)$ in the form $\dfrac{A}{1 - 2x} + \dfrac{B}{2 + x} + \dfrac{C}{(2 + x)^2}$ where A, B and C are constants to be found.
   (b) Show that $f'(1) = -7\tfrac{4}{27}$.

9 $f(x) \equiv \dfrac{x^3 - 6x^2 - x + 28}{(x - 3)(x - 5)}$

   (a) Express $f(x)$ in the form $Ax + B + \dfrac{C}{x - 3} + \dfrac{D}{x - 5}$ where A, B, C and D are constants to be found.
   (b) Show that $f'(x) = 1 - \dfrac{1}{(x - 3)^2} + \dfrac{1}{(x - 5)^2}$.
   (c) Find $f'(4)$.

10 You are given that $f(x) = \dfrac{2x^3 - x^2 - 17x + 2}{(x+3)(x-2)}$.

  (a) Find the quotient and the remainder when $2x^3 - x^2 - 17x + 2$ is divided by $x^2 + x - 6$.
  (b) Express $f(x)$ in partial fractions.
  (c) Find $f'(x)$ and show that $f'(-2) = \tfrac{1}{4}$.

EXERCISE 6F   Examination-style questions

1 $f(x) = \dfrac{x^2 + 7x + 8}{(x+1)(x+2)(x+3)}$

Express $f(x)$ in the form $\dfrac{A}{x+1} + \dfrac{B}{x+2} + \dfrac{C}{x+3}$ where the constants A, B and C are to be found.

2 $f(x) = \dfrac{x+5}{(x-1)(x+1)^2}$

Express $f(x)$ in the form $\dfrac{A}{x-1} + \dfrac{B}{x+1} + \dfrac{C}{(x+1)^2}$ where the constants A, B and C are to be found.

3 $f(x) = \dfrac{4x^2 - 3x + 2}{(x-2)(2x+1)}$

  (a) Show that $f(x) = 2 + \dfrac{3x + 6}{(x-2)(2x+1)}$.
  (b) Express $f(x)$ in terms of partial fractions.
  (c) What happens to $f(x)$ as $x \to \pm\infty$?

4 $f(x) = \dfrac{x^2}{x^2 - 1}$

  (a) Express $f(x)$ in partial fractions.
  (b) Show that $f'(1) = -\tfrac{4}{9}$.

5 Find the values of A, B and C given that $\dfrac{6x^2 - 9x + 2}{x(x-1)^2} \equiv \dfrac{A}{x} + \dfrac{B}{x-1} + \dfrac{C}{(x-1)^2}$.

6 Write $\dfrac{9x^2 + 5x - 6}{(x-1)(2x-1)(3x+1)}$ in partial fractions.

7 $f(x) = \dfrac{4x^2 + 19x + 10}{(x-2)(x+2)^2}$

  (a) Express $f(x)$ in partial fractions.
  (b) Use your answer to part (a) to verify that $f(-1) = \tfrac{5}{3}$.
  (c) Find $f(1)$.

8 Write $\dfrac{x^2 + 3x + 5}{x^2 + x - 6}$ in the form $A + \dfrac{B}{x - 2} + \dfrac{C}{x + 3}$ where the values of A, B and C are to be found.

9 Express $\dfrac{5x + 13}{(x^2 - 1)(x + 2)^2}$ in partial fractions.

10 Find A, B, C and D given that
$$\dfrac{4x^3 - 23x^2 + 52x - 7}{(2x - 1)(x - 3)^2} \equiv A + \dfrac{B}{2x - 1} + \dfrac{C}{x - 3} + \dfrac{D}{(x - 3)^2}.$$

KEY POINTS

Partial fractions

1 A proper algebraic fraction with a denominator which factorises can be decomposed into a sum of proper partial fractions.

2 The following forms of partial fraction should be used:

$$\dfrac{px + q}{(ax + b)(cx + d)} \equiv \dfrac{A}{ax + b} + \dfrac{B}{cx + d}$$

$$\dfrac{px^2 + qx + r}{(ax + b)(cx + d)(ex + f)} \equiv \dfrac{A}{ax + b} + \dfrac{B}{cx + d} + \dfrac{C}{ex + f}$$

$$\dfrac{px^2 + qx + r}{(ax + b)(cx + d)^2} \equiv \dfrac{A}{ax + b} + \dfrac{B}{cx + d} + \dfrac{C}{(cx + d)^2}$$

3 If the order of the numerator is greater than or equal to that of the denominator then the algebraic fraction is improper. In such cases you must first divide the denominator into the numerator.

Chapter seven

Coordinate Geometry in the (x, y) Plane

O tribe of spirits and of men, if you are able to slip through the parameters of the skies and the earth, then do so.

Qur'ān

Parametric equations of curves

Parametric equations are ones in which x and y are given in terms of some *other measure* or *parameter*. The parameter that is often used is t, or, if you are working with trigonometric functions, θ.

Parametric equations are very useful in situations where an otherwise complicated equation may be expressed reasonably in terms of the parameter. Indeed, there are some curves which can be given by parametric equations but cannot be expressed as Cartesian equations (in terms of x and y only).

In this chapter you will learn how to convert parametric equations into Cartesian ones and how to find the area underneath a curve given by its parametric equations.

Although you are not expected in the Core 4 examination to be able to sketch graphs from their parametric equations, it is a useful skill which will be touched on in some of the examples.

Finding the parametric equation by eliminating the parameter

For some pairs of parametric equations it is possible to eliminate the parameter and obtain the Cartesian equation for the curve. This is usually done by making the parameter the subject of one of the equations and substituting this expression into the other.

EXAMPLE 7.1

A curve C is defined by the parametric equations $x = t^2$, $y = 2t$. Eliminate t from the parametric equations and sketch the graph of C.

Solution To eliminate t first make it the subject of the y-equation:

$$y = 2t \implies t = \frac{y}{2}$$

then substitute into the x-equation, which gives

$$x = t^2 \implies x = \left(\frac{y}{2}\right)^2 = \frac{y^2}{4}$$

or
$$y^2 = 4x$$

which is the equation of a parabola. This is drawn in figure 7.1.

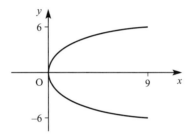

FIGURE 7.1

If you had wished to draw the curve from the original parametric equations then you could have chosen different values of t:

t	−3	−2	−1	0	1	2	3
$x = t^2$	9	4	1	0	1	4	9
$y = 2t$	−6	−4	−2	0	2	4	6

You could then plot the (x, y) coordinates to give the curve shown in figure 7.1.

EXAMPLE 7.2

Eliminate t from the equations $x = t^3 - 2t^2$, $y = \frac{t}{2}$.

Solution $y = \frac{t}{2} \implies t = 2y$

Substituting this in the equation $x = t^3 - 2t^2$ gives

$$x = (2y)^3 - 2(2y)^2 \quad \text{or} \quad x = 8y^3 - 8y^2.$$

Sometimes you need to consider the parametric equations simultaneously. There is often more than one way in which you can do this, and the next example gives two different options.

EXAMPLE 7.3

The parametric equations of a curve are

$$x = t + \frac{1}{t} \quad y = t - \frac{1}{t}.$$

(a) Find the coordinates of the points corresponding to $t = -2, -1, -0.5, 0, 0.5, 1, 2$.
(b) Sketch the curve for $-2 \leq t \leq 2$.
(c) For what values of x is the curve undefined?
(d) Eliminate the parameter by (i) first finding $x + y$ and (ii) first squaring x and y.

Solution

(a)

t	-2	-1	-0.5	0	0.5	1	2
x	-2.5	-2	-2.5	undefined	2.5	2	2.5
y	-1.5	0	1.5	undefined	-1.5	0	1.5

(b)

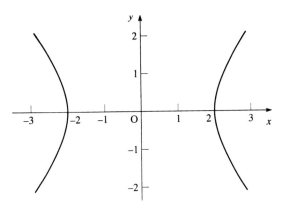

FIGURE 7.2

(c) The curve is undefined for $-2 < x < 2$.

(d) (i) Adding the two equations gives

$$x + y = 2t \quad \text{or} \quad t = \frac{x + y}{2}.$$

Substituting for t in the first equation (it could be either one) gives

$$x = \frac{x + y}{2} + \frac{2}{x + y}.$$

At this point the parameter t has been eliminated, but the equation is not in its neatest form. Multiplying by $2(x + y)$ to eliminate the fractions:

$$2x(x + y) = (x + y)^2 + 4$$
$$\Rightarrow \quad 2x^2 + 2xy = x^2 + 2xy + y^2 + 4$$
$$\Rightarrow \quad x^2 - y^2 = 4$$

(ii) Squaring gives

$$x^2 = t^2 + 2 + \frac{1}{t^2}$$

$$y^2 = t^2 - 2 + \frac{1}{t^2}.$$

Subtracting gives

$$x^2 - y^2 = 4.$$

Note

Figure 7.3 shows that the curve is the rectangular hyperbola $xy = 2$ rotated clockwise through 45°.

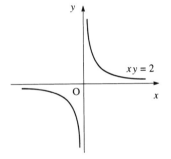
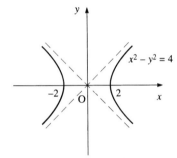

FIGURE 7.3

TRIGONOMETRIC PARAMETRIC EQUATIONS

When trigonometric functions are used in parametric equations, a particular trigonometric identity may help you to eliminate the parameter. The next example illustrates this.

EXAMPLE 7.4

The parametric equations of a curve are $x = 4\cos\theta$, $y = 3\sin\theta$.

(a) Find the Cartesian equation of the curve.
(b) Sketch the curve.

FINDING THE PARAMETRIC EQUATION BY ELIMINATING THE PARAMETER

Solution (a) The identity which connects $\cos\theta$ and $\sin\theta$ is

$$\cos^2\theta + \sin^2\theta = 1. \quad \text{①}$$

$$x = 4\cos\theta \quad \Rightarrow \quad \cos\theta = \frac{x}{4}$$

$$y = 3\sin\theta \quad \Rightarrow \quad \sin\theta = \frac{y}{3}$$

Substituting these in ① gives

$$\left(\frac{x}{4}\right)^2 + \left(\frac{y}{3}\right)^2 = 1.$$

This is usually written as

$$\frac{x^2}{16} + \frac{y^2}{9} = 1$$

and is the equation of an ellipse.

(b) The graph is shown in figure 7.4.

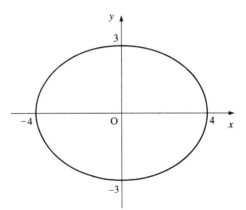

FIGURE 7.4

Note The standard equation of the ellipse is

$$\frac{x^2}{a^2} + \frac{y^2}{b^2} = 1$$

and this crosses the x-axis at (−a, 0) and (a, 0) and the y-axis at (0, b) and (0, −b).

The expansions of $\cos 2\theta$ in terms of either $\sin\theta$ or $\cos\theta$ are also useful in this context.

EXAMPLE 7.5

A curve is defined by the parametric equations

$$x = \cos\theta, \quad y = \cos2\theta.$$

(a) Find the Cartesian equation of the curve.
(b) Sketch the curve.

Solution (a) Using the trigonometric identity

$$\cos2\theta = 2\cos^2\theta - 1$$

gives

$$y = 2x^2 - 1.$$

When you come to sketch the graph you should note that

$$-1 \leqslant \cos\theta \leqslant 1 \quad \Rightarrow \quad -1 \leqslant x \leqslant 1 \quad \text{defines the domain}$$

and

$$-1 \leqslant \cos2\theta \leqslant 1 \quad \Rightarrow \quad -1 \leqslant y \leqslant 1 \quad \text{defines the range.}$$

So the graph is part of the curve $y = 2x^2 - 1$ as shown in figure 7.5.

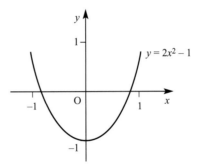

FIGURE 7.5

PARAMETRIC EQUATIONS OF SOME STANDARD CURVES

CIRCLE

The circle with centre $(0, 0)$ and radius 4 units has the equation $x^2 + y^2 = 16$. Alternatively, using the triangle OAB and the angle θ in figure 7.6, we can write the equations

$$x = 4\cos\theta$$
$$y = 4\sin\theta.$$

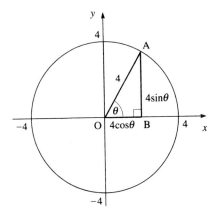

FIGURE 7.6

Generalising, a circle with centre $(0, 0)$ and radius r has the parametric equations

$x = r\cos\theta$
$y = r\sin\theta$.

Translating the centre of the circle to the point (a, b) gives the circle in figure 7.7, whose parametric equations are

$x = a + r\cos\theta$
$y = b + r\sin\theta$.

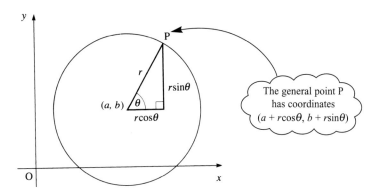

FIGURE 7.7

ELLIPSE

In Example 7.4, we saw that the parametric equations

$x = 4\cos\theta$
$y = 3\sin\theta$.

were equivalent to the Cartesian equation

$$\frac{x^2}{16} + \frac{y^2}{9} = 1.$$

In general, the equations

$$x = a\cos\theta$$
$$y = b\sin\theta$$

correspond to the ellipse

$$\frac{x^2}{a^2} + \frac{y^2}{b^2} = 1$$

(see figure 7.8).

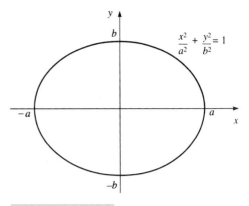

FIGURE 7.8

PARABOLA

The parabola in figure 7.9, whose line of symmetry is the x-axis and whose focus is at the point $(a, 0)$, has the Cartesian equation $y^2 = 4ax$. The corresponding parametric equations are

$$x = at^2$$
$$y = 2at.$$

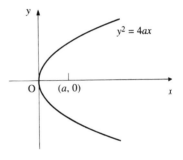

FIGURE 7.9

RECTANGULAR HYPERBOLA

The rectangular hyperbola $xy = c^2$ shown in figure 7.10 has the parametric equations

$$x = ct$$
$$y = \frac{c}{t}.$$

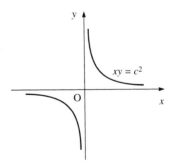

FIGURE 7.10

Exercise 7A

Notes

1 When you convert the equation of a curve from parametric to Cartesian form, you must take care that there are no restrictions on the values of x and y. For example, in figure 7.11, the curve $x = ct^2$, $y = \dfrac{c}{t^2}$ ($c > 0$) is restricted to positive values of x and y (since $t^2 > 0$). However, its Cartesian form, $xy = c^2$, would appear to allow negative values of x and y.

2 When the parametric equations can be recognised as those of a standard curve, the curve can be sketched immediately without the need for further investigation.

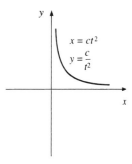

FIGURE 7.11

Exercise 7A

In the following questions, where you are asked to sketch curves, you will find it helpful to use a graphical calculator or a computer graph plotting package to check your sketches.

1 In each of the following questions
   (i) find the Cartesian equation of the curve
   (ii) sketch the curve.

   (a) $x = 2t$
       $y = t^2$

   (b) $x = \cos 2\theta$
       $y = \sin^2 \theta$

   (c) $x = t^2$
       $y = t^3$

   (d) $x = \sin^2 \theta$
       $y = 1 + 2\sin\theta$

   (e) $x = 2\mathrm{cosec}\,\theta$
       $y = 2\cot\theta$

   (f) $x = 2\sin^2\theta$
       $y = 3\cos\theta$

   (g) $x = t$
       $y = (1+t)(1-t)$

   (h) $x = \sin\theta$
       $y = \cos 2\theta$

   (i) $x = \dfrac{t}{1+t}$
       $y = \dfrac{t}{1-t}$

2 For each of the following parametric equations find the Cartesian equation and sketch the curve.

   (a) $x = 5\cos\theta$
       $y = 5\sin\theta$

   (b) $x = 3\cos\theta$
       $y = 2\sin\theta$

   (c) $x = 4 + 3\cos\theta$
       $y = 1 + 3\cos\theta$

   (d) $x = 2\cos\theta - 1$
       $y = 3 + 2\sin\theta$

3 For the two curves with parametric equations

   $x = t,\ y = \dfrac{1}{t}$ and $x = 4t,\ y = \dfrac{4}{t}$

   (a) obtain the Cartesian equations
   (b) on the same axes sketch their graphs
   (c) comment on the relationship between the graphs.

4 (a) Sketch the curve given by the parametric equations $x = 5\cos\theta$, $y = 3\sin\theta$.
   (b) State the parametric equations of the inscribed circle (the largest circle which fits inside the curve) and the circumscribing circle.

5 An ellipse cuts the x-axis at $(-3, 0)$ and $(3, 0)$ and cuts the y-axis at $(0, 4)$ and $(0, -5)$.
   (a) Write down the Cartesian equation of the curve.
   (b) Write down the parametric equations of the curve.

6 A curve has Cartesian equation $xy = 4$. Write down the parametric equations of the curve and sketch the curve.

7 A curve C is defined by the parametric equations
$$x = \frac{t}{2}, \quad y = t^3 - 2t^2.$$
   (a) Obtain the Cartesian equation of the curve in factorised form.
   (b) Sketch the curve C.

8 A curve has parametric equations $x = (t + 1)^2$, $y = t - 1$.
   (a) Find the coordinates of the points corresponding to $t = -4$ to $t = 4$ at intervals of one unit.
   (b) Sketch the curve for $-4 \leqslant t \leqslant 4$.
   (c) State the equation of the line of symmetry of the curve.
   (d) By eliminating the parameter, find the Cartesian equation of the curve.

9 A curve has parametric equations $x = \cos^3\theta$, $y = \sin^3\theta$.
   (a) By eliminating the parameter find the Cartesian equation of the curve.
   (b) Sketch the curve.

10 You are given that a curve has parametric equations $x = \sin^2\theta$, $y = \sin^2 2\theta$.
   (a) Show that $y = 4x - 4x^2$.
   (b) Sketch the curve.

INTEGRATING FUNCTIONS DEFINED PARAMETRICALLY

You now know that a function can be defined in terms of a parameter t using $x = f(t)$ and $y = g(t)$.

You also know from Chapter 12 of *Pure Mathematics: Core 1 and Core 2* that the area under a curve from $x = a$ to $x = b$ is given by

$$\text{area} = \int_a^b y\,dx, \text{ where } y \geqslant 0.$$

INTEGRATING FUNCTIONS DEFINED PARAMETRICALLY

When the curve is defined parametrically then the limits will be given in terms of t (or θ or whatever letter is used for the parameter). Let the limits be t_1 and t_2. In the integral you replace y by $g(t)$. Now, $\frac{dx}{dt} = f'(t)$ and so in the integral you also replace dx by $f'(t)dt$. The integral thus becomes:

$$\int_a^b y\,dx = \int_{t_1}^{t_2} g(t)f'(t)\,dt$$

Clearly, in some cases it may be possible to obtain a Cartesian equation writing y as a function of x. In such cases $\int_a^b y\,dx$ may be used directly but you must ensure that the limits which were defined in terms of the parameter are changed to ones in terms of x-values.

EXAMPLE 7.6

A curve is defined parametrically by $x = t^2$, $y = 2t$.

(a) Sketch the graph of the curve.
(b) Find the area of the region enclosed between the curve and the x-axis from $t = 1$ to $t = 2$.

Solution

(a) The Cartesian equation is

$$y = 2t = \pm 2\sqrt{x} \quad \text{or} \quad y^2 = 4x.$$

Its graph is shown in figure 7.12.

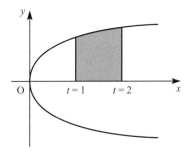

FIGURE 7.12

(b) The required region is shaded on the diagram.

Its area is calculated using $\int_{t=1}^{t=2} y\,dx$.

To solve this, replace y with $2t$.

To rewrite dx use $x = t^2$, so

$$\frac{dx}{dt} = 2t \quad \Rightarrow \quad dx = 2t\,dt.$$

Hence
$$\int_{t=1}^{t=2} y\, dx = \int_1^2 2t \times 2t\, dt$$
$$= \int_1^2 4t^2\, dt$$
$$= \tfrac{4}{3}[t^3]_1^2$$
$$= \tfrac{4}{3}(8-1)$$
$$= \tfrac{28}{3}$$

so the area of the region enclosed is $\tfrac{28}{3}$ square units.

EXAMPLE 7.7

An ellipse has parametric equations $x = 5\cos\theta$, $y = 3\sin\theta$. Sketch the ellipse and find the area of the region within it.

Solution The ellipse is shown in figure 7.13. The required area is twice that of the region shaded in the diagram.

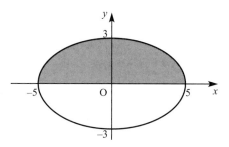

FIGURE 7.13

The shaded area is given by
$$\text{area} = \int_{\theta=0}^{\theta=\pi} y\, dx.$$

Remember to work in radians

But $x = 5\cos\theta$, hence
$$\frac{dx}{d\theta} = -5\sin\theta \quad \Rightarrow \quad dx = -5\sin\theta\, d\theta, \text{ giving}$$

$$\text{area} = \int_0^\pi 3\sin\theta \times -5\sin\theta\, d\theta$$
$$= -15\int_0^\pi \sin^2\theta\, d\theta.$$

But $\cos 2\theta \equiv 1 - 2\sin^2\theta$, so $\sin^2\theta \equiv \tfrac{1}{2}(1 - \cos 2\theta)$, hence
$$\text{area} = -\tfrac{15}{2}\int_0^\pi (1 - \cos 2\theta)\, d\theta$$
$$= -\tfrac{15}{2}[\theta - \tfrac{1}{2}\sin 2\theta]_0^\pi$$
$$= -\tfrac{15}{2}(\pi - 0)$$
$$= -\tfrac{15}{2}\pi.$$

So the magnitude of the area of the top half of the ellipse is $7\frac{1}{2}\pi$ square units, hence the total area is 15π square units.

(The area has a negative sign because as θ increases the value of x decreases.)

EXERCISE 7B

1. Find the area of the region between the following curves, which are defined parametrically, and the x-axis between the given values of t.
   (a) $x = 4t^2, y = 8t,$ between $t = 1$ and $t = 3$
   (b) $x = 1 + t^2, y = 2t,$ between $t = 0$ and $t = 1$
   (c) $x = t + \frac{1}{t}, y = t - \frac{1}{t},$ between $t = 1$ and $t = 2$
   (d) $x = (t + 1)^2, y = (t - 1)^2,$ between $t = 2$ and $t = 4$

2. Find the area of the region between the following curves, which are defined parametrically, and the x-axis between the given values of θ.
   (a) $x = 3\cos\theta, y = 2\sin\theta,$ between $\theta = 0$ and $\theta = \frac{\pi}{2}$
   (b) $x = r\cos\theta, y = r\sin\theta,$ between $\theta = 0$ and $\theta = \pi$
   (c) $x = 3\sin\theta, y = \sin^2\theta,$ between $\theta = \frac{\pi}{6}$ and $\theta = \frac{\pi}{3}$
   (d) $x = \theta - \cos\theta, y = \theta + \sin\theta,$ between $\theta = 0$ and $\theta = \pi$

3. A curve has parametric equations $x = 3t^2, y = t^3$. Its graph is shown below.

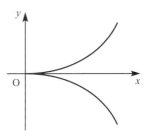

   (a) Copy the graph and on it draw the line $x = 27$. Shade the enclosed region and label it R.
   (b) Calculate the area of R.

4. A curve has parametric equations $x = 1 - t^2, \ y = 2t$.
   (a) Sketch the curve for values of $t \geqslant 0$.
   (b) Show that the curve meets the x-axis at the point where $t = 0$ and state the value of t where the curve cuts the y-axis.
   (c) Find the area of the region enclosed by the curve and the positive x- and y-axes.

5 A curve is defined by $x = \sin\theta$, $y = \sin2\theta$. For values of θ from 0 to $\frac{\pi}{2}$ the curve looks like this:

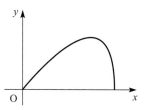

(a) Copy and complete the graph for values of t up to $t = 2\pi$.
(b) Calculate the total area of the region enclosed by the curve.

EXERCISE 7C

1 The curve C is described by the parametric equations

$$x = 3\cos t, \quad y = \cos 2t, \quad 0 \leqslant t \leqslant \pi.$$

(a) Find the Cartesian equation of the curve C.
(b) Draw a sketch of the curve C.

[Edexcel]

2 A curve is given by the parametric equations

$$x = t - 1, \quad y = t^2 + t.$$

(a) By eliminating t find the Cartesian equation of the curve.
(b) Sketch the curve for values of t from $t = -3$ to $t = 2$.

3 A curve is given by the parametric equations

$$x = 2\sin t, \quad y = \cos 2t, \quad -\frac{\pi}{2} \leqslant t \leqslant \frac{\pi}{2}.$$

(a) Find the values of t for which $y = 0$ and the corresponding values of x.
(b) Find the Cartesian equation of the curve, expressing y as a function of x.
(c) Use your Cartesian equation to verify that your answers to part (a) are correct.
(d) Sketch the graph of the curve.

4 (a) The parametric equations of a curve are

$$x = a\cos t, \quad y = b\sin t, \quad -\pi \leqslant t \leqslant \pi.$$

In either order, sketch the curve and find the Cartesian equations of the curves in the cases when

(i) $a = 3$ and $b = 4$
(ii) $a = 4$ and $b = 4$.

What transformation maps the graph in part (i) on to the graph in part (ii)?

(b) Express the Cartesian equation of the circle $x^2 + y^2 = 25$ in parametric form.

EXERCISE 7C

5 (a) A curve, C_1, is given by the parametric equations

$$x = t - 1, \quad y = \frac{t+2}{t-2}.$$

Find y as a function of x and sketch the curve.

(b) A second curve, C_2, has parametric equations

$$x = 5t + 3, \quad y = 2t + 3.$$

Find the Cartesian equation of this curve. Sketch the graph of C_2 on the graph of C_1.

(c) Find the points of intersection of C_1 and C_2.

6

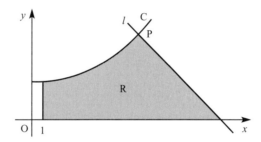

The diagram shows part of the curve C with parametric equations

$$x = (t+1)^2, \quad y = \tfrac{1}{2}t^3 + 3, \quad t \geq -1.$$

P is the point on the curve where $t = 2$. The line l is the normal to C at P.
The normal cuts the x-axis where $t = 3$.
The shaded region R is bounded by C, l, the x-axis and the line with equation $x = 1$.
Using integration and showing all your working, find the area of R.

[Edexcel (adapted)]

7

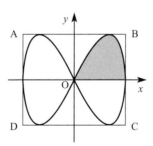

Part of the design of a stained glass window is shown in the diagram. The two loops enclose an area of blue glass. The remaining area within the rectangle ABCD is red glass.

The loops are described by the curve with parametric equations

$$x = 3\cos t, \quad y = 9\sin 2t, \quad 0 \leq t < 2\pi.$$

(a) Find the Cartesian equation of the curve in the form $y^2 = f(x)$.

(b) Show that the shaded area in the diagram enclosed by the curve and the x-axis, is given by

$$\int_0^{\frac{\pi}{2}} A\sin 2t \sin t \, dt,$$ stating the value of the constant A.

(c) Find the value of this integral.

The sides of the rectangle ABCD in the diagram are the tangents to the curve that are parallel to the coordinate axes. Given that 1 unit on each axis represents 1 cm,

(d) find the total area of the red glass.

[Edexcel]

8

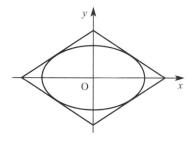

A table top, in the shape of a parallelogram, is made from two types of wood. The design is shown in the diagram. The area inside the ellipse is made from one type of wood, and the surrounding area is made from a second type of wood.

The ellipse has parametric equations,

$$x = 5\cos\theta, \quad y = 4\sin\theta, \quad 0 \leq \theta < 2\pi.$$

The parallelogram consists of four line segments, which are tangents to the ellipse at the points where $\theta = \alpha$, $\theta = -\alpha$, $\theta = \pi - \alpha$, and $\theta = -\pi + \alpha$.

The equation of the tangent ot the ellipse at $(5\cos\alpha, 4\sin\alpha)$ can be written in the form

$$5y\sin\alpha + 4x\cos\alpha = 20.$$

(a) Find by integration, using $\cos 2\theta = 1 - 2\sin^2\theta$, the area enclosed by the ellipse.

(b) Hence show that the area enclosed between the ellipse and the parallelogram is

$$\frac{80}{\sin 2\alpha} - 20\pi.$$

(c) Given that $0 < \alpha < \frac{\pi}{4}$, find the value of α for which the areas of the two types of wood are equal.

[Edexcel (adapted)]

EXERCISE 7C

9 A curve C has parametric equations
$$x = 1 + \frac{1}{t}, \quad y = 1 - \frac{1}{t}, \quad t \geq 1.$$

(a) Show that the integral of y with respect to x from $t = 1$ to $t = 2$ is
$$\int_1^2 \left(\frac{1}{t^3} - \frac{1}{t^2}\right) dt$$
and evaluate the integral.

(b) By eliminating t, find the Cartesian equation of the curve C.

(c) For the defined values of t, sketch the graph of the curve C.

(d) Using your graph and shading the appropriate area show that your answer to part (a) is correct.

10 The diagram shows part of the curve with parametric equations
$x = \sqrt{t}, \quad y = t^2 + 1.$

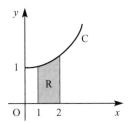

The region R bounded by the curve C, the x-axis and the lines $x = 1$ and $x = 2$ is shaded.

(a) Show that the Cartesian equation of the curve is
$$xy^4 = (1 - y)^4.$$

(b) Calculate the area of the region R.

KEY POINTS

1. In **parametric equations** the relationship between two variables is expressed by writing both of them in terms of a third variable or *parameter*.

2. Eliminating the parameter gives the **Cartesian equation** of the curve.

3. Parametric equations of some well-known curves

    Circle centre $(0, 0)$ and radius r

    $x = r\cos\theta \qquad y = r\sin\theta$

    Circle centre (a, b) and radius r

    $x = a + r\cos\theta \qquad y = b + r\sin\theta$

    Ellipse centre $(0, 0)$ with major axis $2a$ and minor axis $2b$

    $x = a\cos\theta \qquad y = b\sin\theta$

    Parabola whose line of symmetry is the x-axis

    $x = at^2 \qquad y = 2at$

    Rectangular hyperbola $xy = c^2$

    $x = ct \qquad y = \dfrac{c}{t}$

4. **Parametric integration**

    If $x = f(t)$ and $y = g(t)$ then $\int_a^b y\,dx$ becomes

    $$\int_{t_1}^{t_2} g(t)f'(t)\,dt$$

    where t_1 is the value of t when $x = a$ and t_2 is the value of t when $x = b$.

Chapter eight

SEQUENCES AND SERIES

At the age of twenty-one he wrote a treatise upon the Binomial Theorem...
On the strength of it, he won the Mathematical Chair at one of our smaller Universities.

Sherlock Holmes on Professor Moriarty,
'The Final Problem' by Sir Arthur Conan Doyle

How would you find $\sqrt{101}$ correct to 3 decimal places, without using a calculator?

Many people are able to develop a very high degree of skill in mental arithmetic, particularly those, such as bookmakers, whose work calls for quick reckoning. There are also those who have quite exceptional innate skills. M. Hari Prasad, pictured right, is famous for his mathematical speed; on one occasion he found the square root of a six-digit number in just 1 minute 3.8 seconds.

While most mathematicians do not have M. Hari Prasad's high level of talent with numbers they do acquire a sense of when something looks right or wrong. This often involves finding approximate values of numbers, such as $\sqrt{101}$, using methods that are based on series expansions, and these are the subject of this chapter.

Expansion of $(ax + b)^n$ when n is a positive integer

First of all let us recall the work on series from Chapter 8 of *Pure Mathematics: Core 1 and Core 2*. You started by expanding expressions such as $(ax + b)^n$ using Pascal's triangle. The triangle is:

```
                    1     1
                 1     2     1
              1     3     3     1
           1     4     6     4     1
        1     5     10    10    5     1
     1     6     15    20    15    6     1
  1     7     21    35    35    21    7     1
1    8    28    56    70    56    28    8     1
```

EXAMPLE 8.1

Use Pascal's triangle to expand $(2x + 3)^5$.

Solution $(2x)^5 + 5 \times (2x)^4 \times 3 + 10 \times (2x)^3 \times 3^2 + 10 \times (2x)^2 \times 3^3 + 5 \times (2x) \times 3^4 + 3^5$
$= 32x^5 + 240x^4 + 720x^3 + 1080x^2 + 810x + 243$

You then saw how to calculate each term in Pascal's triangle. The rth term of the nth row is given by

$$\binom{n}{r} = {}^nC_r = \frac{n!}{r!(n-r)!},$$

where r can take any value from 0 to n.

You will recall that, by definition, $0! = 1$.

Using this, the expansion of $(ax + b)^n$ when n is an integer is

$$(ax + b)^n = (ax)^n + \binom{n}{1}(ax)^{n-1} b + \binom{n}{2}(ax)^{n-2} b^2 + \ldots + \binom{n}{r}(ax)^{n-r} b^r + \ldots + b^n.$$

This is the binomial expansion of $(ax + b)^n$ when n is a positive integer.

EXERCISE 8A

EXAMPLE 8.2

Find the first three terms in the expansion of $(2x - 3)^{12}$.

Solution
$$(2x - 3)^{12} = (2x)^{12} + \binom{12}{1}(2x)^{11}(-3) + \binom{12}{2}(2x)^{10}(-3)^2 + \ldots$$
$$= 4096x^{12} + 12 \times 2048x^{11} \times -3 + 66 \times 1024x^{10} \times 9 + \ldots$$
$$= 4096x^{12} - 73728x^{11} + 608256x^{10} + \ldots$$

EXERCISE 8A

1. Expand and simplify $(3x - 2)^5$.
2. Expand and simplify $\left(x + \dfrac{1}{x}\right)^8$.
3. Find the first three terms in the expansion of $\left(2x + \dfrac{1}{2}\right)^{10}$.
4. Find the term independent of x (i.e. without x) in the expansion of $\left(\dfrac{4}{x} - \dfrac{x}{2}\right)^{16}$.
5. Expand $(1 + x)^n$ in terms of x and n, writing down the first four terms and the last two terms. Simplify your expression.

THE GENERAL BINOMIAL EXPANSION

In Exercise 8A question 5 you met the binomial expansion in the form

$$(1 + x)^n = 1 + \binom{n}{1}x + \binom{n}{2}x^2 + \binom{n}{3}x^3 + \ldots + \binom{n}{r}x^r + \ldots$$

which holds when n is any positive integer (or zero), that is $n \in \mathbb{N}$.

Since, for example,

$$\binom{n}{1} = \frac{n!}{1!(n-1)!} = \frac{n \times (n-1)!}{1 \times (n-1)!} = n$$

and
$$\binom{n}{2} = \frac{n!}{2!(n-2)!} = \frac{n \times (n-1) \times (n-2)!}{2 \times (n-2)!} = \tfrac{1}{2}n(n-1)$$

then this may also be written as

$$(1 + x)^n = 1 + nx + \frac{1}{2!}n(n-1)x^2 + \frac{1}{3!}n(n-1)(n-2)x^3 + \ldots$$
$$+ \frac{1}{r!}n(n-1)(n-2)\ldots(n-r+1)x^r + \ldots$$

and, being the same expansion as above, holds when $n \in \mathbb{N}$.

The general binomial theorem states that this second form is true when *n is any real number* but there are two important differences to note when $n \notin \mathbb{N}$:

- the series is infinite (or non-terminating);
- the expansion of $(1 + x)^n$ is valid only if $|x| < 1$.

Proving this result is not done until the Further Pure 3 unit, but you can assume that it is true!

Consider now the coefficients in the binomial expansion:

$$1 \quad n \quad \frac{1}{2!}n(n-1) \quad \frac{1}{3!}n(n-1)(-2) \quad \frac{1}{4!}n(n-1)(n-2)(n-3) \ldots$$

When $n = 0$, we get 1 0 0 0 0 ...   (infinitely many zeros)
$n = 1$ 1 1 0 0 0 ...   ditto
$n = 2$ 1 2 1 0 0 ...   ditto
$n = 3$ 1 3 3 1 0 ...   ditto
$n = 4$ 1 4 6 4 1 ...   ditto

so that, for example

$$(1 + x)^2 = 1 + 2x + x^2 + 0x^3 + 0x^4 + 0x^5 + \ldots$$
$$(1 + x)^3 = 1 + 3x + 3x^2 + x^3 + 0x^4 + 0x^5 + \ldots$$
$$(1 + x)^4 = 1 + 4x + 6x^2 + 4x^3 + x^4 + 0x^5 + \ldots$$

Of course, it is usual to discard all the zeros and write these binomial coefficients in the familiar form of Pascal's triangle:

$$\begin{array}{ccccccccc}
 & & & & 1 & & & & \\
 & & & 1 & & 1 & & & \\
 & & 1 & & 2 & & 1 & & \\
 & 1 & & 3 & & 3 & & 1 & \\
1 & & 4 & & 6 & & 4 & & 1
\end{array}$$

and the expansions as

$$(1 + x)^2 = 1 + 2x + x^2$$
$$(1 + x)^3 = 1 + 3x + 3x^2 + x^3$$
$$(1 + x)^4 = 1 + 4x + 6x^2 + 4x^3 + x^4$$

However, for other values of n (where $n \notin \mathbb{N}$) there are no zeros in the row of binomial coefficients and so we obtain an infinite sequence of non-zero terms. For example:

THE GENERAL BINOMIAL EXPANSION

$n = -3$ gives $\quad 1 \quad -3 \quad \frac{1}{2!}(-3)(-4) \quad \frac{1}{3!}(-3)(-4)(-5) \quad \frac{1}{4!}(-3)(-4)(-5)(-6) \ldots$

that is $\quad 1 \quad -3 \quad 6 \quad -10 \quad 15 \ldots$

$n = \frac{1}{2}$ gives $\quad 1 \quad \frac{1}{2} \quad \frac{1}{2!}(\frac{1}{2})(-\frac{1}{2}) \quad \frac{1}{3!}(\frac{1}{2})(-\frac{1}{2})(-\frac{3}{2}) \quad \frac{1}{4!}(\frac{1}{2})(-\frac{1}{2})(-\frac{3}{2})(-\frac{5}{2}) \ldots$

that is $\quad 1 \quad \frac{1}{2} \quad -\frac{1}{8} \quad \frac{1}{16} \quad -\frac{5}{128} \ldots$

so that $(1 + x)^{-3} = 1 - 3x + 6x^2 - 10x^3 + 15x^4 + \ldots$

and $(1 + x)^{\frac{1}{2}} = 1 + \frac{1}{2}x - \frac{1}{8}x^2 + \frac{1}{16}x^3 - \frac{5}{128}x^4 + \ldots$

Note Remember: these two expansions are valid only if $|x| < 1$.

These examples confirm that there will be an infinite sequence of non-zero coefficients when $n \notin \mathbb{N}$. You can also see that, after a certain stage, the remaining terms of the sequence will alternate in sign.

When $|x| < 1$, the magnitudes of $x, x^2, x^3, x^4, x^5, \ldots$ form a decreasing geometric sequence. In this case, the binomial expansion converges (just as a geometric series converges for $-1 < r < 1$, where r is the common ratio) and has a sum to infinity.

To summarise: when n is not a positive integer or zero, the binomial expansion of $(1 + x)^n$ becomes an infinite series, and is only valid when some restriction is placed on the values of x.

The binomial theorem states that for any value of n:

$$(1 + x)^n = 1 + nx + \frac{1}{2!}n(n-1)x^2 + \frac{1}{3!}n(n-1)(n-2)x^3 + \ldots$$

where

- if $n \in \mathbb{N}$, x may take any value;
- if $n \notin \mathbb{N}$, $|x| < 1$.

Notes

1 The full statement is the binomial *theorem*, and the right-hand side is referred to as the binomial *expansion*.

2 The term being expanded must be in the form $(1 + \ldots)$. *Must be 1*

3 If $n \notin \mathbb{N}$ then the expansion of $(1 + ax)^n$ is valid only if $|ax| < 1$ or $|x| < 1$.

EXAMPLE 8.3

Expand $(1 - x)^{-2}$ as a series of ascending powers of x up to and including the term in x^3, stating the set of values of x for which the expansion is valid.

Solution $(1 + x)^n = 1 + nx + \frac{1}{2!}n(n - 1)x^2 + \frac{1}{3!}n(n - 1)(n - 2)x^3 + \ldots$

Replacing n by -2, and x by $(-x)$ gives

$$(1 + (-x))^{-2} = 1 + (-2)(-x) + \frac{1}{2!}(-2)(-3)(-x)^2$$
$$+ \frac{1}{3!}(-2)(-3)(-4)(-x)^3 + \ldots \quad \text{when } |-x| < 1$$

> It is important to put brackets round the term $-x$, since, for example, $(-x)^2$ is not the same as $-x^2$

which leads to

$$(1 - x)^{-2} \approx 1 + 2x + 3x^2 + 4x^3 \quad \text{when } |x| < 1.$$

Note

In this example the coefficients of the powers of x form a recognisable sequence, and it would be possible to write down a general term in the expansion. The coefficient is always one more than the power, so the rth term would be rx^{r-1}. Using sigma notation, the infinite series could be written as

$$\sum_{r=1}^{\infty} rx^{r-1}.$$

EXAMPLE 8.4

Find a quadratic approximation for $\dfrac{1}{\sqrt{1 + 2t}}$ and state for which values of t the expansion is valid.

Solution $\dfrac{1}{\sqrt{1 + 2t}} = \dfrac{1}{(1 + 2t)^{\frac{1}{2}}} = (1 + 2t)^{-\frac{1}{2}}$

The binomial theorem states that

$$(1 + x)^n = 1 + nx + \frac{1}{2!}n(n - 1)x^2 + \frac{1}{3!}n(n - 1)(n - 2)x^3 + \ldots$$

Replacing n by $-\frac{1}{2}$ and x by $2t$ gives

$$(1 + 2t)^{-\frac{1}{2}} = 1 + (-\tfrac{1}{2})(2t) + \frac{1}{2!}(-\tfrac{1}{2})(-\tfrac{3}{2})(2t)^2 + \ldots \text{ when } |2t| < 1$$

$\Rightarrow \quad (1 + 2t)^{-\frac{1}{2}} \approx 1 - t + \tfrac{3}{2}t^2 \quad \text{when } |t| < \tfrac{1}{2}.$

THE GENERAL BINOMIAL EXPANSION

The equivalent binomial expansion of $(a + x)^n$ when n is not a positive integer is rather unwieldy. It is easier to start by taking a outside the brackets:

$$(a + x)^n = a^n\left(1 + \frac{x}{a}\right)^n.$$

The first entry inside the bracket is now 1 and so the first few terms of the expansion are

$$(a + x)^n = a^n\left[1 + n\left(\frac{x}{a}\right) + \frac{1}{2!}n(n-1)\left(\frac{x}{a}\right)^2 + \frac{1}{3!}n(n-1)(n-2)\left(\frac{x}{a}\right)^3 + \ldots\right]$$

for $\left|\frac{x}{a}\right| < 1$.

Note

Since the bracket is raised to the power n, any quantity you take out must be raised to the power n too, as in the following example.

EXAMPLE 8.5

Expand $(2 + x)^{-3}$ as a series of ascending powers of x up to and including the term in x^2, stating the values of x for which the expansion is valid.

Solution

$$(2 + x)^{-3} = \frac{1}{(2 + x)^3}$$

$$= \frac{1}{2^3\left(1 + \frac{x}{2}\right)^3}$$

$$= \frac{1}{8}\left(1 + \frac{x}{2}\right)^{-3}$$

Notice that this is the same as $2^{-3}\left(1 + \frac{x}{2}\right)^{-3}$

Take the binomial expansion

$$(1 + x)^n = 1 + nx + \frac{1}{2!}n(n-1)x^2 + \frac{1}{3!}n(n-1)(n-2)x^3 + \ldots$$

and replace n by -3 and x by $\frac{x}{2}$ to give

$$\frac{1}{8}\left(1 + \frac{x}{2}\right)^{-3} = \frac{1}{8}\left[1 + (-3)\left(\frac{x}{2}\right) + \frac{1}{2!}(-3)(-4)\left(\frac{x}{2}\right)^2 + \ldots\right] \quad \text{when } \left|\frac{x}{2}\right| < 1$$

$$\approx \frac{1}{8} - \frac{3x}{16} + \frac{3x^2}{16} \quad \text{when } |x| < 2.$$

EXAMPLE 8.6

(a) Expand $\sqrt{100 + x}$ as far as the term in x^3.
(b) Use the expansion to find $\sqrt{101}$ to 9 decimal places without the use of a calculator.

Solution This example refers to the introductory remarks at the beginning of this chapter.

(a) Start with the expansion of $\sqrt{100 + x}$:

$$\sqrt{100 + x} = \sqrt{100} \times \sqrt{1 + \frac{x}{100}}$$

$$= 10 \times \left(1 + \frac{x}{100}\right)^{\frac{1}{2}}$$

$$= 10 \times \left(1 + \frac{1}{2} \times \left(\frac{x}{100}\right) + \frac{1}{2!}\left(\frac{1}{2}\right)\left(-\frac{1}{2}\right)\left(\frac{x}{100}\right)^2 + \frac{1}{3!}\left(\frac{1}{2}\right)\left(-\frac{1}{2}\right)\left(-\frac{3}{2}\right)\left(\frac{x}{100}\right)^3 + \ldots\right)$$

$$= 10 \times \left(1 + \frac{x}{200} - \frac{x^2}{80\,000} + \frac{x^3}{16\,000\,000} + \ldots\right)$$

$$= 10 + \frac{x}{20} - \frac{x^2}{8000} + \frac{x^3}{1\,600\,000} - \ldots$$

(b) To find $\sqrt{101}$, put $\sqrt{101} = \sqrt{100 + x}$, from which it follows that $x = 1$.

Substituting into the expansion:

$$\sqrt{101} = 10 + \frac{1}{20} - \frac{1}{8000} + \frac{1}{1\,600\,000}$$

$$= 10 + 0.05 - 0.000\,125 + 0.000\,000\,625$$

$$= 10.049\,875\,625.$$

This is more figures than your calculator gives and taking more terms would give an answer to even more decimal places.

Note that if we had taken the first two terms only the answer of 10.05 would be correct to 2 decimal places.

EXAMPLE 8.7

Find a quadratic approximation for $\frac{(2 + x)}{(1 - x^2)}$, stating the values of x for which the expansion is valid.

Solution $\frac{(2 + x)}{(1 - x^2)} = (2 + x)(1 - x^2)^{-1}$

So you need to expand $(1 - x^2)^{-1}$ as far as the term in x^2 and multiply the result by $(2 + x)$.

Take the binomial expansion

$$(1 + x)^n = 1 + nx + \frac{1}{2!}n(n - 1)x^2 + \frac{1}{3!}n(n - 1)(n - 2)x^3 + \ldots$$

and replace n by -1 and x by $(-x^2)$ to give

$$(1 + (-x^2))^{-1} = 1 + (-1)(-x^2) + \frac{1}{2!}(-1)(-2)(-x^2)^2 + \ldots \quad \text{when } |-x^2| < 1$$

$$(1 - x^2)^{-1} = 1 + x^2 + \ldots \quad \text{when } |x^2| < 1, \text{ i.e. } |x| < 1.$$

THE GENERAL BINOMIAL EXPANSION

Pure Mathematics: Core 4

Multiply both sides by $(2 + x)$ to obtain $(2 + x)(1 - x^2)^{-1}$:

$$(2 + x)(1 - x^2)^{-1} \approx (2 + x)(1 + x^2)$$
$$\approx 2 + x + 2x^2 \quad \text{when } |x| < 1.$$

The term in x^3 has been omitted because the question asked for a quadratic approximation

Sometimes two or more binomial expansions may be used together. If these impose different restrictions on the values of x, you need to decide which is the strictest.

EXAMPLE 8.8

Find a and b such that

$$\frac{1}{(1 - 2x)(1 + 3x)} \approx a + bx$$

and state the values of x for which the expansion is valid.

Solution

$$\frac{1}{(1 - 2x)(1 + 3x)} = (1 - 2x)^{-1}(1 + 3x)^{-1}$$

Using the binomial expansion:

$$(1 - 2x)^{-1} \approx 1 + (-1)(-2x) \quad \text{for } |-2x| < 1$$
$$\text{and} \quad (1 + 3x)^{-1} \approx 1 + (-1)(3x) \quad \text{for } |3x| < 1$$
$$\Rightarrow (1 - 2x)^{-1}(1 + 3x)^{-1} \approx (1 + 2x)(1 - 3x)$$
$$\approx 1 - x \quad \text{(ignoring higher powers of } x\text{)}$$

giving $a = 1$ and $b = -1$.

For the result to be valid, both $|2x| < 1$ and $|3x| < 1$ need to be satisfied.

$$|2x| < 1 \quad \Rightarrow \quad -\tfrac{1}{2} < x < \tfrac{1}{2}$$
$$\text{and} \quad |3x| < 1 \quad \Rightarrow \quad -\tfrac{1}{3} < x < \tfrac{1}{3}.$$

Both of these restrictions are satisfied if $-\tfrac{1}{3} < x < \tfrac{1}{3}$. This is the stricter restriction.

Note

The binomial expansion may also be used when the first term is the variable. For example:

$(x + 2)^{-1}$ may be written as $(2 + x)^{-1} = 2^{-1}(1 + \tfrac{x}{2})^{-1}$

or $(2x - 1)^{-3}$ may be written as $[(-1)(1 - 2x)]^{-3}$
$$= (-1)^{-3}(1 - 2x)^{-3}$$
$$= -(1 - 2x)^{-3}.$$

EXERCISE 8B For each of the functions in questions 1 to 12:

(a) write down the first three non-zero terms in their expansions as a series of ascending powers of x;

(b) state the values of x for which the expansion is valid;

(c) if you have access to a graphical calculator or suitable computer package, draw the graph of each function and the first three terms of its binomial expansion on the same axes. In each case, notice how the graphs illustrate the need for some restriction on the values of x.

1. $(1 + x)^{-2}$
2. $\dfrac{1}{1 + 2x}$
3. $\sqrt{1 - x^2}$
4. $\dfrac{1 + 2x}{1 - 2x}$
5. $(3 + x)^{-1}$
6. $(1 - x)\sqrt{4 + x}$
7. $\dfrac{x + 2}{x - 3}$
8. $\dfrac{1}{\sqrt{3x + 4}}$
9. $\dfrac{1 + 2x}{(2x - 1)^2}$
10. $\dfrac{1 + x^2}{1 - x^2}$
11. $\sqrt[3]{1 + 2x^2}$
12. $\dfrac{1}{(1 + 2x)(1 + x)}$

13. (a) Write down the expansion of $(1 + x)^3$.

    (b) Find the first four terms in the expansion of $(1 - x)^{-4}$ in ascending powers of x. For what values of x is this expansion valid?

    (c) When the expansion is valid
    $$\dfrac{(1 + x)^3}{(1 - x)^4} = 1 + 7x + ax^2 + bx^3 + \ldots$$
    Find the values of a and b. [MEI]

14. (a) Write down the expansion of $(2 - x)^4$.

    (b) Find the first four terms in the expansion of $(1 + 2x)^{-3}$ in ascending powers of x. For what range of values of x is this expansion valid?

    (c) When the expansion is valid
    $$\dfrac{(2 - x)^4}{(1 + 2x)^3} = 16 + ax + bx^2 + \ldots$$
    Find the values of a and b. [MEI]

15. Write down the expansions of the following expressions in ascending powers of x as far as the term containing x^3. In each case state the values of x for which the expansion is valid.

    (a) $(1 - x)^{-1}$
    (b) $(1 + 2x)^{-2}$
    (c) $\dfrac{1}{(1 - x)(1 + 2x)^2}$

EXERCISE 8B

16 (a) Show that $\dfrac{1}{\sqrt{4-x}} = \dfrac{1}{2}\left(1 - \dfrac{x}{4}\right)^{-\frac{1}{2}}$.

(b) Write down the first three terms in the binomial expansion of $\left(1 - \dfrac{x}{4}\right)^{-\frac{1}{2}}$ in ascending powers of x stating the range of values of x for which this expansion is valid.

(c) Find the first three terms in the expansion of $\dfrac{2(1+x)}{\sqrt{4-x}}$ in ascending powers of x, for small values of x.

[MEI]

17 (a) Expand $(1 + y)^{-1}$, where $-1 < y < 1$, as a series in powers of y giving the first four terms.

(b) Hence find the first four terms of the expansion of $\left(1 + \dfrac{2}{x}\right)^{-1}$ where $-1 < \dfrac{2}{x} < 1$.

(c) Show that $\left(1 + \dfrac{2}{x}\right)^{-1} = \dfrac{x}{x+2} = \dfrac{x}{2}\left(1 + \dfrac{x}{2}\right)^{-1}$.

(d) Find the first four terms of the expansion of $\dfrac{x}{2}\left(1 + \dfrac{x}{2}\right)^{-1}$ where $-1 < \dfrac{x}{2} < 1$.

(e) State the conditions on x under which your expansions for $\left(1 + \dfrac{2}{x}\right)^{-1}$ and $\dfrac{x}{2}\left(1 + \dfrac{x}{2}\right)^{-1}$ are valid and explain briefly why your expansions are different.

[MEI]

18 (a) Expand $(1 - 3x)^{\frac{1}{3}}$, $|x| < \frac{1}{3}$, in ascending powers of x up to and including the term in x^3.

(b) By substituting $x = 10^{-3}$ in your expansion, find, to 9 significant figures, the cube root of 997.

USING PARTIAL FRACTIONS WITH THE BINOMIAL EXPANSION

One of the most common reasons for writing an expression in partial fractions is to enable binomial expansions to be applied, as in the following example.

EXAMPLE 8.9

Express $\dfrac{2x+7}{(x-1)(x+2)}$ in partial fractions and hence find the first three terms of its binomial expansion, stating the values of x for which this is valid.

Solution $\dfrac{2x+7}{(x-1)(x+2)} = \dfrac{A}{(x-1)} + \dfrac{B}{(x+2)}$

Multiplying both sides by $(x - 1)(x + 2)$ gives

$$2x + 7 \equiv A(x + 2) + B(x - 1).$$
$$x = 1 \implies 9 = 3A \implies A = 3$$
$$x = -2 \implies 3 = -3B \implies B = -1$$

This gives

$$\frac{2x + 7}{(x - 1)(x + 2)} \equiv \frac{3}{(x - 1)} - \frac{1}{(x + 2)}.$$

In order to obtain the binomial expansion, each bracket must be of the form $(1 \pm \ldots)$, giving

$$\frac{2x + 7}{(x - 1)(x + 2)} \equiv \frac{-3}{(1 - x)} - \frac{1}{2\left(1 + \frac{x}{2}\right)}$$

$$\equiv -3(1 - x)^{-1} - \frac{1}{2}\left(1 + \frac{x}{2}\right)^{-1}. \qquad ①$$

The two binomial expansions are

$$(1 - x)^{-1} = 1 + (-1)(-x) + \frac{(-1)(-2)}{2!}(-x)^2 + \ldots \quad \text{for } |x| < 1$$

$$\approx 1 + x + x^2$$

and $\left(1 + \frac{x}{2}\right)^{-1} = 1 + (-1)\left(\frac{x}{2}\right) + \frac{(-1)(-2)}{2!}\left(\frac{x}{2}\right)^2 + \ldots \quad \text{for } \left|\frac{x}{2}\right| < 1$

$$\approx 1 - \frac{x}{2} + \frac{x^2}{4}.$$

Substituting these in ① gives

$$\frac{2x + 7}{(x - 1)(x + 2)} \approx -3(1 + x + x^2) - \frac{1}{2}\left(1 - \frac{x}{2} + \frac{x^2}{4}\right)$$

$$\approx -\frac{7}{2} - \frac{11x}{4} - \frac{25x^2}{8}.$$

The expansion is valid when $|x| < 1$ and $\left|\frac{x}{2}\right| < 1$. The stricter of these is $|x| < 1$.

EXERCISE 8C

1. (a) Express $\dfrac{7 - 4x}{(2x - 1)(x + 2)}$ in partial fractions as $\dfrac{A}{(2x - 1)} + \dfrac{B}{(x + 2)}$ where A and B are to be found.

   (b) Find the expansion of $\dfrac{1}{(1 - 2x)}$ in the form $a + bx + cx^2 + \ldots$ where a, b and c are to be found. Give the range of values of x for which this expansion is valid.

EXERCISE 8D

(c) Find the expansion of $\dfrac{1}{(2+x)}$ as far as the term containing x^2. Give the range of values of x for which this expansion is valid.

(d) Hence find a quadratic approximation for $\dfrac{7-4x}{(2x-1)(x+2)}$ when $|x|$ is small.

[MEI, part]

In questions 2 to 8 find the first three terms in the binomial expansion of the function and state the values of x for which the expansion is valid.

2 $\dfrac{1}{(1+x)(1-x)}$

3 $\dfrac{2+x}{(2-x)(3+x)}$

4 $\dfrac{4}{(1+x)(1+2x)(2-x)}$

5 $\dfrac{2x^2+3x-1}{(3+x)(1-x)}$

6 $\dfrac{4}{(1-3x)(1-x)^2}$

7 $\dfrac{5+2x}{(2x-1)(x+1)^2}$

8 $\dfrac{5-2x}{(x-1)^2(x+2)}$

EXERCISE 8D **Examination-style questions**

1 (a) Expand $(1+4x)^{\frac{1}{4}}$, $|x| < \frac{1}{4}$, in ascending powers of x up to and including the term in x^3.

(b) By substituting a suitable value for x in the expansion, find $\sqrt[4]{1.04}$ to 6 decimal places.

2 (a) Find the expansion of $\dfrac{1}{(1+2x)^3}$, $|x| < \frac{1}{2}$, in ascending powers of x up to and including the term in x^3.

(b) The first four terms in the expansion of $\dfrac{1-x}{(1+2x)^3}$, $|x| < \frac{1}{2}$, in ascending powers of x are

$$1 + ax + 30x^2 + bx^3.$$

Find the values of a and b.

3 (a) Expand $\sqrt[3]{1000 + x}$ in ascending powers of x up to and including the term in x^2.

   (b) Use your expansion to evaluate $\sqrt[3]{1003}$ without using a calculator.

   (c) Show that the percentage error between the answer to part **(b)** and the exact value of $\sqrt[3]{1003}$ is approximately 1.7×10^{-7}.

4 (a) Expand $\sqrt{1 - x}$, $|x| < 1$, in ascending powers of x up to and including the term in x^3.

   (b) By putting $x = 0.02$ show that $\sqrt{\frac{98}{100}}$ is approximately $0.989\,949\,5$.

   (c) Use your answer to find $\sqrt{2}$ to 5 decimal places.

5 (a) Expand $\frac{1}{2 - x}$ in ascending powers of x up to and including the term in x^3.

   (b) For what values of x is the expansion valid?

6 (a) Given that $|x| < 1$, expand $\sqrt[3]{1 + x}$ as a series of ascending powers of x up to and including the term in x^2.

   (b) Show that, if x is small,

   $$(3 - x)\sqrt[3]{1 + x} \approx a + bx^2$$

   stating the exact values of a and b.

7 (a) Expand $\frac{1 + x}{1 - x}$ in ascending powers of x up to and including the term in x^3.

   (b) For what values of x is the expansion valid?

8 (a) Find the first three terms in the series expansion, in ascending powers of x, of $(1 + 2x)^{-2}$, stating the values of x for which the expansion is valid.

   (b) Show that, when x is small,

   $$\left(\frac{1 - 2x}{1 + 2x}\right)^2 \approx 1 - 8x + 32x^2.$$

9 Given that $(1 + 3x)^n = 1 + ax + \left(\frac{3}{2}\right)^3 x^2$, neglecting terms of order x^3 and higher, find

   (a) the values of n;

   (b) the corresponding values of a.

10 $f(x) = (1 + 3x)^{-1}$, $|x| < \frac{1}{3}$

   (a) Expand $f(x)$ in ascending powers of x up to and including the term in x^3.

   (b) Hence show that, for small x,

   $$\frac{1 + x}{1 + 3x} \approx 1 - 2x + 6x^2 - 18x^3.$$

   (c) Taking a suitable value for x, which should be stated, use the series expansion in part **(b)** to find an approximate value for $\frac{101}{103}$, giving your answer to 5 decimal places.

   [Edexcel]

Exercise 8D

11 **(a)** Express $\dfrac{x-8}{(x+1)(x-2)}$ in partial fractions.

   (b) Using the partial fractions show that the expansion of $\dfrac{x-8}{(x+1)(x-2)}$ as far as the term in x^3 is

   $$4 - \dfrac{5}{2}x + \dfrac{13}{4}x^2 - \dfrac{23}{8}x^3.$$

12 Given that

   $$\dfrac{10(2-3x)}{(1-2x)(2+x)} \equiv \dfrac{A}{1-2x} + \dfrac{B}{2+x}.$$

   (a) find the values of the constants A and B.

   (b) Hence, or otherwise, find the series expansion in ascending powers of x, up to and including the term in x^3, of $\dfrac{10(2-3x)}{(1-2x)(2+x)}$, for $|x| < \tfrac{1}{2}$.

   [Edexcel]

13 Given that $\dfrac{2x^3 - x^2 - 9x - 3}{x^2 - x - 6} \equiv Ax + B + \dfrac{C}{x-3} + \dfrac{D}{x+2}$

   (a) Find the values of the constants A, B, C and D.

   (b) Find the series expansion in ascending powers of x up to and including the term in x^2 of $\dfrac{2x^3 - x^2 - 9x - 3}{x^2 - x - 6}$.

   (c) For what values of x is the expansion valid?

14 **(a)** Expand $\dfrac{1}{2+x}$ in ascending powers of x up to and including the term in x^5.

   (b) Write down the expansion of $\dfrac{1}{2-x}$ up to the term in x^5.

   (c) Express $\dfrac{1}{4-x^2}$ in partial fractions.

   (d) Expand $\dfrac{1}{4-x^2}$ in ascending powers of x up to and including the term in x^4.

15 **(a)** Given that $\dfrac{2x^2 + 6x + 12}{(3-x)(1+x)^2} \equiv \dfrac{A}{3-x} + \dfrac{B}{1+x} + \dfrac{C}{(1+x)^2}$, find the values of A, B and C.

   (b) The first four terms in the expansion of $\dfrac{2x^2 + 6x + 12}{(3-x)(1+x)^2}$ in ascending powers of x are

   $$4 + ax - \dfrac{64}{9}x^2 + bx^3.$$

   Find the values of a and b.

   (c) For what values of x is the expansion valid?

KEY POINTS

1 The general binomial expansion for $n \in \mathbb{R}$ is

$$(1 + x)^n = 1 + nx + \frac{1}{2!}n(n-1)x^2 + \frac{1}{3!}n(n-1)(n-2)x^3 + \ldots$$

2 If n is a positive integer or zero (i.e. $n \in \mathbb{N}$) then

- x may take any value;
- the expansion has $n + 1$ terms.

3 If n is not a positive integer or zero (i.e. $n \notin \mathbb{N}$) then

- the expansion is valid if $|x| < 1$;
- the expansion has an infinite number of terms.

4 The expansion must be of a function in the form $(1 \pm \ldots)$.

5 When $n \notin \mathbb{N}$, $(a + x)^n$ must be written as $a^n\left(1 + \frac{x}{a}\right)^n$ before obtaining the binomial expansion.

6 The expansion of $(1 + ax)^n$ is valid for $|ax| < 1$, i.e. $|x| < \frac{1}{|a|}$.

7 Rational functions such as $\dfrac{1 + 2x}{(1 - x)(1 - 2x)}$ must be expressed in terms of partial fractions before their binomial expansion can be found.

Chapter nine

DIFFERENTIATION

An education isn't how much you have committed to memory, or even how much you know. It's being able to differentiate between what you do know and what you do not know.

Anatole France

Below is a summary of what you have learnt so far.

y	$\dfrac{dy}{dx}$	y	$\dfrac{dy}{dx}$
x^n	nx^{n-1}	$\sin x$	$\cos x$
$e^{f(x)}$	$f'(x)\, e^{f(x)}$	$\cos x$	$-\sin x$
$\ln(f(x))$	$\dfrac{f'(x)}{f(x)}$	$\tan x$	$\sec^2 x$

You should also know the following rules:

derivative: $\dfrac{dx}{dy} = \dfrac{1}{\frac{dy}{dx}}$ product rule: $\dfrac{d}{dx}(uv) = u\dfrac{dv}{dx} + v\dfrac{du}{dx}$

chain rule: $\dfrac{dy}{dx} = \dfrac{dy}{du} \times \dfrac{du}{dx}$ quotient rule: $\dfrac{d}{dx}\left(\dfrac{u}{v}\right) = \dfrac{v\dfrac{du}{dx} - u\dfrac{dv}{dx}}{v^2}$

In this chapter you will use this knowledge with implicit and parametric functions. You will look at the derivative of a^x and finally learn how to form simple differential equations.

DIFFERENTIATING FUNCTIONS DEFINED IMPLICITLY

All the functions you have differentiated so far have been of the form $y = f(x)$. However, many functions cannot be arranged in this way at all, for example $x^3 + y^3 = xy$, and others can look clumsy when you try to make y the subject.

An example of this is the semicircle $x^2 + y^2 = 4$, $y \geq 0$, illustrated in figure 9.1

The curve is much more easily recognised in this form than in the equivalent $y = \sqrt{4 - x^2}$.

When a function is specified by an equation connecting x and y which does not have y as the subject it is called an *implicit function*.

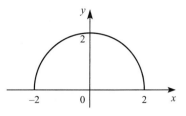

FIGURE 9.1

Note   Although restrictions on x or y are often necessary to make the function unambiguous, we frequently assume such restrictions but do not mention them.

The chain rule $\frac{dy}{dx} = \frac{dy}{du} \times \frac{du}{dx}$ and the product rule $\frac{d}{dx}(uv) = u\frac{dv}{dx} + v\frac{du}{dx}$ are used extensively to help in the differentiation of implicit functions.

EXAMPLE 9.1

Differentiate each of the following with respect to x.

(a) y^2
(b) xy
(c) $3x^2y^3$
(d) $\sin y$

Solution (a) $\frac{d}{dx}(y^2) = \frac{d}{dy}(y^2) \times \frac{dy}{dx}$ (chain rule)

$$= 2y\frac{dy}{dx}$$

(b) $\frac{d}{dx}(xy) = x\frac{dy}{dx} + 1 \times y = x\frac{dy}{dx} + y$ (product rule)

(c) $\frac{d}{dx}(3x^2y^3) = 3(x^2\frac{d}{dx}(y^3) + y^3\frac{d}{dx}(x^2))$ (product rule)

$$= 3(x^2 \times 3y^2\frac{dy}{dx} + y^3 \times 2x)$$ (chain rule)

$$= 3xy^2(3x\frac{dy}{dx} + 2y)$$

(d) $\frac{d}{dx}(\sin y) = \frac{d}{dy}(\sin y) \times \frac{dy}{dx}$ (chain rule)

$$= \cos y \frac{dy}{dx}$$

DIFFERENTIATING FUNCTIONS DEFINED IMPLICITLY

EXAMPLE 9.2

The equation of a curve is given by $y^3 + xy = 2$.

(a) Find an expression for $\frac{dy}{dx}$ in terms of x and y.
(b) Hence find the gradient of the curve at $(1, 1)$ and the equation of the tangent to the curve at that point.

Solution

(a) $$y^3 + xy = 2$$
$$\Rightarrow \quad 3y^2\frac{dy}{dx} + (x\frac{dy}{dx} + y) = 0$$
$$\Rightarrow \quad (3y^2 + x)\frac{dy}{dx} = -y$$
$$\Rightarrow \quad \frac{dy}{dx} = \frac{-y}{(3y^2 + x)}$$

(b) At $(1, 1)$, $\frac{dy}{dx} = -\frac{1}{4}$. Using $y - y_1 = m(x - x_1)$, the equation of the tangent is

$$(y - 1) = -\tfrac{1}{4}(x - 1)$$
$$\Rightarrow \quad 4y - 4 = -x + 1$$
$$\Rightarrow \quad x + 4y - 5 = 0.$$

Note

For interest, the graph of $y^3 + xy = 2$ is shown in figure 9.2.

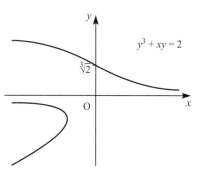

FIGURE 9.2

For this particular equation it is easy to make x the subject. You can then swap x with y to give you an equation which you can enter into a graphical calculator. You must then reflect the image on your calculator in the line $y = x$ to give the correct graph of $y^3 + xy = 2$.

STATIONARY POINTS

As before these occur where $\frac{dy}{dx} = 0$. Putting $\frac{dy}{dx} = 0$ will not usually give values of x directly, but will give a relationship between x and y. This needs to be solved simultaneously with the equation of the curve to find the coordinates.

EXAMPLE 9.3

(a) Differentiate $x^3 + y^3 = 3xy$ with respect to x.

(b) Hence find the coordinates of any stationary points.

Solution

(a) $\dfrac{d}{dx}(x^3) + \dfrac{d}{dx}(y^3) = \dfrac{d}{dx}(3xy)$

$\Rightarrow 3x^2 + 3y^2\dfrac{dy}{dx} = 3\left(x\dfrac{dy}{dx} + y\right)$

Notice how it is not necessary to find an expression for $\dfrac{dy}{dx}$ unless you are told to.

(b) At stationary points, $\dfrac{dy}{dx} = 0$

$\Rightarrow \quad 3x^2 = 3y$

$\Rightarrow \quad x^2 = y.$

To find the coordinates of the stationary points, solve

$\left. \begin{array}{l} x^2 = y \\ x^3 + y^3 = 3xy \end{array} \right\}$ simultaneously.

Substituting for y we obtain

$x^3 + (x^2)^3 = 3x(x^2)$

$\Rightarrow \qquad x^3 + x^6 = 3x^3$

$\Rightarrow \qquad \qquad x^6 = 2x^3$

$\Rightarrow \qquad x^3(x^3 - 2) = 0$

$\Rightarrow \qquad \qquad x = 0 \quad \text{or} \quad x = \sqrt[3]{2}.$

$y = x^2 \Rightarrow$ stationary points are $(0, 0)$ and $(\sqrt[3]{2}, \sqrt[3]{4})$.

The stationary points are A and B in figure 9.3.

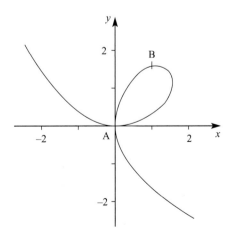

FIGURE 9.3

Types of stationary points

As with explicit functions, the type of stationary point can usually be determined by considering the sign of the second derivative $\frac{d^2y}{dx^2}$ at the stationary point.

EXAMPLE 9.4

The curve with equation $\sin x + \sin y = 1$ for $0 \leqslant x \leqslant \pi$, $0 \leqslant y \leqslant \pi$ is shown in figure 9.4.

(a) Differentiate the equation of the curve with respect to x and hence find the coordinates of any stationary points.

(b) Differentiate the equation again with respect to x to determine the types of stationary points.

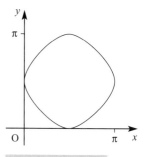

FIGURE 9.4

Solution

$$\sin x + \sin y = 1 \qquad ① $$

(a) $\Rightarrow \quad \cos x + (\cos y)\dfrac{dy}{dx} = 0$

$\Rightarrow \quad \dfrac{dy}{dx} = -\dfrac{\cos x}{\cos y}$

At any stationary points $\dfrac{dy}{dx} = 0$

$\Rightarrow \quad \cos x = 0$
$\Rightarrow \quad x = \dfrac{\pi}{2}$ (only solution).

Substitute in $\sin x + \sin y = 1$.

When $x = \dfrac{\pi}{2}$, $\sin x = 1$

$\Rightarrow \quad \sin y = 0$
$\Rightarrow \quad y = 0 \quad \text{or} \quad y = \pi$

$\Rightarrow \quad$ turning points at $\left(\dfrac{\pi}{2}, 0\right)$ and $\left(\dfrac{\pi}{2}, \pi\right)$.

(b) Differentiating equation ① again with respect to x:

$$\cos x + (\cos y)\dfrac{dy}{dx} = 0$$

$\Rightarrow \quad -\sin x + \left[(\cos y)\dfrac{d^2y}{dx^2} + \dfrac{dy}{dx}\left((-\sin y)\dfrac{dy}{dx}\right)\right] = 0$

At $\left(\dfrac{\pi}{2}, 0\right)$, $\dfrac{dy}{dx} = 0$

$\Rightarrow \quad -\sin\dfrac{\pi}{2} + (\cos 0)\dfrac{d^2y}{dx^2} = 0$

$\Rightarrow \quad \dfrac{d^2y}{dx^2} = 1 \quad \Rightarrow \quad$ minimum turning point.

At $(\frac{\pi}{2}, \pi)$, $\frac{dy}{dx} = 0$

$$\Rightarrow \quad -\sin\frac{\pi}{2} + (\cos\pi)\frac{d^2y}{dx^2} = 0$$

$$\Rightarrow \quad -1 - \frac{d^2y}{dx^2} = 0$$

$$\Rightarrow \quad \frac{d^2y}{dx^2} = -1 \quad \Rightarrow \quad \text{maximum turning point.}$$

These points are confirmed by considering the sketch in figure 9.4.

EXERCISE 9A

1 Differentiate each of the following with respect to x.
   (a) y^4
   (b) $x^2 + y^3 - 5$
   (c) $xy + x + y$
   (d) $\cos y$
   (e) $e^{(y+2)}$
   (f) xy^3
   (g) $2x^2y^5$
   (h) $x + \ln y - 3$
   (i) $xe^y - \cos y$
   (j) $x^2 \ln y$
   (k) $xe^{\sin y}$
   (l) $x\tan y - y\tan x$

2 Find the gradient of the curve $xy^3 = 5\ln y$ at the point $(0, 1)$.

3 Find the gradient of the curve $e^{\sin x} + e^{\cos y} = e + 1$ at the point $(\frac{\pi}{2}, \frac{\pi}{2})$.

4 (a) Find the gradient of the curve $x^2 + 3xy + y^2 = x + 3y + 2$ at the point $(2, 0)$.
   (b) Hence find the equation of the tangent to the curve at this point.

5 Find the coordinates of all the stationary points on the curve $x^2 + y^2 + xy = 3$.

6 (a) Show that the graph of $xy + 48 = x^2 + y^2$ has stationary points at $(4, 8)$ and $(-4, -8)$.
   (b) By differentiating with respect to x a second time, determine the nature of these stationary points.

7 A curve has equation $(x - 6)(y + 4) = 2$.
   (a) Find an expression for $\frac{dy}{dx}$ in terms of x and y.
   (b) Find the equation of the normal to the curve at the point $(7, -2)$.
   (c) Find the coordinates of the point where the normal meets the curve again.
   (d) By rewriting the equation in the form $y - a = \frac{b}{x - c}$, identify any asymptotes and sketch the curve.

8 A curve has equation $x^2 + 3xy + y^2 = 11$.
   (a) Show that the gradient of the curve at the point $(1, 2)$ is $-\frac{8}{7}$.
   (b) Find the equation of the tangent to the curve at $(1, 2)$.
   (c) Show that the normal to the curve at $(1, 2)$ goes through the point $(9, 9)$.

9 A curve has equation $x^2 + 2xy + 3y^2 = 33$.
   (a) Find an expression for $\frac{dy}{dx}$ in terms of x and y.
   (b) Find the gradient of the curve at the points where $x = 3$.

10 The curve C has equation $x^2 + y^2 - 3xy + 5 = 0$.
   (a) Find an expression for $\frac{dy}{dx}$ in terms of x and y.
   (b) Find the equation of the tangent to the curve at the point where $x = 2$.
   (c) Find the points P and Q on the curve where $x = 3$.
   (d) Show that the normal to the curve at one of the points P and Q is parallel to the y-axis and find the equation of the normal to the curve at the other point.

DIFFERENTIATING FUNCTIONS DEFINED PARAMETRICALLY

In Chapter 7 you were introduced to parametric equations. Example 7.1 illustrated the parametric equations

$$x = t^2, \quad y = 2t.$$

You learnt how to eliminate the parameter to obtain a Cartesian equation connecting x and y and how to sketch graphs of these functions.

This next section deals with differentiating parametric equations. To differentiate a function which is defined in terms of a parameter t, you need to use the chain rule:

$$\frac{dy}{dx} = \frac{dy}{dt} \times \frac{dt}{dx}$$

Since

$$\frac{dt}{dx} = \frac{1}{\frac{dx}{dt}}$$

it follows that

$$\frac{dy}{dx} = \frac{\frac{dy}{dt}}{\frac{dx}{dt}}$$

provided that $\frac{dx}{dt} \neq 0$.

EXAMPLE 9.5

A curve has the parametric equations $x = t^2$, $y = 2t$. Find
(a) $\frac{dy}{dx}$ in terms of the parameter t;
(b) the equation of the tangent to the curve at the general point $(t^2, 2t)$;
(c) the equation of the tangent at the point where $t = 3$.
(d) Eliminate the parameter, and hence sketch the curve and the tangent at the point where $t = 3$.

Solution (a) $x = t^2 \implies \dfrac{dx}{dt} = 2t$

$y = 2t \implies \dfrac{dy}{dt} = 2$

$\dfrac{dy}{dx} = \dfrac{\frac{dy}{dt}}{\frac{dx}{dt}} = \dfrac{2}{2t} = \dfrac{1}{t}$ The gradient of the curve at $(t^2, 2t)$

(b) Using $y - y_1 = m(x - x_1)$ and taking the point (x_1, y_1) as $(t^2, 2t)$, the equation of the tangent at the point $(t^2, 2t)$ is

$$y - 2t = \dfrac{1}{t}(x - t^2)$$
$$\implies ty - 2t^2 = x - t^2$$
$$\implies x - ty + t^2 = 0.$$

This equation still contains the parameter and is called the equation of the tangent at the general point

(c) Substituting $t = 3$ into this equation gives the equation of the tangent at the point where $t = 3$.

The tangent is $x - 3y + 9 = 0$.

(d) Eliminating t from $x = t^2$, $y = 2t$ gives

$$x = \left(\dfrac{y}{2}\right)^2 \quad \text{or} \quad y^2 = 4x.$$

This is a parabola whose line of symmetry is the x-axis.

The point where $t = 3$ has coordinates $(9, 6)$.

The tangent $x - 3y + 9 = 0$ crosses the axes at $(0, 3)$ and $(-9, 0)$.

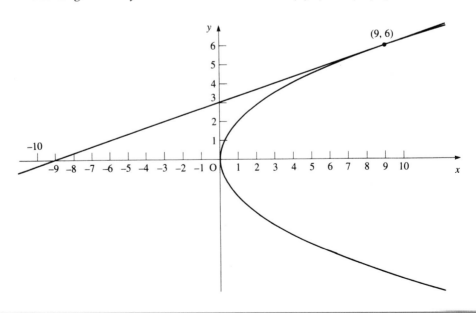

FIGURE 9.5

EXAMPLE 9.6

An ellipse has parametric equations $x = 4\cos\theta$, $y = 3\sin\theta$. Find

(a) $\frac{dy}{dx}$ at the point with parameter θ;
(b) the equation of the normal at the general point $(4\cos\theta, 3\sin\theta)$;
(c) the equation of the normal at the point where $\theta = \frac{\pi}{4}$;
(d) the coordinates of the point where $\theta = \frac{\pi}{4}$.
(e) Show the ellipse and the normal on a sketch.

Solution

(a) $x = 4\cos\theta \implies \frac{dx}{d\theta} = -4\sin\theta$

$y = 3\sin\theta \implies \frac{dy}{d\theta} = 3\cos\theta$

$$\frac{dy}{dx} = \frac{\frac{dy}{d\theta}}{\frac{dx}{d\theta}} = \frac{3\cos\theta}{-4\sin\theta}$$

$$= -\frac{3\cos\theta}{4\sin\theta}$$

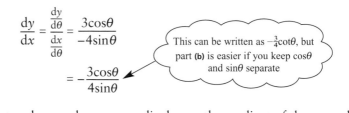

This can be written as $-\frac{3}{4}\cot\theta$, but part (b) is easier if you keep $\cos\theta$ and $\sin\theta$ separate

(b) The tangent and normal are perpendicular, so the gradient of the normal is

$$-\frac{1}{\frac{dy}{dx}} \text{ which is } +\frac{4\sin\theta}{3\cos\theta}.$$

$m_1 m_2 = -1$ for perpendicular lines

Using $y - y_1 = m(x - x_1)$ and taking the point (x_1, y_1) as $(4\cos\theta, 3\sin\theta)$, the equation of the normal at the point $(4\cos\theta, 3\sin\theta)$ is

$$y - 3\sin\theta = \frac{4\sin\theta}{3\cos\theta}(x - 4\cos\theta)$$

$\implies \quad 3y\cos\theta - 9\sin\theta\cos\theta = 4x\sin\theta - 16\sin\theta\cos\theta$

$\implies \quad 4x\sin\theta - 3y\cos\theta - 7\sin\theta\cos\theta = 0.$

(c) When $\theta = \frac{\pi}{4}$, $\cos\theta = \frac{1}{\sqrt{2}}$ and $\sin\theta = \frac{1}{\sqrt{2}}$, so the equation of the normal is

$$4x \times \frac{1}{\sqrt{2}} - 3y \times \frac{1}{\sqrt{2}} - 7 \times \frac{1}{\sqrt{2}} \times \frac{1}{\sqrt{2}} = 0$$

$\implies \quad 4\sqrt{2}x - 3\sqrt{2}y - 7 = 0$

or $\quad 4x - 3y - 4.95 = 0$ (to 2 decimal places).

(d) The coordinates of the point where $\theta = \frac{\pi}{4}$ are

$$(4\cos\tfrac{\pi}{4}, 3\sin\tfrac{\pi}{4}) = \left(4 \times \frac{1}{\sqrt{2}}, 3 \times \frac{1}{\sqrt{2}}\right)$$

$$= (2\sqrt{2}, \tfrac{3}{2}\sqrt{2}) \text{ or } (2.83, 2.12) \text{ (to 2 decimal places)}.$$

(See figure 9.6.)

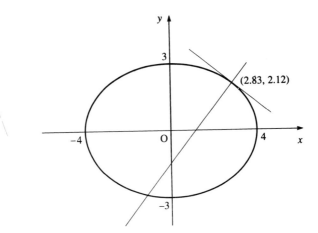

FIGURE 9.6

EXERCISE 9B

1 For each of the following curves, find $\frac{dy}{dx}$ in terms of the parameter.

(a) $x = 3t^2$
$y = 2t^3$

(b) $x = \theta - \cos\theta$
$y = \theta + \sin\theta$

(c) $x = t + \dfrac{1}{t}$
$y = t - \dfrac{1}{t}$

(d) $x = 3\cos\theta$
$y = 2\sin\theta$

(e) $x = (t + 1)^2$
$y = (t - 1)^2$

(f) $x = \theta\sin\theta + \cos\theta$
$y = \theta\cos\theta - \sin\theta$

(g) $x = e^{2t} + 1$
$y = e^t$

(h) $x = \dfrac{t}{1 + t}$
$y = \dfrac{t}{1 - t}$

2 A curve has the parametric equations $x = \tan\theta$, $y = \tan 2\theta$. Find
   (a) the value of $\frac{dy}{dx}$ when $\theta = \frac{\pi}{6}$;
   (b) the equation of the tangent to the curve at the point where $\theta = \frac{\pi}{6}$;
   (c) the equation of the normal to the curve at the point where $\theta = \frac{\pi}{6}$.

3 A curve has the parametric equations $x = t^2$, $y = 1 - \dfrac{1}{2t}$ for $t > 0$. Find
   (a) the coordinates of the point P where the curve cuts the x-axis;
   (b) the gradient of the curve at this point;
   (c) the equation of the tangent to the curve at P;
   (d) the coordinates of the point where the tangent cuts the y-axis.

4 A curve has parametric equations $x = at^2$, $y = 2at$, where a is constant. Find
   (a) the equation of the tangent to the curve at the point with parameter t;
   (b) the equation of the normal to the curve at the point with parameter t;
   (c) the coordinates of the points where the normal cuts the x- and y-axes.

EXERCISE 9B

5 A curve has parametric equations $x = \cos\theta$, $y = \cos 2\theta$.
   (a) Show that $\frac{dy}{dx} = 4\cos\theta$.
   (b) By writing $\frac{dy}{dx}$ in terms of x, show that $\frac{d^2y}{dx^2} - 4 = 0$.

6 The parametric equations of a curve are $x = at$, $y = \frac{b}{t}$, where a and b are constant. Find in terms of a, b and t
   (a) $\frac{dy}{dx}$;
   (b) the equation of the tangent to the curve at the general point $\left(at, \frac{b}{t}\right)$;
   (c) the coordinates of the points X and Y where the tangent cuts the x- and y-axes.
   (d) Show that the area of triangle OXY is constant, where O is the origin.

7 The equation of a curve is given in terms of the parameter t by the equations $x = 4t$ and $y = 2t^2$ where t takes positive and negative values.

   (a) Sketch the curve.

   P is the point on the curve with parameter t.

   (b) Show that the gradient at P is t.
   (c) Find and simplify the equation of the tangent at P.

   The tangents at two points Q (with parameter t_1) and R (with parameter t_2) meet at S.

   (d) Find the coordinates of S.
   (e) In the case when $t_1 + t_2 = 2$, show that S lies on a straight line. Give the equation of the line.

   [MEI]

8 The diagram shows a sketch of the curve given parametrically in terms of t by the equations $x = 1 - t^2$, $y = 2t + 1$.
   (a) Show that the point Q (0, 3) lies on the curve, stating the value of t corresponding to this point.
   (b) Show that, at the point with parameter t,
   $$\frac{dy}{dx} = -\frac{1}{t}.$$
   (c) Find the equation of the tangent at Q.
   (d) Verify that the tangent at Q passes through the point R (4, −1).
   (e) The other tangent from R to the curve touches the curve at the point S and has equation $3y - x + 7 = 0$. Find the coordinates of S.

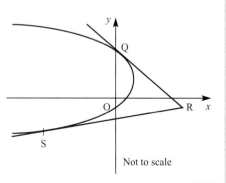

Not to scale

[MEI]

9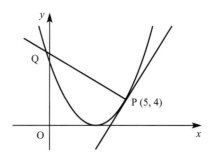

The diagram shows a sketch of the curve with parametric equations $x = 1 - 2t$, $y = t^2$. The tangent and normal at P are also shown.

(a) Show that the point P (5, 4) lies on the curve by stating the value of t corresponding to this point.

(b) Show that, at the point with parameter t, $\frac{dy}{dx} = -t$.

(c) Find the equation of the tangent at P.

(d) The normal at P cuts the curve again at Q. Find the coordinates of Q.

[MEI]

10 A point P moves in a plane so that at time t its coordinates are given by $x = 4\cos t$, $y = 3\sin t$. Find

(a) $\frac{dy}{dx}$ in terms of t;

(b) the equation of the tangent to its path at time t;

(c) the value of t for which the particle is travelling parallel to the line $x + y = 0$.

11 A circle has parametric equations $x = 3 + 2\cos\theta$, $y = 3 + 2\sin\theta$.

(a) Find the equation of the tangent at the point with parameter θ.

(b) Show that this tangent will pass through the origin provided that $\sin\theta + \cos\theta = -\frac{2}{3}$.

(c) By writing $\sin\theta + \cos\theta$ in the form $R\sin(\theta + \alpha)$, solve the equation $\sin\theta + \cos\theta = -\frac{2}{3}$ for $0 \leq \theta \leq 2\pi$.

(d) Illustrate the circle and tangents on a sketch, showing clearly the values of θ which you found in part (c).

12 The parametric equations of the circle centre (2, 5) and radius 3 units are $x = 2 + 3\cos\theta$, $y = 5 + 3\sin\theta$.

(a) Find the gradient of the circle at the point with parameter θ.

(b) Find the equation of the normal to the circle at this point.

(c) Show that the normal at any point on the circle passes through the centre. (This is an alternative proof of the result 'tangent and radius are perpendicular'.)

13 The parametric equations of a curve are

$$x = 3\cos\theta, \quad y = 2\sin\theta \quad \text{for } 0 \leq \theta < 2\pi.$$

(a) By eliminating θ between these two equations, find the Cartesian equation of the curve.

(b) Draw a sketch of the curve, giving the coordinates of the points where it cuts the axes. On your sketch show the pair of tangents which pass through the point (6, 2).

(c) Use the parametric equations to calculate $\frac{dy}{dx}$ in terms of θ.

You are given that the equation of the tangent to the curve at $(3\cos\theta, 2\sin\theta)$ is

$$2x\cos\theta + 3y\sin\theta = 6.$$

(d) Show that, for tangents to the curve which pass through the point (6, 2),

$$2\cos\theta + \sin\theta = 1.$$

(e) Solve the equation in (d) to find the two values of θ (in radians correct to 2 decimal places) corresponding to the two tangents.

[MEI]

14 A curve C is given by the equations

$$x = 2\cos t + \sin 2t, \quad y = \cos t - 2\sin 2t, \quad 0 \leq t < \pi,$$

where t is a parameter.

(a) Find $\frac{dx}{dt}$ and $\frac{dy}{dt}$ in terms of t.

(b) Find the value of $\frac{dy}{dx}$ at the point P on C where $t = \frac{\pi}{4}$.

(c) Find an equation of the normal to the curve at P.

[Edexcel]

15 The curve C has parametric equations

$$x = t^3, \quad y = t^2, \quad t > 0.$$

(a) Find an equation of the tangent to C at A (1, 1).

Given that the line l with equation $3y - 2x + 4 = 0$ cuts the curve C at point B,

(b) find the coordinates of B;

(c) prove that the line l only cuts C at the point B.

[Edexcel]

Exponential growth and decay, and the derivative of a^x

You will recall from Chapter 10 of *Pure Mathematics: Core 1 and Core 2* that the general equation for exponential growth (or exponential decay) is

$$y = a^x.$$

If $a > 1$ the graph of this function takes the form shown in figure 9.7.

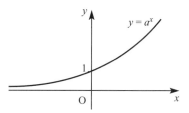

FIGURE 9.7

We are interested in the gradient of this function, that is, we would like to determine its derivative. For example, if instead of x we have time t, we might have the population P given by the equation

$$P = P_0 2^{0.1t}$$

where P_0 is the initial population. The derivative of this function at time t will give us the rate of change of population. So we need to know how to differentiate a^x.

You have seen that the derivative of e^x is e^x and, more generally,

$$\text{if } y = e^{f(x)} \text{ then } \frac{dy}{dx} = f'(x)e^{f(x)}.$$

However, it is more complicated when you have $y = a^x$. One way to find the derivative is to take natural logarithms of both sides of the equation and then differentiate as an implicit function.

$$\ln y = \ln a^x = x \ln a$$

Differentiating gives:

$$\frac{1}{y} \times \frac{dy}{dx} = \ln a$$

$$\therefore \quad \frac{dy}{dx} = y \ln a$$

so $\quad \dfrac{dy}{dx} = a^x \ln a$

or $\quad \dfrac{d(a^x)}{dx} = a^x \ln a$

EXAMPLE 9.7

Differentiate $y = 5^x$ with respect to x.

Solution Using the above rule

$$\frac{dy}{dx} = 5^x \ln 5.$$

EXAMPLE 9.8

Find the derivative of $a^{f(x)}$.

Solution Let $y = a^{f(x)}$.
Let $u = f(x)$, then $y = a^u$.

Applying the chain rule:

$$\frac{dy}{dx} = \frac{dy}{du} \times \frac{du}{dx} = a^u \ln a \times f'(x)$$

so $\quad \dfrac{dy}{dx} = a^{f(x)} \ln a \times f'(x)$

EXAMPLE 9.9

The area A (m²) of weed covering a pond is given by

$$A = 2 \times 1.5^t$$

where t is the time measured in days.

(a) What was the initial area of the weed?

(b) What is the rate at which the weed is growing after 5 days?

(c) After how many days is the rate at which the weed is growing more than 100 m² per day?

Solution (a) The initial area is found by putting $t = 0$.

$$\Rightarrow \quad A_0 = 2 \times 1.5^0 = 2 \text{ m}^2$$

(b) The rate of growth of the weed is given by the derivative of A with respect to t, so

$$\frac{dA}{dt} = 2 \times 1.5^t \times \ln 1.5$$

and when $t = 5$

$$\frac{dA}{dt} = 2 \times 1.5^5 \times \ln 1.5 = 6.16 \text{ m}^2 \text{ per day (to 3 significant figures)}.$$

(c) If the rate of growth is $100 \, \text{m}^2$ per day then

$$\frac{dA}{dt} = 2 \times 1.5^t \times \ln 1.5 = 100$$

$$1.5^t = \frac{100}{2 \times \ln 1.5} = 123.3.$$

Taking natural logarithms of both sides gives

$$t = \frac{\ln 123.3}{\ln 1.5} = 11.87.$$

During and after the twelfth day the model predicts that the growth will be greater than $100 \, \text{m}^2$ per day.

EXERCISE 9C

1 Differentiate the following with respect to x.
 (a) 3^x
 (b) b^x
 (c) $2^{\sin x}$
 (d) $3^x \times 4^x$
 (e) 10^{x^2}
 (f) x^x
 (g) $(\tan x)^x$
 (h) $e^{\sec x}$
 (i) $\dfrac{8^x}{\ln 2}$

2 A fungus grows exponentially such that its area A (mm^2) at time t (days) is given by

$$A = 100 \times 2^t.$$

 (a) Find the rate of growth when $t = 10$.
 (b) When is the rate of growth of the fungus $10\,000 \, \text{mm}^2/\text{day}$?

3 A student predicts that the population P of a group of people grows such that after t years it is given by

$$P = P_0 \times 4^{0.2t}$$

where P_0 is the initial population.

 (a) If the population after 5 years is 3000, what was the initial population?
 (b) What is the rate of growth of the population when $t = 5$?
 (c) Explain why this is not a sensible model for calculating the population.

4 A manufacturer claims that *Edcel* eradicates unwanted cells to cure a disease. If the initial number of cells is C_0 then after t days use of *Edcel* there should be C cells, where

$$C = C_0 \times 2.5^{-0.25t}.$$

If the initial number of cells is estimated to be $100\,000$,
 (a) how many cells will there be after 12 days?
 (b) what is the rate of change of cells per day after 7 days?
 (c) does the rate of change increase or decrease as the days go by? Illustrate your answer on a graph.

5 Linda attempts to climb a 1000 m high mountain. The height s (metres) she has gained from her starting point is given in terms of the time t (hours) by the equation

$$s = 1000(1 - 2000^{-0.1t}).$$

(a) Sketch the graph of s against t.
(b) Does Linda reach the top of the mountain?
(c) Find Linda's speed (i.e. the rate of change of distance with time) after 1 hour.
(d) Linda stops after 3 hours. What was her speed when she stopped climbing the mountain and how far had she gone?

DIFFERENTIAL EQUATIONS

The idea of a differential equation was first considered in Chapter 5 of *Pure Mathematics: Core 1* where integration was introduced as the reverse process of differentiation.

The first illustration was the differential equation

$$\frac{dy}{dx} = 2x$$

giving rise to $y = x^2 + c$.

In general, differential equations are ones that involve derivatives such as $\frac{dy}{dx}$, $\frac{d^2y}{dx^2}$, $\frac{dx}{dt}$ or $\frac{d\theta}{dt}$. A differential equation which only involves a first derivative, such as $\frac{dy}{dx}$, is called a *first-order differential equation*. One which involves a second derivative, such as $\frac{d^2y}{dx^2}$, is called a *second-order differential equation*. A *third-order differential equation* involves a third derivative, and so on.

In *Pure Mathematics: Core 4* you will look at first-order differential equations only. Second-order equations appear in the Further Pure Mathematics 1 unit of the specification.

FORMING DIFFERENTIAL EQUATIONS FROM RATES OF CHANGE

If you are given sufficient information about the rate of change of a quantity, such as temperature or velocity, you can work out a differential equation to model the situation. It is important to look carefully at the wording of the problem which you are studying in order to write an equivalent mathematical statement. For example, if the altitude of an aircraft is being considered, the phrase 'the rate of change of height' might be used. This actually means 'the rate of change of height *with respect to time*' and could be written as $\frac{dh}{dt}$. However, we might be more interested in how the height of the aircraft changes according to the horizontal distance it has travelled. In this case, we would talk about 'the rate of change of height *with respect to horizontal distance*' and could write this as $\frac{dh}{dx}$, where x is the horizontal distance travelled.

Some of the situations you meet in this chapter involve motion along a straight line, and so you will need to know the meaning of the associated terms.

FIGURE 9.8

The position of an object (+5 in figure 9.8) is its distance from the origin O in the direction you have chosen to define as being positive.

The rate of change of position of the object with respect to time is its velocity v and this can take positive or negative values according to whether the object is moving away from the origin or towards it:

$$v = \frac{ds}{dt}$$ where s is the displacement from the origin at time t.

The rate of change of an object's velocity with respect to time is called its acceleration, a:

$$a = \frac{dv}{dt}$$

Velocity and acceleration are vector quantities but in one-dimensional motion there is no choice in direction, only in sense (i.e. whether positive or negative). Consequently we have chosen not to use the conventional bold type for vectors in this chapter.

EXAMPLE 9.10

An object is moving through a liquid so that the rate at which its velocity decreases is proportional to its velocity at any given instant. When it enters the liquid, it has a velocity of 5 m s^{-1} and the velocity is decreasing at a rate of 1 m s^{-2}. Find the differential equation to model this situation.

Solution   The rate of change of velocity means the rate of change of velocity with respect to time and so can be written as $\frac{dv}{dt}$. As it is decreasing, the rate of change must be negative, so

$$\frac{dv}{dt} \propto -v$$

or

$$\frac{dv}{dt} = -kv$$

where k is a positive constant.

When the object enters the liquid its velocity is $5\,\text{m s}^{-1}$, so $v = 5$, and the velocity is decreasing at the rate of $1\,\text{m s}^{-2}$, so

$$\frac{dv}{dt} = -1.$$

Putting this information into the equation gives

$$-1 = -k \times 5 \quad \Rightarrow \quad k = \tfrac{1}{5}.$$

So the situation is modelled by the differential equation

$$\frac{dv}{dt} = -\frac{v}{5}.$$

EXAMPLE 9.11

A model is proposed for the temperature gradient within a star, in which the temperature decreases with respect to the distance from the centre of the star at a rate which is inversely proportional to the square of the distance from the centre. Express this model as a differential equation.

Solution In this example the rate of change of temperature is not with respect to time but with respect to distance. If θ represents the temperature of the star and r the distance from the centre of the star, the rate of change of temperature with respect to distance may be written as $\frac{d\theta}{dr}$, so

$$\frac{d\theta}{dr} \propto -\frac{1}{r^2} \quad \text{or} \quad \frac{d\theta}{dr} = -\frac{k}{r^2}$$

where k is a positive constant.

Note

This model must break down at the centre of the star, otherwise it would be infinitely hot there.

EXAMPLE 9.12

The area A of a square is increasing at a rate proportional to the length of its side s. Find an expression for $\frac{ds}{dt}$.

Solution The rate of increase of A with respect to time may be written as $\frac{dA}{dt}$. As this is proportional to s, we can write

$$\frac{dA}{dt} = ks$$

where k is a positive constant.

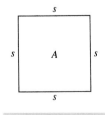

FIGURE 9.9

We can use the chain rule to write down an expression for $\frac{ds}{dt}$ in terms of $\frac{dA}{dt}$.

$$\frac{ds}{dt} = \frac{ds}{dA} \times \frac{dA}{dt}$$

We now need an expression for $\frac{ds}{dA}$. Because A is a square,

$$A = s^2$$

$$\Rightarrow \frac{dA}{ds} = 2s$$

$$\Rightarrow \frac{ds}{dA} = \frac{1}{2s}.$$

Substituting the expressions for $\frac{ds}{dA}$ and $\frac{dA}{dt}$ into the expression for $\frac{ds}{dt}$:

$$\frac{ds}{dt} = \frac{1}{2s} \times ks$$

$$\Rightarrow \frac{ds}{dt} = \tfrac{1}{2}k.$$

You will learn how to solve first-order differential equations of this type in the next chapter.

Exercise 9D

1. The differential equation

$$\frac{dv}{dt} = 5v^2$$

   models the motion of a particle, where v is the velocity of the particle in m s^{-1} and t the time in seconds. Explain the meaning of $\frac{dv}{dt}$ and what the differential equation tells us about the motion of the particle.

2. A spark from a Roman candle is moving in a straight line at a speed which is inversely proportional to the square of the distance which the spark has travelled from the candle. Find an expression for the speed (i.e. the rate of change of distance travelled) of the spark.

3. The rate at which a sunflower increases in height is proportional to the natural logarithm of the difference between its final height H and its height h at a particular time. Find a differential equation to model this situation.

4. In a chemical reaction in which substance A is converted into substance B, the rate of increase of the mass of substance B is inversely proportional to the mass of substance B present. Find a differential equation to model this situation.

5. After a major advertising campaign, an engineering company finds that its profits are increasing at a rate proportional to the square root of the profits at any given time. Find an expression to model this situation.

Exercise 9D

6 The coefficient of restitution e of a squash ball increases with respect to the ball's temperature θ at a rate proportional to the temperature, for typical playing temperatures. (The coefficient of restitution is a measure of how elastic, or bouncy, the ball is. Its value lies between 0 and 1, 0 meaning that the ball is not at all elastic and 1 meaning that it is perfectly elastic.) Find a differential equation to model this situation.

7 A cup of tea cools at a rate proportional to the temperature of the tea above that of the surrounding air. Initially, the tea is at a temperature of 95°C and is cooling at a rate of 0.5°C s^{-1}. The surrounding air is at 15°C. Find a differential equation to model this situation.

8 The rate of increase of bacteria is modelled as being proportional to the amount of bacteria at any time during their initial growth phase. When the bacteria number 2×10^6 they are increasing at a rate of 10^5 per day. Find a differential equation to model this situation.

9 The acceleration (i.e. the rate of change of velocity) of a moving object under a particular force is inversely proportional to the square root of its velocity. When the speed is $4\,\text{m s}^{-1}$ the acceleration is $2\,\text{m s}^{-2}$. Find a differential equation to model this situation.

10 The radius of a circular ink blot is increasing at a rate inversely proportional to its area A. Find an expression for $\dfrac{dA}{dt}$.

11 A poker, 80 cm long, has one end in a fire. The temperature of the poker decreases with respect to the distance from that end at a rate proportional to that distance. Half-way along the poker, the temperature is decreasing at a rate of 10°C cm^{-1}. Find a differential equation to model this situation.

12 A conical egg timer, shown in the diagram, is letting sand through from top to bottom at a rate of $0.02\,\text{cm}^3\,\text{s}^{-1}$. Find an expression for the rate of change of height $\left(\dfrac{dh}{dt}\right)$ of the sand in the top of the timer.

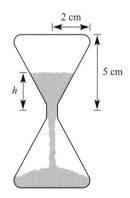

13 A spherical balloon is allowed to deflate. The rate at which air is leaving the balloon is proportional to the volume V of air left in the balloon. When the radius of the balloon is 15 cm, air is leaving at a rate of $8\,\text{cm}^3\,\text{s}^{-1}$. Find an expression for $\dfrac{dV}{dt}$.

14 A tank is shaped as a cuboid with a square base of side 10 cm. Water runs out through a hole in the base at a rate proportional to the square root of the height, h cm, of water in the tank. At the same time, water is pumped into the tank at a constant rate of $2\,\text{cm}^3\,\text{s}^{-1}$. Find an expression for $\frac{dh}{dt}$.

Exercise 9E

Examination-style questions

1 A curve has equation $2x^2 + 5xy + y^2 = 16$.
  (a) Show that the point $(1, 2)$ lies on the curve and find the other point on the curve where $x = 1$.
  (b) Find an expression for $\frac{dy}{dx}$ in terms of x and y.
  (c) Find the equation of the normal to the curve at the point $(1, 2)$.

2 Find the equation of the normal to
$$x^3 - 4xy^2 + y^3 + 1 = 0$$
at the point $(2, -1)$.

3 A curve has equation $x^2 + 4xy + 2y^2 = 41$.
  (a) When $x = 3$ there are two distinct points on the curve. Show that one has coordinates $(3, 2)$ and find the coordinates of the other point.
  (b) Find the equation of the tangent to the curve at the point $(3, 2)$.

4 Find the equation of the normal to the curve $x^3 - y^3 = 72$ at the point where $x = 4$.

5 A curve has equation $7x^2 + 48xy - 7y^2 + 75 = 0$.
A and B are two distinct points on the curve. At each of these points the gradient of the curve is equal to $\frac{2}{11}$.
  (a) Use implicit differentiation to show that $x + 2y = 0$ at the points A and B.
  (b) Find the coordinates of the points A and B.
  [Edexcel]

6 A curve, C, is given by
$$x = 2t + 3, \qquad y = t^3 - 4t$$
where t is a parameter. The point A has parameter $t = -1$ and the line l is the tangent to C at A. The line l also intersects the curve at B.
  (a) Show that an equation for l is $2y + x = 7$.
  (b) Find the value of t at B.
  [Edexcel]

7 A curve is given by the parametric equations
$$x = 4\sin^3 t, \quad y = \cos 2t, \qquad 0 \leq t \leq \tfrac{\pi}{4}.$$
  (a) Show that $\frac{dx}{dy} = -3\sin t$.
  (b) Find an equation of the normal to the curve at the point where $t = \tfrac{\pi}{6}$.
  [Edexcel]

EXERCISE 9E

8 The curve C has parametric equations

$$x = 4\cos 2t, \quad y = 3\sin t, \quad -\tfrac{\pi}{2} < t < \tfrac{\pi}{2}.$$

A is the point $(2, 1\tfrac{1}{2})$ and lies on C.

(a) Find the value of t at the point A.
(b) Find $\dfrac{dy}{dx}$ in terms of t.
(c) Show that an equation of the normal to C at A is

$$6y - 16x + 23 = 0.$$

The normal at A cuts C again at the point B.

(d) Find the y-coordinate of the point B.

[Edexcel]

9 A curve is given by the parametric equations

$$x = t + \frac{1}{t}, \quad y = t - \frac{1}{t}, \quad t \neq 0.$$

(a) Show that $\dfrac{dy}{dx} = \dfrac{t^2 + 1}{t^2 - 1}$.

(b) Find the equation of the normal to the curve at the point where $t = \tfrac{1}{2}$.

10 A curve is given parametrically by the equations

$$x = 5\cos t, \quad y = -2 + 4\sin t, \quad 0 \leq t \leq 2\pi.$$

(a) Find the coordinates of all the points at which C intersects the coordinate axes, giving your answers in surd form where appropriate.
(b) Sketch the graph of C.

P is the point where $t = \tfrac{\pi}{6}$.

(c) Show that the normal to C at P has equation

$$8\sqrt{3}y = 10x - 25\sqrt{3}.$$

[Edexcel]

11 (a) Given that $y = 3^x$ prove that $\dfrac{dy}{dx} = 3^x \ln 3$.
(b) Show that the graph of $y = x \times 3^x$ has only one turning point and that its coordinates are $\left(-\dfrac{1}{\ln 3}, -\dfrac{1}{e\ln 3}\right)$.

12 Myles noticed that the area of weed on the lake where he goes sailing was doubling every day. He guessed that when he first noticed the problem the weed had covered 5 m².

(a) Find an expression for the area A covered after t days.
(b) Show that the rate of change of growth of the area with respect to time is $5\ln 2 \times 2^t$.
(c) Use your result to show that the ratio of the rate of growth of the weed between any day and the next day is 1 : 2.

13 (a) Given that $a^x = e^{kx}$, where a and k are constants, $a > 0$ and $x \in \mathbb{R}$, prove that $k = \ln a$.

  (b) Hence, using the derivative of e^{kx}, prove that when $y = 2^x$,
  $$\frac{dy}{dx} = 2^x \ln 2.$$

  (c) Hence deduce that the gradient of the curve with equation $y = 2^x$ at the point $(2, 4)$ is $\ln 16$.

  [Edexcel]

14 An egg timer is made from two cones as shown. The diameter of the base of each cone is 4 cm and the height of each cone is 6 cm.

The timer has the lower half full of sand and is then turned upside down so that all the sand falls from the upper half into the lower half. If it takes $4\tfrac{1}{2}$ minutes for this to happen, show that

$$\frac{dV}{dt} = \frac{16\pi}{9} \text{ cm}^3 \text{ per minute.}$$

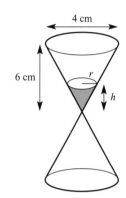

In general, the sand in the upper half is at a height h cm with radius r cm, as shown in the diagram. Find an expression for the volume V of the sand in the upper half in terms of h only, and show that

$$\frac{dV}{dh} = \tfrac{1}{9}\pi h^2.$$

Using the chain rule find an expression for $\frac{dh}{dt}$ and show that when $h = 2$ cm, $\frac{dh}{dt} = 4$ cm per minute.

15 In an experiment the volume V of a spherical balloon is measured every t hours. Show that

$$\frac{dV}{dt} = 4\pi r^2 \frac{dr}{dt}$$

where r is the radius.

It is assumed that $r = 2^{-t}$. Find $\frac{dV}{dt}$ when

(a) $t = 0$

(b) $t = 3$.

Key points

1. **Stationary points**

   (a) $\dfrac{dy}{dx} = 0$, $\dfrac{d^2y}{dx^2} < 0$ Maximum

   (b) $\dfrac{dy}{dx} = 0$, $\dfrac{d^2y}{dx^2} > 0$ Minimum

   (c) $\dfrac{d^2y}{dx^2} = 0$ Inflection (but could be a maximum or minimum)

   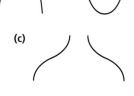

2. **Chain rule**

   $\dfrac{dy}{dx} = \dfrac{dy}{du} \times \dfrac{du}{dx}$

3. **Product rule**

   $y = uv \implies \dfrac{dy}{dx} = u\dfrac{dv}{dx} + v\dfrac{du}{dx}$

4. **Quotient rule**

   $y = \dfrac{u}{v} \implies \dfrac{v\dfrac{du}{dx} - u\dfrac{dv}{dx}}{v^2}$

5. **Inverses**

   $\dfrac{dy}{dx} = \dfrac{1}{\dfrac{dx}{dy}}$

6. **Functions**

y	$\dfrac{dy}{dx}$	y	$\dfrac{dy}{dx}$
$e^{f(x)}$	$f'(x)e^{f(x)}$	$\ln(f(x))$	$\dfrac{f'(x)}{f(x)}$
a^x	$a^x \ln a$	$a^{f(x)}$	$a^{f(x)} \ln a \times f'(x)$
$\sin x$	$\cos x$	$\sec x$	$\sec x \tan x$
$\cos x$	$-\sin x$	$\text{cosec } x$	$-\text{cosec } x \cot x$
$\tan x$	$\sec^2 x$	$\cot x$	$-\text{cosec}^2 x$

7. **Implicit**

   e.g. $\dfrac{d}{dx}(y^2) = 2y\dfrac{dy}{dx}$ $\dfrac{d}{dx}(xy) = x\dfrac{dy}{dx} + y$

8. **Parametric**

   If $x = f(t)$ and $y = g(t)$ then $\dfrac{dy}{dx} = \dfrac{\dfrac{dy}{dt}}{\dfrac{dx}{dt}}$.

9. **Differential equations**

   A first-order differential equation is one that contains a term of the type $\dfrac{dy}{dx}$ without powers and without higher derivatives.

Chapter ten

INTEGRATION

I have been ever of opinion that revolutions are not to be evaded.

Benjamin Disraeli

REVIEW OF INTEGRATION

You will recall from Chapters 5 and 12 of *Pure Mathematics: Core 1 and Core 2* that the general rule for integrating the nth power of x is

$$\int x^n \, dx = \frac{1}{n+1} x^{n+1} + c \quad \text{provided } n \neq -1.$$

Further, you could find the area between the curve $y = f(x)$ and the x-axis from $x = a$ to $x = b$, such as that shown in figure 10.1, using

$$\text{area} = \int_a^b f(x) \, dx, \text{ if } f(x) \geq 0.$$

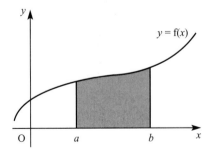

FIGURE 10.1

Do you recall how to find areas when the curve cuts the x-axis? Do you recall how to find areas between curves and the y-axis? If not, a review of the integration chapters in *Pure Mathematics: Core 1 and Core 2* would be beneficial.

Here are some examples to remind you of some of the various problems. After Exercise 10A we shall extend the methods and techniques of integration.

EXAMPLE 10.1

Find the indefinite integral $\int \dfrac{4}{x^2}\,dx$.

Solution

$$\int \dfrac{4}{x^2}\,dx = \int 4x^{-2}\,dx$$

$$= \dfrac{4x^{-2+1}}{-1} + c \qquad (-2+1 = -1)$$

$$= -4x^{-1} + c$$

$$= -\dfrac{4}{x} + c$$

It is usual to put the final answer in the same form as the original function.

EXAMPLE 10.2

Evaluate the definite integral $\int_4^9 x^{\frac{3}{2}}\,dx$.

Solution

$$\int_4^9 x^{\frac{3}{2}}\,dx = \left[\dfrac{x^{\frac{5}{2}}}{\frac{5}{2}}\right]_4^9 \qquad \left(\tfrac{3}{2} + 1 = \tfrac{5}{2}\right)$$

$$= \tfrac{2}{5}\left[x^{\frac{5}{2}}\right]_4^9$$

$$= \tfrac{2}{5}\left(9^{\frac{5}{2}} - 4^{\frac{5}{2}}\right)$$

$$= \tfrac{2}{5}(243 - 32)$$

$$= 84\tfrac{2}{5}$$

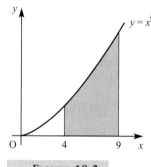

FIGURE 10.2

This gives the area of the shaded region in figure 10.2.

EXAMPLE 10.3

Find the area represented by $\int_1^2 \left(\dfrac{3}{x^4} - \dfrac{1}{x^2} + 4\right)dx$.

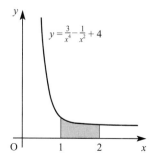

FIGURE 10.3

Solution

$$\int_1^2 \left(\frac{3}{x^4} - \frac{1}{x^2} + 4\right) dx = \int_1^2 (3x^{-4} - x^{-2} + 4) \, dx$$

$$= \left[\frac{3x^{-3}}{-3} - \frac{x^{-1}}{-1} + 4x\right]_1^2$$

$$= \left[-\frac{1}{x^3} + \frac{1}{x} + 4x\right]_1^2$$

$$= \left(-\frac{1}{8} + \frac{1}{2} + 8\right) - (-1 + 1 + 4)$$

$$= 4\tfrac{3}{8}$$

A *differential equation* is any equation which involves a derivative such as $\frac{dy}{dx}$.

The example below reminds you how to solve simple differential equations.

EXAMPLE 10.4

Given that $\frac{dy}{dx} = \sqrt{x} + \frac{1}{x^2}$:

(a) find the general solution of the differential equation;
(b) find the equation of the curve with this gradient function which passes through (1, 5).

Solution (a) $\frac{dy}{dx} = \sqrt{x} + \frac{1}{x^2}$

$$= x^{\frac{1}{2}} + x^{-2}$$

$$\Rightarrow y = \frac{x^{\frac{3}{2}}}{\frac{3}{2}} + \frac{x^{-1}}{-1} + c$$

$$= \frac{2}{3}x^{\frac{3}{2}} - \frac{1}{x} + c$$

> Integrate $\frac{dy}{dx}$ to obtain y

(b) Since the curve passes through (1, 5)

$$5 = \tfrac{2}{3} - 1 + c$$
$$c = 5\tfrac{1}{3}$$
$$\Rightarrow y = \tfrac{2}{3}x^{\frac{3}{2}} - \frac{1}{x} + 5\tfrac{1}{3}.$$

EXERCISE 10A

EXAMPLE 10.5

Find the area of the region between $y = \sqrt{x-1}$, $y = 1$, $y = 2$ and the y-axis.

Solution   First sketch the region to ensure that it does not cross the y-axis. This is shown in figure 10.4.

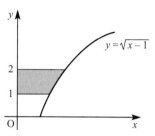

FIGURE 10.4

The area is

$$A = \int_a^b x \, dy$$
$$= \int_1^2 (y^2 + 1) \, dy$$
$$= \left[\tfrac{1}{3}y^3 + y\right]_1^2$$
$$= \frac{10}{3}.$$

Since $y = \sqrt{x-1}$
$x - 1 = y^2$
$x = y^2 + 1$

EXERCISE 10A

1 In each of the following questions, find the indefinite integral. Remember to include the constant of integration.

(a) $\int 10x^{-4} \, dx$

(b) $\int (2x - 3x^{-4}) \, dx$

(c) $\int (2 + x^3 + 5x^{-3}) \, dx$

(d) $\int (6x^2 - 7x^{-2}) \, dx$

(e) $\int 5x^{\frac{1}{4}} \, dx$

(f) $\int \frac{1}{x^4} \, dx$

(g) $\int \sqrt{x} \, dx$

(h) $\int \left(2x^4 - \frac{4}{x^2}\right) dx$

2 Evaluate the following definite integrals. Give answers as fractions or to 3 significant figures as appropriate.

(a) $\int_1^4 3x^{-2} \, dx$

(b) $\int_2^4 8x^{-3} \, dx$

(c) $\int_2^4 12x^{\frac{1}{2}} \, dx$

(d) $\int_{-3}^{-1} \frac{6}{x^3} \, dx$

(e) $\int_1^8 \left(\frac{x^2 + 3x + 4}{x^4}\right) dx$

(f) $\int_4^9 \left(\sqrt{x} - \frac{1}{\sqrt{x}}\right) dx$

3 The graph of $y = \sqrt{x} + \frac{1}{\sqrt{x}}$ for $x > 0$ is shown.

The shaded region is bounded by the curve, the x-axis and the lines $x = 1$ and $x = 9$. Find its area.

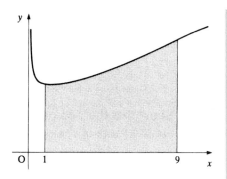

4 The graph of $y = 4 - \frac{16}{x^2}$ is shown below, together with the line $y = 3$.

(a) Find the coordinates of the points P, Q, R and S.
(b) Find the area of the shaded regions.

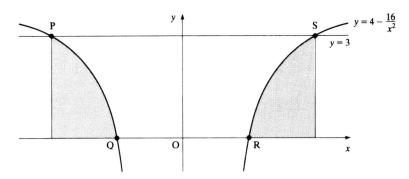

5 (a) Find the point of intersection of the graphs of $y = \sqrt{x}$ and $y = \frac{128}{x^3}$.
   (b) Sketch, on the same axes, the graphs of $y = \sqrt{x}$ and $y = \frac{128}{x^3}$ for $0 \leqslant x \leqslant 6$.
   (c) Shade the region bounded by the curves, the x-axis and the line $x = 5$.
   (d) Find the area of the shaded region.

6 The graphs of $y = 1 - \frac{3}{x^2}$ and $y = -\frac{2}{x^3}$ are shown below.

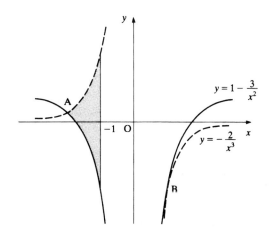

(a) Show that the x-coordinates of the points where the curves meet are the roots of the equation:

$$x^3 - 3x + 2 = 0.$$

(b) Hence show that the coordinates of A are $(-2, \tfrac{1}{4})$, and find the coordinates of point B.

(c) Find the area of the shaded region.

7 Given that $\dfrac{dy}{dx} = \dfrac{2}{x^2} - 3$

(a) find the general solution of the differential equation;

(b) find the equation of the curve with this gradient function which passes through $(2, 10)$.

8 Given that $\dfrac{dy}{dx} = \sqrt{x}$, find the general solution of this differential equation. Find the equation of the curve with this gradient function which passes through $(9, 20)$.

9 Given that $\dfrac{dy}{dx} = \dfrac{5x^2 - 1}{\sqrt{x}}$

(a) find the general solution of the differential equation;

(b) find the equation of the curve which passes through $(4, 68)$.

10 (a) Sketch the graph of $y = \sqrt[3]{x - 2}$.

(b) Find the area enclosed between $y = \sqrt[3]{x - 2}$, $y = 1$, $y = 2$ and the y-axis.

INTEGRAL OF e^x

In Chapter 4 you learnt that

if $y = e^x$ then $\dfrac{dy}{dx} = e^x$.

You also know from *Pure Mathematics: Core 1 and Core 2* that integration may be considered as the reverse process of differentiation, from which it follows that

$$\int e^x \, dx = e^x + c.$$

Remember: for indefinite integrals you need the constant c which disappears when you differentiate

EXAMPLE 10.6

Find $\int 2e^x \, dx$.

Solution $\int 2e^x \, dx = 2 \int e^x \, dx$

$= 2e^x + c$

EXAMPLE 10.7

Work out $\int_0^1 5e^x\,dx$ leaving your answer in terms of e.

Solution
$$\int_0^1 5e^x\,dx = 5\int_0^1 e^x\,dx$$
$$= 5\,[e^x]_0^1$$
$$= 5(e^1 - e^0)$$
$$= 5(e - 1)$$

You will recall that the integral of a function with respect to x gives the area between the curve and the x-axis between the two limits. Figure 10.5 shows the area defined by the integral in Example 10.7.

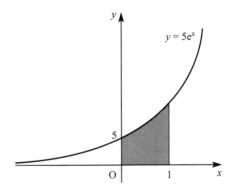

FIGURE 10.5

EXAMPLE 10.8

Find the area enclosed between the curve $y = e^x - 2$, the x-axis and the lines $x = -1$ and $x = 1$. Leave your answer in terms of e and natural logarithms.

Solution First sketch the curve as shown in figure 10.6.

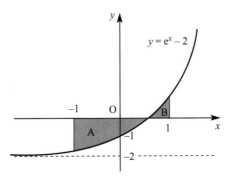

FIGURE 10.6

It cuts the x-axis when $y = 0$,
i.e. when $e^x - 2 = 0$
so $e^x = 2$
and $x = \ln 2$.

To find the area enclosed consider the parts below and above the x-axis separately.

$$\text{Area A} = \int_{-1}^{\ln 2} (e^x - 2) \, dx = [e^x - 2x]_{-1}^{\ln 2}$$

Note that $e^{\ln 2} = 2$

$$= e^{\ln 2} - 2\ln 2 - (e^{-1} + 2)$$
$$= -2\ln 2 - e^{-1}$$

Areas under the x-axis are negative

Magnitude of the area A is $2\ln 2 + e^{-1}$.

$$\text{Area B} = \int_{\ln 2}^{1} (e^x - 2) \, dx = [e^x - 2x]_{\ln 2}^{1}$$

$$= e - 2 - (e^{\ln 2} - 2\ln 2)$$
$$= e + 2\ln 2 - 4$$

So the total area enclosed is

$$e + e^{-1} + 4\ln 2 - 4.$$

EXERCISE 10B

1 Find the following indefinite integrals:

(a) $\int e^x \, dx$

(b) $\int 2e^x \, dx$

(c) $3 \int e^x \, dx$

(d) $\int (e^x + 2) \, dx$

(e) $5 \int (4 + 3e^x) \, dx$

(f) $\int (4 - e^x) \, dx$

(g) $\int (2x + 3e^x) \, dx$

(h) $\int (\tfrac{1}{2}e^x + \sqrt{x}) \, dx$

(i) $\int \left(4e^x + \dfrac{2}{x^3}\right) dx$

(j) $\int e^{x+2} \, dx$ (Hint: $a^m \times a^n = a^{m+n}$.)

2 Find the following definite integrals, giving your answers to 3 significant figures. You may have a calculator that can compute these integrals. If so, use it to check your answers but ensure that you have shown sufficient working to prove that you can find the answers without such a facility.

(a) $\int_0^1 e^x \, dx$

(b) $\int_0^2 2e^x \, dx$

(c) $\int_{-1}^0 (1 + e^x) \, dx$

(d) $\int_1^4 (\sqrt{x} - e^x) \, dx$

(e) $\int_{.5}^{.75} \left(e^x - \dfrac{1}{x^2}\right) dx$

(f) $\int_3^5 (1 - 2e^x) \, dx$

(g) $4\int_1^2 (2e^x - \sqrt[3]{x}) \, dx$

(h) $\int_{-1}^3 (e^x + x(1 + x)) \, dx$

(i) $\int_2^4 e^{x-2} \, dx$

(Hint: multiply out the bracket.)

3 Prove that $\int_{-1}^{1} e^x \, dx = \dfrac{e^2 - 1}{e}$.

4 A curve has equation $y = e^x - 3$.
  (a) Sketch the curve.
  (b) Show that it cuts the x-axis at $(\ln 3, 0)$.
  (c) Show that the total area enclosed between the curve, the x-axis and the lines $x = 0$ and $x = 2$ is

  $$e^2 + 6\ln 3 - 11.$$

5 Prove that the area enclosed between $y = e^x$, the x-axis and the lines $x = 1$ and $x = 2$ is $e(e - 1)$.

INTEGRAL OF $\frac{1}{x}$

In Chapter 4 you saw that $y = \ln x \Rightarrow \dfrac{dy}{dx} = \dfrac{1}{x}$, so it follows that when you reverse the process

$$\int \frac{1}{x} dx = \ln x + c.$$

The function $y = \ln x$ is defined for the domain $x > 0$ so it follows that the integral is also valid for $x > 0$.

If you have a multiplier, e.g. $y = \dfrac{k}{x}$, then take k out of the integral to give

$$\int \frac{k}{x} dx = k \int \frac{1}{x} dx = k \ln x + c.$$

EXAMPLE 10.9

Find the indefinite integral $\int \dfrac{1}{4x} dx$.

Solution $\int \dfrac{1}{4x} dx = \dfrac{1}{4} \int \dfrac{1}{x} dx$

$= \dfrac{1}{4} \ln x + c$

EXAMPLE 10.10

Find the exact value of $\int_1^3 \dfrac{2}{x} dx$.

Solution $\int_1^3 \dfrac{2}{x} dx = 2 \int_1^3 \dfrac{1}{x} dx = 2[\ln x]_1^3$

$= 2(\ln 3 - \ln 1)$ ⟵ $\ln 1 = 0$

$= 2\ln 3$

INTEGRAL OF $\frac{1}{x}$

EXAMPLE 10.11

Evaluate $\int_2^4 \left(\frac{x+1}{x}\right) dx$.

Solution You must first remove the quotient by splitting it into two fractions and simplifying:

$$\int_2^4 \left(\frac{x+1}{x}\right) dx = \int_2^4 \left(\frac{x}{x} + \frac{1}{x}\right) dx$$

$$= \int_2^4 \left(1 + \frac{1}{x}\right) dx$$

$$= [x + \ln x]_2^4$$

$$= (4 + \ln 4) - (2 + \ln 2)$$

$$= 2 + \ln 4 - \ln 2$$

$$= 2 + \ln \tfrac{4}{2}$$

$$= 2 + \ln 2.$$

EXAMPLE 10.12

Sketch the graph of $y = \frac{1}{x}$ for all values of x.

Find the area between the curve, the x-axis and
(a) the lines $x = 1$ and $x = 2$
(b) the lines $x = -2$ and $x = -1$.

Solution

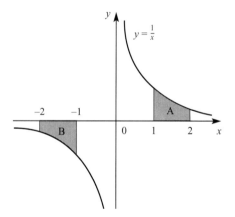

FIGURE 10.7

(a) Area A $= \int_1^2 \frac{1}{x} dx = [\ln x]_1^2 = \ln 2 - \ln 1 = \ln 2.$

(b) Area B $= \int_{-2}^{-1} \frac{1}{x} dx = [\ln x]_{-2}^{-1} = \ln(-1) - \ln(-2) = \ln \frac{-1}{-2} = \ln \frac{1}{2} = -\ln 2.$

But $\ln(-1)$ and $\ln(-2)$ do not exist – something very dubious here!

Example 10.12 illustrates a problem. It is clear from figure 10.7 that the areas A and B have equal magnitude and opposite sign. This allows us to drop the restriction $x > 0$ and to avoid dubious mathematics with logarithms of negative numbers by generalising the integral of $\frac{1}{x}$ to

$$\int \frac{1}{x} dx = \ln|x| + c \qquad x \neq 0$$

and

$$\int \frac{k}{x} dx = k\ln|x| + c \qquad x \neq 0.$$

Note

$x \neq 0$ since $\frac{1}{x}$ is undefined if $x = 0$.

EXAMPLE 10.13

Evaluate $\int_{-4}^{-2} \frac{1}{2x} dx$.

Solution

$\int_{-4}^{-2} \frac{1}{2x} dx = \frac{1}{2}\left[\ln|x|\right]_{-4}^{-2} = \frac{1}{2}(\ln 2 - \ln 4) = \frac{1}{2}\ln\left(\frac{2}{4}\right) = \frac{1}{2}\ln\frac{1}{2}.$

EXERCISE 10C

1 Find the following indefinite integrals.

(a) $\int \frac{2}{x} dx$

(b) $\int \frac{1}{2x} dx$

(c) $\int \left(\frac{3}{4x} + 5\right) dx$

(d) $\int \left(\frac{1}{x} + e^x\right) dx$

(e) $\int \left(\frac{1}{2x} + \frac{1}{2\sqrt{x}}\right) dx$

(f) $\int \left(\frac{1}{x^2} - \frac{1}{x}\right) dx$

(g) $\int \left(\frac{4+x}{x}\right) dx$

(h) $\int \left(\frac{5}{x} + \frac{x}{5}\right) dx$

(i) $\int \left(\frac{(x+1)^2}{x}\right) dx$

(j) $\int \left(\frac{x^2-1}{x}\right) dx$

2 Evaluate the following definite integrals.

(a) $\int_1^2 \frac{1}{x} dx$

(b) $\int_1^4 \frac{3}{x} dx$

(c) $\int_1^4 \left(\frac{1}{2x} - \sqrt{x}\right) dx$

(d) $\int_{-6}^{-2} \frac{1}{x} dx$

(e) $\int_4^9 \left(\frac{1}{\sqrt{x}} + \frac{1}{x}\right) dx$

(f) $\int_2^3 \left(\frac{x-2}{x}\right) dx$

(g) $2\int_1^2 \left(\frac{1}{x} + e^x\right) dx$

(h) $\int_1^2 \left(\frac{(1+x)^3}{x}\right) dx$

(i) $\int_2^4 \left(1 + \frac{1}{x}\right)^2 dx$

(j) $\int_{-2}^{-1} \left(\frac{4}{x} - \frac{x^2}{4}\right) dx$

EXERCISE 10C

3 Sketch the graph of $y = \frac{1}{x} + 2$.
   (a) Find the area between the graph, the x-axis and the lines $x = 2$ and $x = 3$.
   (b) Show that the area between the graph, the x-axis and the lines $x = -3$ and $x = -2$ is $2 - \ln\left(\frac{3}{2}\right)$.

4 The diagram shows part of the curve $y = \frac{4x^2 + 1}{x}$.

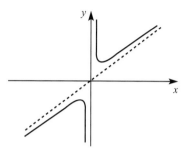

Show that the area between the curve, the x-axis and the lines $x = 1$ and $x = e$ is $2e^2 - 1$.

5 Evaluate $\int_2^3 \left(\frac{1}{2x} - \frac{1}{4}e^x + \sqrt[3]{x}\right) dx$ correct to 4 significant figures.

FURTHER INTEGRATION

INTEGRAL OF $(ax + b)^n$

In Chapter 4 you saw that if $y = (ax + b)^n$ then $\frac{dy}{dx} = an(ax + b)^{n-1}$. Reversing the process gives

$$\int (ax + b)^n \, dx = \frac{1}{a} \times \frac{1}{n+1}(ax + b)^{n+1} + c.$$

This allows you to integrate functions of this type. It is probably easier to think 'What do I have to differentiate to obtain $(ax + b)^n$?' rather than learn this integral.

EXAMPLE 10.14

Find $\int (2x + 3)^4 \, dx$.

Solution Using the rule, or better still thinking 'What do I have to differentiate to give $(2x + 3)^4$?' gives the answer

$$\int (2x + 3)^4 \, dx = \tfrac{1}{2} \times \tfrac{1}{5} (2x + 3)^5 + c$$
$$= \tfrac{1}{10} (2x + 3)^5 + c.$$

EXAMPLE 10.15

Evaluate $\int_0^1 \dfrac{1}{\sqrt{4-3x}}\, dx$.

Solution

$$\int_0^1 \dfrac{1}{\sqrt{4-3x}}\, dx = \int_0^1 (4-3x)^{-\frac{1}{2}}\, dx$$
$$= -\tfrac{1}{3} \times 2\left[(4-3x)^{\frac{1}{2}}\right]_0^1$$
$$= -\tfrac{2}{3}(1-2)$$
$$= \tfrac{2}{3}$$

INTEGRAL OF e^{ax+b}

You saw in Chapter 4 that if $y = e^{f(x)}$ then $\dfrac{dy}{dx} = f'(x)e^{f(x)}$. Putting $f(x) = ax + b$, where a and b are constants, gives

$$y = e^{ax+b} \quad \Rightarrow \quad \dfrac{dy}{dx} = ae^{ax+b}.$$

When you reverse the process it follows that

$$\int e^{ax+b}\, dx = \dfrac{1}{a} e^{ax+b} + c.$$

EXAMPLE 10.16

Find $\int e^{2x+3}\, dx$.

Solution

Using the rule

$$\int e^{2x+3}\, dx = \tfrac{1}{2} e^{2x+3} + c.$$

EXAMPLE 10.17

Evaluate $\int_1^2 4e^{5-2x}\, dx$, leaving your answer in terms of e.

Solution

$$\int_1^2 4e^{5-2x}\, dx = \left[-2e^{5-2x}\right]_1^2$$
$$= -2e - (-2e^3)$$
$$= 2e(e^2 - 1).$$

Integral of $\frac{1}{ax+b}$

You saw in Chapter 4 that if $y = \ln(f(x))$ then $\frac{dy}{dx} = \frac{f'(x)}{f(x)}$. Putting $f(x) = ax + b$, where a and b are constants, gives

$$y = \ln(ax + b) \implies \frac{dy}{dx} = \frac{a}{ax + b}.$$

When you reverse the process it follows that

$$\int \frac{1}{ax + b} \, dx = \frac{1}{a} \ln|ax + b| + c.$$

The reason for the modulus sign is explained on page 249.

EXAMPLE 10.18

Find $\int \frac{1}{2x + 3} \, dx$.

Solution Using the rule

$$\int \frac{1}{2x + 3} \, dx = \tfrac{1}{2} \ln|2x + 3| + c.$$

This can be written slightly differently by putting $c = \tfrac{1}{2} \ln k$, in which case

$$\int \frac{1}{2x + 3} \, dx = \tfrac{1}{2} \ln|2x + 3| + \tfrac{1}{2} \ln k$$
$$= \tfrac{1}{2} \ln k|2x + 3|.$$

Using k as part of the logarithm can often make further steps in problems easier.

EXAMPLE 10.19

Find the exact value of $\int_1^2 \frac{1}{7 - 2x} \, dx$.

Solution
$$\int_1^2 \frac{1}{7 - 2x} \, dx = -\tfrac{1}{2}[\ln|7 - 2x|]_1^2$$
$$= -\tfrac{1}{2}(\ln 3 - \ln 5)$$
$$= -\tfrac{1}{2} \ln \tfrac{3}{5}$$
$$= \tfrac{1}{2} \ln \tfrac{5}{3}$$

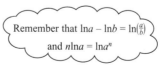
Remember that $\ln a - \ln b = \ln(\tfrac{a}{b})$ and $n \ln a = \ln a^n$

Exercise 10D

1 Integrate the following with respect to x.
  (a) e^{5x}
  (b) e^{3-4x}
  (c) $2e^{3x}$
  (d) $4e^{\frac{x}{4}}$
  (e) $\dfrac{3}{e^{x-2}}$
  (f) $5e^{3x-1}$

2 Find the exact value of the following integrals.
  (a) $\displaystyle\int_0^1 e^x\, dx$
  (b) $\displaystyle\int_1^2 2e^{2x}\, dx$
  (c) $\displaystyle\int_2^5 e^{\frac{1}{3}(x+1)}\, dx$
  (d) $\displaystyle\int_0^1 \dfrac{2}{e^{4x}}\, dx$
  (e) $\displaystyle\int_{-1}^1 4e^{3-2x}\, dx$
  (f) $\displaystyle\int_{-2}^{-1} \dfrac{3}{e^{1-4x}}\, dx$

3 Integrate the following with respect to x.
  (a) $\dfrac{1}{x+3}$
  (b) $\dfrac{1}{3x+2}$
  (c) $\dfrac{4}{3-5x}$
  (d) $1 - \dfrac{1}{1+x}$
  (e) $\dfrac{1}{5(2+3x)}$
  (f) $(5+6x)^{-1}$

4 Find the exact value of the following integrals.
  (a) $\displaystyle\int_1^2 \dfrac{1}{x}\, dx$
  (b) $\displaystyle\int_{-1}^3 \dfrac{1}{x+2}\, dx$
  (c) $\displaystyle\int_{-1}^1 \dfrac{2}{3-x}\, dx$
  (d) $\displaystyle\int_2^7 \dfrac{1}{2(3x+4)}\, dx$
  (e) $\displaystyle\int_1^5 \dfrac{6}{2x-1}\, dx$
  (f) $\displaystyle\int_{-2}^{-1} (5-3x)^{-1}\, dx$

5 The curve C has equation $y = \dfrac{2}{x-3}$.

  (a) Sketch the graph of C.
  (b) Find the area of the region enclosed between C, the x-axis and the lines $x = 4$ and $x = 6$.
  (c) Explain why you cannot calculate $\displaystyle\int_2^4 \dfrac{2}{x-3}\, dx$.

6 Show that area of the region enclosed between the curve $y = e^{2x-1}$, the x-axis, the y-axis and the line $x = 1$ is $\dfrac{e^2 - 1}{2e}$.

7 Given that $f(x) = \dfrac{3x}{(x-1)(2x+1)}$:

  (a) express $f(x)$ in partial fractions;
  (b) prove that $\displaystyle\int_2^3 f(x)\, dx = \tfrac{1}{2}\ln\tfrac{28}{5}$.

8 It is believed that the population P of a colony of ants grows at a rate proportional to $e^{0.02t}$, where t is the number of days. Initially there are 1000 ants and numbers are increasing at a rate of 100 per day.

  (a) Explain why $\dfrac{dP}{dt} = 100e^{0.02t}$.
  (b) Show that the number of ants after one week is approximately 1750.

9 Using partial fractions prove that
$$\int \dfrac{1}{1-x^2}\, dx = \tfrac{1}{2}\ln\dfrac{k(1+x)}{(1-x)}.$$

EXERCISE 10D

10 The curve C has equation $y = \ln(x - 2)$.
  (a) Sketch the graph of C, indicating where it crosses the x-axis and clearly drawing any asymptotes.
  (b) On your sketch shade the region R bounded by the curve C, the x- and y-axes and the line $y = 2$.

INTEGRAL OF $\dfrac{f'(x)}{f(x)}$

You have just been reminded that if $y = \ln(f(x))$ then $\dfrac{dy}{dx} = \dfrac{f'(x)}{f(x)}$, from which it follows that

$$\int \frac{f'(x)}{f(x)} \, dx = \ln(f(x)) + c.$$

This is valid provided $f(x) > 0$. However,

$$\int \frac{f'(x)}{f(x)} \, dx = \int \frac{-f'(x)}{-f(x)} \, dx = \ln(-f(x)) + c$$

which is valid for $f(x) < 0$.

These two conditions can be combined into a single statement by using the modulus of $f(x)$, that is

$$\int \frac{f'(x)}{f(x)} \, dx = \ln|f(x)| + c.$$

Putting $f(x) = ax + b$ is a special case of this integral. More generally, if the numerator is the derivative of the denominator (or a multiple of the derivative) then the integral can be written down, as illustrated in the next examples.

EXAMPLE 10.20

Find $\displaystyle\int \frac{x}{x^2 + 1} \, dx$.

Solution If $f(x) = x^2 + 1$ then $f'(x) = 2x$, so in this integral the numerator is half of the derivative of the denominator, hence

$$\int \frac{x}{x^2 + 1} \, dx = \tfrac{1}{2}\ln|x^2 + 1| + c \qquad \text{or} \qquad \int \frac{x}{x^2 + 1} \, dx = \tfrac{1}{2}\ln k|x^2 + 1|.$$

EXAMPLE 10.21

Evaluate $\int_1^2 \dfrac{6x^2 + 6}{x^3 + 3x}\, dx$.

Solution If $f(x) = x^3 + 3x$ then $f'(x) = 3x^2 + 3$, so in this integral the numerator is twice the derivative of the denominator, hence

$$\int_1^2 \dfrac{6x^2 + 6}{x^3 + 3x}\, dx = 2\left[\ln|x^3 + 3x|\right]_1^2$$
$$= 2(\ln 14 - \ln 4) = 2\ln\tfrac{7}{2}.$$

EXERCISE 10E

1 Find the following integrals.

(a) $\displaystyle\int \dfrac{x}{x^2 - 1}\, dx$
(b) $\displaystyle\int \dfrac{4x}{3 - x^2}\, dx$
(c) $\displaystyle\int \dfrac{x + 2}{x^2 + 4x + 1}\, dx$

(d) $\displaystyle\int \dfrac{\cos x}{1 + \sin x}\, dx$
(e) $\displaystyle\int \dfrac{e^{3x}}{1 + e^{3x}}\, dx$
(f) $\displaystyle\int \dfrac{1}{x \ln x}\, dx$

2 Find the exact value of the following integrals.

(a) $\displaystyle\int_0^1 \dfrac{x^2}{x^3 + 1}\, dx$
(b) $\displaystyle\int_0^1 \dfrac{2x + 2}{(x + 1)^2}\, dx$
(c) $\displaystyle\int_3^4 \dfrac{4x - 6}{x^2 - 3x + 1}\, dx$

(d) $\displaystyle\int_0^{\frac{\pi}{9}} \dfrac{\sin 3x}{\cos 3x}\, dx$
(e) $\displaystyle\int_1^2 \dfrac{e^x}{e^x - 1}\, dx$
(f) $\displaystyle\int_0^2 \dfrac{3}{4x + 1}\, dx$

3 Show that $\dfrac{4x}{x^2 + 1} + \dfrac{1}{2x + 1} \equiv \dfrac{9x^2 + 4x + 1}{(x^2 + 1)(2x + 1)}$.

Hence prove that $\displaystyle\int_0^2 \dfrac{9x^2 + 4x + 1}{(x^2 + 1)(2x + 1)}\, dx = \tfrac{5}{2}\ln 5$.

4 The diagram shows part of the graph of $y = \dfrac{x}{x^2 + 1}$.

Find the area bounded by the curve, the x-axis and the lines $x = 2$ and $x = 7$.

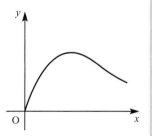

5 A first-order differential equation is

$$\dfrac{dy}{dx} = \dfrac{x^3}{x^4 + 1}.$$

(a) Find the general solution of the differential equation.
(b) Find the particular solution given that $y = \ln 2$ when $x = 1$.
(c) Show that your answer to part (b) may be written in the form
$$x^4 = \tfrac{1}{8}(e^{4y} - 8).$$

INTEGRATION OF TRIGONOMETRIC FUNCTIONS

In Chapter 4 you learnt the derivatives of the six trigonometric functions. Reversing the process gives the integrals.

Since $\frac{d}{dx}(\sin x) = \cos x$ it follows that

$$\int \cos x \, dx = \sin x + c.$$

Similarly, since $\frac{d}{dx}(\cos x) = -\sin x$ it follows that

$$\int \sin x \, dx = -\cos x + c.$$

The derivative of $\tan x$ is $\sec^2 x$ so

$$\int \sec^2 x \, dx = \tan x + c.$$

Exercise 10E question **2(d)** gives you a hint for finding the integral of $\tan x$:

$$\int \tan x \, dx = \int \frac{\sin x}{\cos x} \, dx$$
$$= -\ln|\cos x| + c$$
$$= \ln|\cos x|^{-1} + c$$

so

$$\int \tan x \, dx = \ln|\sec x| + c.$$

By a similar argument

$$\int \cot x \, dx = \ln|\sin x| + c.$$

Since, for example, you also know that $\frac{d}{dx}(\sin ax) = a\cos ax$, you can find the integrals of functions such as $\cos 3x$ and $\tan 5x$. You will see how to integrate some of the more complicated functions involving trigonometric functions later in the chapter.

EXAMPLE 10.22

Find $\int \sin 8x \, dx$.

Solution Thinking of the reverse process of differentiating, then

$$\int \sin 8x \, dx = -\tfrac{1}{8}\cos 8x + c.$$

EXAMPLE 10.23

Find the area of the region enclosed between $y = \cos x$ and the x-axis from $x = 0$ to $x = \frac{\pi}{2}$.

Solution Sketching the function (figure 10.8) gives

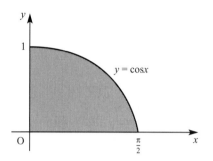

$$\text{area} = \int_0^{\frac{\pi}{2}} \cos x \, dx = [\sin x]_0^{\frac{\pi}{2}}$$
$$= 1 - 0$$
$$= 1.$$

FIGURE 10.8

Note

When integrating (and differentiating) trigonometric functions you must work in radians.

EXERCISE 10F

1 Find the following integrals.

(a) $\int \sin 2x \, dx$ (b) $\int 2\cos 3x \, dx$

(c) $\int 5\sec x \tan x \, dx$ (d) $4\int \cosec^2 x \, dx$

(e) $\int \tan\tfrac{1}{2}x \, dx$ (f) $\int 6\sec^2 3x \, dx$

2 Evaluate the following integrals.

(a) $\int_{\frac{\pi}{6}}^{\frac{\pi}{3}} \cos x \, dx$ (b) $\int_0^{\frac{\pi}{2}} \sin\tfrac{1}{2}x \, dx$

(c) $\int_{\frac{\pi}{6}}^{\frac{\pi}{2}} \cosec x \cot x \, dx$ (d) $\int_0^{\frac{\pi}{12}} 3\cos 6x \, dx$

(e) $\int_{\frac{\pi}{4}}^{\frac{\pi}{2}} \cot x \, dx$ (f) $\int_{\frac{2\pi}{3}}^{\pi} \tfrac{1}{4}\sec^2 \tfrac{1}{4}x \, dx$

3 Find the area of the region enclosed between the curve $y = \sin x$ and the x-axis from $x = 0$ to $x = \pi$.

4 The function f(x) is defined by f(x) = $3\sin x + 4\cos x$.

(a) Find $\int f(x) \, dx$.

(b) Evaluate $\int_{\frac{\pi}{4}}^{\frac{\pi}{2}} f(x) \, dx$.

(c) Express f(x) in the form $r\sin(x + \alpha)$ where $r > 0$ and $0 \leq \alpha \leq \frac{\pi}{2}$.

(d) Find $\int_{\frac{\pi}{4}}^{\frac{\pi}{2}} r\sin(x + \alpha) \, dx$, using $r\sin(x + \alpha)$ from part **(c)**, and show that your answer is the same as that in part **(b)**.

5 (a) Prove that $\int \tan x \, dx = \ln|\sec x| + c$.

(b) Show that
$$\frac{1 + \sin 2\theta}{1 - \sin^2 \theta} \equiv \sec^2 \theta + 2\tan\theta.$$

(c) Prove that
$$\int_0^{\frac{\pi}{4}} \frac{1 + \sin 2x}{1 - \sin^2 x} \, dx = 1 + \ln 2.$$

INTEGRATION BY SUBSTITUTION

The graph of $y = \sqrt{x - 1}$ is shown in figure 10.9.

The shaded area is given by

$$\int_1^5 \sqrt{x - 1} \, dx = \int_1^5 (x - 1)^{\frac{1}{2}} \, dx$$
$$= \left[\tfrac{2}{3}(x - 1)^{\frac{3}{2}}\right]_1^5 = \tfrac{2}{3}(8 - 0) = 5\tfrac{1}{3}.$$

An alternative method to find $\int_1^5 \sqrt{x - 1} \, dx$ is to make the substitution $u = x - 1$.

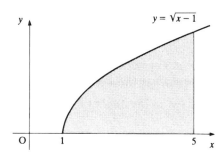

FIGURE 10.9

When you make this substitution it means that you are now integrating with respect to a new variable, namely u. The limits of the integral, and the 'dx', must be written in terms of u.

The new limits are given by $\quad x = 1 \quad \Rightarrow \quad u = 1 - 1 = 0$
and $\quad x = 5 \quad \Rightarrow \quad u = 5 - 1 = 4$.

Since $u = x - 1$, $\frac{du}{dx} = 1 \quad \Rightarrow \quad du = dx$.

Even though $\frac{du}{dx}$ is not a fraction, it is usual to treat it as one in this situation (see the note on page 254).

The integral now becomes

$$\int_{u=0}^{u=4} u^{\frac{1}{2}} \, du = \left[\frac{u^{\frac{3}{2}}}{\frac{3}{2}} \right]_{0}^{4}$$

$$= \left[\frac{2u^{\frac{3}{2}}}{3} \right]_{0}^{4}$$

$$= 5\tfrac{1}{3}.$$

This method by integration is known as *integration by substitution*. It is a very powerful method which allows you to integrate many more functions. Since you are changing the variable from x to u, the method is also referred to as *integration by change of variable*.

Note

The last example included the statement '$du = dx$'. Some mathematicians are reluctant to write such statements on the grounds that du and dx may only be used in the form $\frac{du}{dx}$, i.e. as a gradient. This is not in fact true; there is a well defined branch of mathematics which justifies such statements. It may help you to think of it as shorthand for 'in the limit as $\delta x \to 0$, $\frac{\delta u}{\delta x} \to 1$, and so $\delta u = \delta x$'.

EXAMPLE 10.24

Evaluate $\int_{1}^{3} (x + 1)^3 \, dx$ by making a suitable substitution.

Solution Let $u = x + 1$.

Converting the limits: $x = 1 \implies u = 1 + 1 = 2$
$x = 3 \implies u = 3 + 1 = 4.$

Converting dx to du:

$$\frac{du}{dx} = 1 \implies du = dx.$$

$$\int_{1}^{2} (x + 1)^3 \, dx = \int_{2}^{4} u^3 \, du$$

$$= \left[\frac{u^4}{4} \right]_{2}^{4}$$

$$= \frac{4^4}{4} - \frac{2^4}{4}.$$

$$= 60.$$

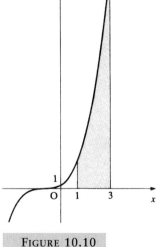

FIGURE 10.10

Integration by substitution may be described as the reverse of the chain rule.

INTEGRATION BY SUBSTITUTION

EXAMPLE 10.25

Evaluate $\int_3^4 2x(x^2 - 4)^{\frac{1}{2}} \, dx$ by making a suitable substitution.

Solution Notice that $2x$ is the derivative of the function in the brackets, $x^2 - 4$, and so $u = x^2 - 4$ is a natural substitution to try.

This gives $\dfrac{du}{dx} = 2x \quad \Rightarrow \quad du = 2x \, dx.$

Converting the limits: $\quad x = 3 \quad \Rightarrow \quad u = 9 - 4 = 5$
$\quad\quad\quad\quad\quad\quad\quad\quad\quad\; x = 4 \quad \Rightarrow \quad u = 16 - 4 = 12.$

So the integral becomes

$$\int_3^4 (x^2 - 4)^{\frac{1}{2}} 2x \, dx = \int_5^{12} u^{\frac{1}{2}} \, du$$

$$= \left[\tfrac{2}{3} u^{\frac{3}{2}}\right]_5^{12}$$

$$= 20.3 \text{ (to 3 significant figures)}.$$

Note

In the last example there were two functions of x multiplied together, the second function being an expression in brackets raised to a power. The two functions are in this case related, since the first function, $2x$, is the derivative of the expression in brackets, $x^2 - 4$. It was this relationship that made the integration possible.

EXAMPLE 10.26

Find $\int x(x^2 + 2)^3 \, dx$ by making an appropriate substitution.

Solution Since this is an indefinite integral there are no limits to change, and the final answer will be a function of x.

Let $u = x^2 + 2$, then:

$\dfrac{du}{dx} = 2x \quad \Rightarrow \quad \tfrac{1}{2} du = x \, dx.$

> You only have $x\,dx$ in the integral, not $2x\,dx$

So $\quad \int x(x^2 + 2)^3 \, dx = \int (x^2 + 2)^3 x \, dx$

$$= \int u^3 \times \tfrac{1}{2} \, du$$

$$= \tfrac{1}{8} u^4 + c$$

$$= \tfrac{1}{8}(x^2 + 2)^4 + c.$$

Note: Always remember, when finding an indefinite integral by substitution, to substitute back at the end. The original integral was in terms of x, so your final answer must be too.

EXAMPLE 10.27

By making a suitable substitution, find $\int x\sqrt{x-2}\, dx$.

Solution

This question is not of the same type as the previous ones since x is not the derivative of $(x-2)$. However, by making the substitution $u = x - 2$ you can still make the integral into one you can do.

Let $u = x - 2$, then:

$$\frac{du}{dx} = 1 \quad \Rightarrow \quad du = dx.$$

There is also an x in the integral so you need to write down an expression for x in terms of u. Since $u = x - 2$ it follows that $x = u + 2$.

In the original integral you can now replace $\sqrt{x-2}$ by $u^{\frac{1}{2}}$, dx by du, and x by $u + 2$:

$$\int x\sqrt{x-2}\, dx = \int (u+2)u^{\frac{1}{2}}\, du$$

$$= \int (u^{\frac{3}{2}} + 2u^{\frac{1}{2}})\, du$$

$$= \tfrac{2}{5}u^{\frac{5}{2}} + \tfrac{4}{3}u^{\frac{3}{2}} + c.$$

Replacing u by $x - 2$ and tidying up gives $\tfrac{2}{15}(3x+4)(x-2)^{\frac{3}{2}} + c$.

EXAMPLE 10.28

By making a suitable substitution, find $\int_0^4 2xe^{x^2}\, dx$.

Solution

$\int_0^4 2xe^{x^2}\, dx = \int_0^4 e^{x^2} 2x\, dx$.

Since $2x$ is the derivative of x^2, let $u = x^2$.

$$\frac{du}{dx} = 2x \quad \Rightarrow \quad du = 2x\, dx.$$

The new limits are given by $\quad x = 0 \quad \Rightarrow \quad u = 0$
and $\quad x = 4 \quad \Rightarrow \quad u = 16$.

The integral can now be written as

$$\int_0^{16} e^u\, du = [e^u]_0^{16}$$

$$= e^{16} - e^0$$

$$= 8.89 \times 10^6 \quad \text{(to 3 significant figures)}.$$

INTEGRATION BY SUBSTITUTION

Pure Mathematics: Core 4

EXAMPLE 10.29

Evaluate $\int_1^5 \frac{2x}{x^2 + 3} \, dx$.

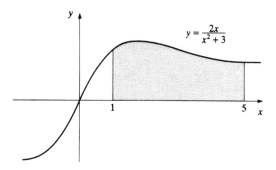

FIGURE 10.11

Solution   In this case, substitute $u = x^2 + 3$, so that

$$\frac{du}{dx} = 2x \quad \Rightarrow \quad du = 2x \, dx.$$

The new limits are given by $\quad x = 1 \quad \Rightarrow \quad u = 4$
and $\quad x = 5 \quad \Rightarrow \quad u = 28$.

$$\int_1^5 \frac{2x}{x^2 + 3} \, dx = \int_4^{28} \frac{1}{u} \, du$$

$$= [\ln |u|]_4^{28}$$
$$= \ln 28 - \ln 4$$
$$= 1.95 \qquad \text{(to 3 significant figures)}$$

The last example is of the form $\int \frac{f'(x)}{f(x)} \, dx$, where $f(x) = x^2 + 3$. In such cases the substitution $u = f(x)$ transforms the integral into $\int \frac{1}{u} \, du$. The answer is then $\ln |u| + c$ or $\ln |f(x)| + c$. This result was stated as the working rule

$$\int \frac{f'(x)}{f(x)} \, dx = \ln |f(x)| + c$$

earlier in this chapter.

EXAMPLE 10.30

Find $\int 2x \cos(x^2 + 1) \, dx$

Solution Make the substitution $u = x^2 + 1$. Then differentiate:

$$\frac{du}{dx} = 2x \implies 2x \, dx = du$$

$$\int 2x \cos(x^2 + 1) \, dx = \int \cos u \, du$$

$$= \sin u + c$$

$$= \sin(x^2 + 1) + c$$

EXAMPLE 10.31

Find $\int_0^{\frac{\pi}{2}} \cos x \sin^2 x \, dx$.

(Remember that $\sin^2 x$ means the same as $(\sin x)^2$.)

Solution Now $(\sin x)^2$ is a function of $\sin x$, and $\cos x$ is the derivative of $\sin x$, so you should use the substitution $u = \sin x$.

Differentiating:

$$\frac{du}{dx} = \cos x \implies du = \cos x \, dx$$

The limits of integration need to be changed as well:

$$x = \frac{\pi}{2} \implies u = 1$$
$$x = 0 \implies u = 0.$$

Therefore $\int_0^{\frac{\pi}{2}} \cos x \sin^2 x \, dx = \int_0^1 u^2 \, du$

$$= \left[\frac{u^3}{3}\right]_0^1$$

$$= \frac{1}{3}.$$

Note

You may find that as you gain practice in this type of integration you become able to work out the integral without writing down the substitution. In this case you will appreciate the general rule

$$\int f'(g(x)) \, g'(x) \, dx = f(g(x)) + c.$$

However, if you are unsure or if the question asks for an integration by substitution, it is best to write down the whole process.

Exercise 10G

1. Find the following indefinite integrals by making the suggested substitution. Remember to give your final answer in terms of x.

   (a) $\int (x+1)^3 \, dx$, $u = x+1$

   (b) $\int 2\sqrt{2x-1} \, dx$, $u = 2x-1$

   (c) $\int 3x^2(x^3+1)^7 \, dx$, $u = x^3+1$

   (d) $\int 2x(x^2+1)^5 \, dx$, $u = x^2+1$

   (e) $\int 3x^2(x^3-2)^4 \, dx$, $u = x^3-2$

   (f) $\int x\sqrt{2x^2-5} \, dx$, $u = 2x^2-5$

   (g) $\int x\sqrt{2x+1} \, dx$, $u = 2x+1$

   (h) $\int \dfrac{x}{\sqrt{x+9}} \, dx$, $u = x+9$

   (i) $\int 12x e^{3x^2} \, dx$, $u = 3x^2$

   (j) $\int x^2 \sin(x^3) \, dx$, $u = x^3$

2. Evaluate each of the following definite integrals by using a suitable substitution. Give your answer to 3 significant figures where appropriate.

   (a) $\int_{-1}^{4} (x-3)^4 \, dx$

   (b) $\int_{0}^{3} (3x+2)^6 \, dx$

   (c) $\int_{5}^{9} \sqrt{x-5} \, dx$

   (d) $\int_{2}^{15} \sqrt[3]{2x-3} \, dx$

   (e) $\int_{1}^{5} x^2(x^3+1)^2 \, dx$

   (f) $\int_{-1}^{2} 2x(x-3)^5 \, dx$

   (g) $\int_{1}^{5} x\sqrt{x-1} \, dx$

   (h) $\int_{2}^{6} 2xe^{-x^2} \, dx$

   (i) $\int_{2}^{4} \dfrac{x+2}{x^2+4x-3} \, dx$

   (j) $\int_{0.2}^{0.5} x \sec^2(x^2+1) \, dx$

3. The graph of $y = xe^{x^2}$ is shown here.
   (a) Using the substitution $u = x^2$, find the area of region A.
   (b) Find the area of region B.
   (c) Hence write down the total area of the shaded region.

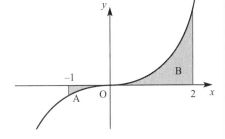

4. The graph of $y = xe^{-x^2}$ is shown below.

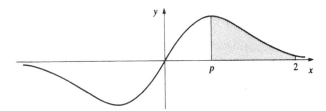

   (a) Find $\dfrac{dy}{dx}$ using the product rule.
   (b) Find the x-coordinate, p, of the maximum point. (You do not need to prove that it corresponds to a maximum.)
   (c) Use your answer from (b) to find the area of the shaded region.

5 The graph of $\dfrac{x+2}{x^2+4x+3}$ is shown below.

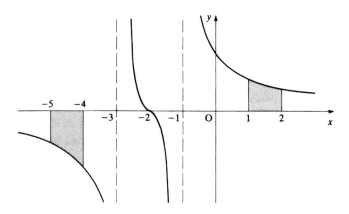

Find, to 3 significant figures, the area of each shaded region using the substitution $u = x^2 + 4x + 3$.

6 The sketch shows part of the graph of $y = x\sqrt{1+x}$.

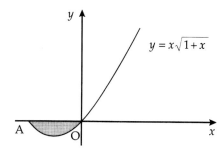

(a) Find the coordinates of point A and the range of values of x for which the function is defined.

(b) Show that the shaded area is $\tfrac{4}{15}$. You may find the substitution $u = 1 + x$ useful.

[MEI]

7 (a) By substituting $u = 1 + x$ or otherwise, find:

(i) $\displaystyle\int (1+x)^3\,dx$;

(ii) $\displaystyle\int_{-1}^{1} x(1+x)^3\,dx$.

(b) By substituting $t = 1 + x^2$ or otherwise, evaluate $\displaystyle\int_0^1 x\sqrt{1+x^2}\,dx$.

[MEI]

8 The curve with equation $y = \dfrac{e^x}{e^x + 1}$ is shown for values of x between 0 and 2.

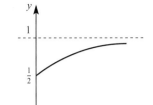

(a) Find the area of the region enclosed by this curve, the axes and the line $x = 2$.

(b) Find the value of $\int_0^e \dfrac{2t}{t^2 + 1}\,dt$.

(c) Compare your answers to (a) and (b). Explain this result.

9 The diagram shows a sketch of the curve $y = \dfrac{x}{x^2 + 1}$.

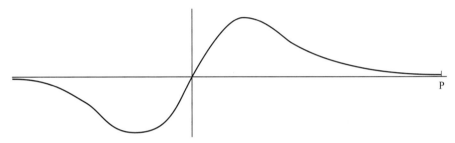

(a) Differentiate $y = \dfrac{x}{x^2 + 1}$.

(b) Hence find the coordinates of the two turning points of the curve.

(c) By substituting $t = x^2 + 1$, or otherwise, find $\int \dfrac{x}{x^2 + 1}\,dx$.

(d) Hence find the x-coordinate of the point P, given that the area of the region between the curve and the x-axis, from the origin to P is 3 units^2.

[MEI, adapted]

10 (a) Show that $\int_5^{10} \dfrac{1}{u}\,du = \ln 2$.

The function f(x) is defined by

$$f(x) = \dfrac{x}{x^2 + 1}.$$

The graph of $y = f(x)$ for positive values of x is shown below.

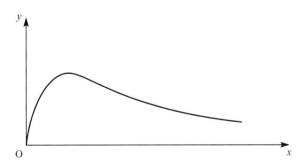

(b) Calculate $\int_2^3 f(x)\,dx$. (You may wish to use the substitution $u = x^2 + 1$.)

(c) Show that $f(x)$ is an odd function.

Write down the value of $\int_{-3}^{-2} f(x)\,dx$.

(d) State a transformation which will transform the graph of $y = f(x)$ into the graph of $y = f(x + 1)$.

(e) Using **(c)** and **(d)**, or otherwise, calculate the value of

$$\int_{-4}^{-3} \frac{x + 1}{x^2 + 2x + 2}\,dx.$$

[MEI]

11 Use the substitution $u = \sin x$ to show that

$$\int_0^{\frac{\pi}{2}} \sin^6 x \cos x\,dx = \tfrac{1}{7}.$$

Hence find the answer to

$$\int_0^{\frac{\pi}{4}} \sin^6 2x \cos 2x\,dx.$$

12 Use the substitution $u = 2x + 1$ to express $\int x(2x + 1)^4\,dx$ as an integral in terms of u.

Hence find $\int x(2x + 1)^4\,dx$, giving your answer in terms of x.

13 Use the substitution $t = \ln x$ to find

$$\int_1^{e^{\frac{\pi}{2}}} \frac{\sin(\ln x)}{x}\,dx.$$

14 The diagram shows part of the curve $y = x\sec(x^2)$.

(a) Find the area of the region enclosed between the curve, the x-axis and the line $x = 0.2$.

(b) The curve passes through the point $(0.2, h)$. Find the value of h.

(c) Without further integration find the area of the region between the curve, the y-axis and the line $y = h$.

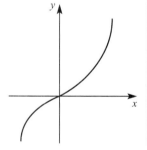

INTEGRATION BY PARTS

Suppose you wish to find $\int x\cos x\,dx$.

The function to be integrated is clearly a product of two simpler functions, x and $\cos x$, so your first thought may be to look for a substitution to enable you to perform the integration. However, there are some functions which are products but which cannot be integrated by substitution. This is one of them. You need a new technique to integrate such functions.

Take the function $x\sin x$ and differentiate it, using the product rule:

$$\frac{d}{dx}(x\sin x) = x\cos x + \sin x.$$

Now integrate both sides. This has the effect of 'undoing' the differentiation, so

$$x\sin x = \int x\cos x \, dx + \int \sin x \, dx.$$

Rearranging this gives

$$\int x\cos x \, dx = x\sin x - \int \sin x \, dx$$
$$= x\sin x - (-\cos x) + c$$
$$= x\sin x + \cos x + c.$$

This has enabled you to find the integral of $x\cos x$.

The work in this example can be generalised into the method of integration by parts.

If in the above we use u to stand for the function x, and v to stand for the function $\sin x$, then using the product rule to differentiate the function uv:

$$\frac{d}{dx}(uv) = v\frac{du}{dx} + u\frac{dv}{dx}.$$

Integrating gives

$$uv = \int v\frac{du}{dx} \, dx + \int u\frac{dv}{dx} \, dx.$$

Rearranging gives the general result for integration by parts:

$$\int u\frac{dv}{dx} \, dx = uv - \int v\frac{du}{dx} \, dx.$$

In order to use this, you have to split the function you want to integrate into two simpler functions. One of these functions will be called u, and the other $\frac{dv}{dx}$, to fit the left-hand side of the expression. You will need to decide which will be which. Two considerations will help you:

1 As you want to use $\frac{du}{dx}$ on the right-hand side of the expression, u should be a function which becomes a simpler function after differentiation. So in this case, u will be the function x.

2 As you need v to work out the right-hand side of the expression, it must be possible to integrate the function $\frac{dv}{dx}$ to obtain v.

EXAMPLE 10.32

Find $\int x\cos x \, dx$.

Solution Put $\quad u = x \quad \Rightarrow \quad \dfrac{du}{dx} = 1$

and $\quad \dfrac{dv}{dx} = \cos x \quad \Rightarrow \quad v = \sin x$.

Substituting in

$$\int u \frac{dv}{dx} \, dx = uv - \int v \frac{du}{dx} \, dx$$

gives

$$\int x\cos x \, dx = x\sin x - \int 1 \times \sin x \, dx$$
$$= x\sin x - (-\cos x) + c$$
$$= x\sin x + \cos x + c.$$

This is the result we obtained earlier.

EXAMPLE 10.33

Find $\int 2xe^x \, dx$.

Solution First split $2xe^x$ into the two simpler functions, $2x$ and e^x. Both can be integrated easily, but as $2x$ becomes a simpler function after differentiation and e^x does not, take u to be $2x$.

$$u = 2x \quad \Rightarrow \quad \frac{du}{dx} = 2$$
$$\frac{dv}{dx} = e^x \quad \Rightarrow \quad v = e^x$$

Substituting in

$$\int u \frac{dv}{dx} \, dx = uv - \int v \frac{du}{dx} \, dx$$

gives

$$\int 2xe^x \, dx = 2xe^x - \int 2e^x \, dx$$
$$= 2xe^x - 2e^x + c.$$

Sometimes it is necessary to use integration by parts twice or more to complete the integration successfully.

INTEGRATION BY PARTS

EXAMPLE 10.34

Find $\int x^2 \sin x \, dx$.

Solution First split $x^2 \sin x$ into the two functions x^2 and $\sin x$. As x^2 becomes a simpler function after differentiation, take u to be x^2.

$$u = x^2 \quad \Rightarrow \quad \frac{du}{dx} = 2x$$

$$\frac{dv}{dx} = \sin x \quad \Rightarrow \quad v = -\cos x$$

Substituting in

$$\int u \frac{dv}{dx} \, dx = uv - \int v \frac{du}{dx} \, dx$$

gives

$$\int x^2 \sin x \, dx = -x^2 \cos x - \int -2x \cos x \, dx$$
$$= -x^2 \cos x + \int 2x \cos x \, dx. \qquad ①$$

Now the integral of $2x \cos x$ cannot be found without using integration by parts again. It has to be split into the functions $2x$ and $\cos x$, and as $2x$ becomes a simpler function after differentiation, take u to be $2x$.

$$u = 2x \quad \Rightarrow \quad \frac{du}{dx} = 2$$

$$\frac{dv}{dx} = \cos x \quad \Rightarrow \quad v = \sin x$$

Substituting in

$$\int u \frac{dv}{dx} \, dx = uv - \int v \frac{du}{dx} \, dx$$

gives

$$\int 2x \cos x \, dx = 2x \sin x - \int 2 \sin x \, dx$$
$$= 2x \sin x - (-2 \cos x) + c$$
$$= 2x \sin x + 2 \cos x + c.$$

Substituting in ①

$$\int x^2 \sin x \, dx = -x^2 \cos x + 2x \sin x + 2 \cos x + c.$$

In some cases, the choice of u and v may be less obvious.

EXAMPLE 10.35

Find $\int x\ln x \, dx$.

Solution It might seem at first that u should be taken as x, because it becomes a simpler function after differentiation.

$$u = x \implies \frac{du}{dx} = 1$$
$$\frac{dv}{dx} = \ln x \implies v = ?$$

Now you need to integrate $\ln x$ to obtain v. Although it is possible to integrate $\ln x$, it has to be done by parts, as you will see in Example 10.36. The wrong choice has been made for u and v, resulting in a more complicated integral.

So instead, let $u = \ln x$.

$$u = \ln x \implies \frac{du}{dx} = \frac{1}{x}$$
$$\frac{dv}{dx} = x \implies v = \tfrac{1}{2}x^2$$

Substituting in

$$\int u \frac{dv}{dx} \, dx = uv - \int v \frac{du}{dx} \, dx$$

gives

$$\int x\ln x \, dx = \tfrac{1}{2}x^2 \ln x - \int \tfrac{1}{2}x^2 \times \tfrac{1}{x} \, dx$$
$$= \tfrac{1}{2}x^2 \ln x - \int \tfrac{1}{2}x \, dx$$
$$= \tfrac{1}{2}x^2 \ln x - \tfrac{1}{4}x^2 + c.$$

EXAMPLE 10.36

Find $\int \ln x \, dx$

Solution At first sight the way to do the integral is not obvious. The 'trick' is to multiply by 1 and integrate by parts,

i.e
$$\int \ln x \, dx = \int \ln x \times 1 \, dx.$$

let
$$u = \ln x \implies \frac{du}{dx} = \frac{1}{x}$$
$$\frac{dv}{dx} = 1 \implies v = x$$

Substituting in
$$\int u \frac{dv}{dx} dx = uv - \int v \frac{du}{dx} dx$$

gives
$$\int \ln x \, dx = \ln x \times x - \int x \times \frac{1}{x} dx$$

$$\therefore \int \ln x \, dx = x \ln x - x + c.$$

The technique of integration by parts is usually used when the two functions are of different types: polynomials, trigonometric functions, exponentials, logarithms. There are, however, some exceptions, as in questions 11 and 12 of Exercise 10H.

Integration by parts is a very important technique which is needed in many other branches of mathematics. For example, integrals of the form $\int x f(x) \, dx$ are used in statistics to find the mean for a given probability density function, and in mechanics to find the centre of mass for a given shape. Integrals of the form $\int x^2 f(x) \, dx$, requiring integration by parts to be used twice, are used in statistics in finding variance and in mechanics to find moments of inertia.

Exercise 10H

In questions 1 to 6:

(a) write down the function to be taken as u and function to be taken as $\frac{dv}{dx}$;
(b) use the formula for integration by parts to complete the integration.

1. $\int x e^x \, dx$
2. $\int x \cos 3x \, dx$
3. $\int (2x + 1) \cos x \, dx$
4. $\int x e^{-2x} \, dx$
5. $\int x e^{-x} \, dx$
6. $\int x \sin 2x \, dx$

In questions 7 to 10, use integration by parts to integrate the functions given.

7. $x^3 \ln x$
8. $x^2 e^x$
9. $(2 - x)^2 \cos x$
10. $x^2 \ln 2x$

11. Find $\int x \sqrt{1 + x} \, dx$
    (a) by using integration by parts;
    (b) by using the substitution $u = 1 + x$.

12. Find $\int 2x(x - 2)^4 \, dx$
    (a) by using integration by parts;
    (b) by using the substitution $u = x - 2$.

13. Find $\int \ln 3x \, dx$.

Definite integration by parts

When you use the method of integration by parts on a definite integral, it is important to remember that the term uv on the right-hand side of the expression has already been integrated and so should be written in square brackets with the limits indicated.

$$\int_a^b u \frac{dv}{dx} dx = \left[uv \right]_a^b - \int_a^b v \frac{du}{dx} dx$$

EXAMPLE 10.37

Evaluate $\int_0^2 xe^x \, dx$.

Solution Put $u = x \implies \frac{du}{dx} = 1$

and $\frac{dv}{dx} = e^x \implies v = e^x$.

Substituting in

$$\int_a^b u \frac{dv}{dx} dx = \left[uv \right]_a^b - \int_a^b v \frac{du}{dx} dx$$

gives

$$\int_0^2 xe^x \, dx = \left[xe^x \right]_0^2 - \int_0^2 e^x \, dx$$

$$= \left[xe^x \right]_0^2 - \left[e^x \right]_0^2$$

$$= (2e^2 - 0) - (e^2 - e^0)$$

$$= 2e^2 - e^2 + 1$$

$$= e^2 + 1.$$

EXAMPLE 10.38

Find the area in the region between the curve $y = x\cos x$ and the x-axis, between $x = 0$ and $x = \frac{\pi}{2}$.

Solution Figure 10.12 shows the region whose area is to be found.

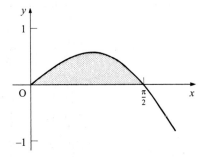

FIGURE 10.12

To find the required area you need to integrate the function $x\cos x$ between the limits 0 and $\frac{\pi}{2}$. You therefore need to work out

$$\int_0^{\frac{\pi}{2}} x\cos x \, dx.$$

Put $u = x \implies \dfrac{du}{dx} = 1$

and $\dfrac{dv}{dx} = \cos x \implies v = \sin x.$

$$\int_a^b u \dfrac{dv}{dx} dx = \left[uv\right]_a^b - \int_a^b v \dfrac{du}{dx} dx$$

$$\int_0^{\frac{\pi}{2}} x\cos x \, dx = \left[x\sin x\right]_0^{\frac{\pi}{2}} - \int_0^{\frac{\pi}{2}} \sin x \, dx$$

$$= \left[x\sin x\right]_0^{\frac{\pi}{2}} - \left[-\cos x\right]_0^{\frac{\pi}{2}}$$

$$= \left[x\sin x + \cos x\right]_0^{\frac{\pi}{2}}$$

$$= \left(\tfrac{\pi}{2} + 0\right) - \left(0 + 1\right)$$

$$= \tfrac{\pi}{2} - 1$$

So the required area is $\tfrac{\pi}{2} - 1$ square units.

EXERCISE 10I

Evaluate the definite integrals in questions 1 to 6.

1 $\displaystyle\int_0^1 xe^{3x} \, dx$

2 $\displaystyle\int_0^{\pi} (x-1)\cos x \, dx$

3 $\displaystyle\int_0^2 (x+1)e^x \, dx$

4 $\displaystyle\int_1^2 \ln 2x \, dx$

5 $\displaystyle\int_0^{\frac{\pi}{2}} x^2 \sin 2x \, dx$

6 $\displaystyle\int_1^4 x^2 \ln x \, dx$

7 (a) Find the coordinates of the points where the graph of $y = (2-x)e^{-x}$ cuts the x- and y-axes.
  (b) Hence sketch the graph of $y = (2-x)e^{-x}$.
  (c) Use integration by parts to find the area of the region between the x-axis, the y-axis and the graph $y = (2-x)e^{-x}$.

8 (a) Sketch the graph of $y = x\sin x$ from $x = 0$ to $x = \pi$ and shade the region between the curve and the x-axis.
  (b) Find the area of this region using integration by parts.

9 Find the area of the region between the x-axis, the line $x = 5$ and the graph $y = \ln x$.

10 Find the area of the region between the x-axis and the graph $y = x^2\cos x$ from $x = -\tfrac{\pi}{2}$ to $x = \tfrac{\pi}{2}$.

11 Find the area of the region between the negative x-axis and the graph $y = x\sqrt{x+1}$
  (a) using integration by parts;
  (b) using the substitution $u = x + 1$.

12 The sketch shows the curve with equation $y = x^2\ln 2x$.

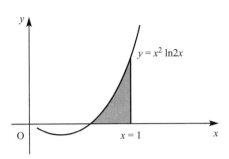

Find the x-coordinate of the point where the curve cuts the x-axis. Hence calculate the area of the shaded region using the method of integration by parts applied to the product of $\ln 2x$ and x^2. Give your answer correct to 3 decimal places.

[MEI]

13 Show that $\int_0^1 x^2 e^x \, dx = e - 2$.

Show that the use of the trapezium rule with five strips (six ordinates) gives an estimate that is about 3.8% too high. Explain why approximate evaluation of this integral using the trapezium rule will always result in an overestimate, however many strips are used.

(You will recall from *Pure Mathematics: Core 2* that the trapezium rule is $A = \tfrac{1}{2}h\{y_0 + y_n + 2(y_1 + y_2 + \ldots + y_{n-1})\}$.)

[MEI]

14 Part of the graph of $y = x\sin x$ from $x = 0$ to $x = \pi$ is shown below.

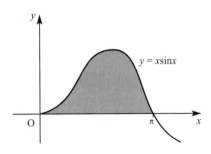

Calculate the area of the shaded region.

15 (a) Use integration by parts to show that
$$\int x\cos x \, dx = x\sin x + \cos x + c.$$

(b) Find $\int x^2 \sin x \, dx$.

(c) Solve $\dfrac{dy}{dx} = x^2 \sin x$ given that $y = 0$ when $x = \pi$.

16 The diagram shows a cross-section R of a dam.

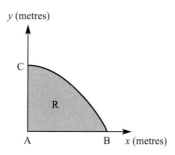

The line AC is the vertical face of the dam, AB is the horizontal base and the curve BC is the profile. Taking x and y to be the horizontal and vertical axes, then A, B and C have coordinates $(0, 0)$, $(3\pi^2, 0)$ and $(0, 30)$ respectively. The area of the cross-section is to be calculated.

Initially the profile BC is approximated by a straight line.

(a) Find an estimate for the area of the cross-section R using this approximation.

The profile BC is actually described by the parametric equations

$$x = 16t^2 - \pi^2, \quad y = 30\sin 2t, \quad \tfrac{\pi}{4} \leqslant t \leqslant \tfrac{\pi}{2}.$$

(b) Find the exact area of the cross-section R.
(c) Calculate the percentage error in the estimate of the area of the cross-section R that you found in part **(a)**.

[Edexcel]

USING PARTIAL FRACTIONS IN INTEGRATION

You will recall from Chapter 1 how to express algebraic fractions in partial fractions.

EXAMPLE 10.39

Find $\int \dfrac{2}{x^2 - 1}\, dx$.

Solution Since $x^2 - 1$ can be factorised to give $(x + 1)(x - 1)$, we can put the function to be integrated into partial fractions:

$$\dfrac{2}{x^2 - 1} \equiv \dfrac{A}{x - 1} + \dfrac{B}{x + 1}$$

> This is true for all values of x. It is an identity and to emphasise this point we use the identity symbol $\equiv$

$$2 \equiv A(x + 1) + B(x - 1).$$

Let $x = 1$ $2 = 2A$ $\Rightarrow$ $A = 1$
Let $x = -1$ $2 = -2B$ $\Rightarrow$ $B = -1$

Substituting these values for A and B gives

$$\frac{2}{x^2 - 1} \equiv \frac{1}{x - 1} - \frac{1}{x + 1}.$$

The integral then becomes

$$\int \frac{2}{x^2 - 1} \, dx = \int \frac{1}{x - 1} \, dx - \int \frac{1}{x + 1} \, dx.$$

Now the two integrals on the right can be recognised as giving logarithms. So

$$\int \frac{2}{x^2 - 1} \, dx = \ln|x - 1| - \ln|x + 1| + c$$

$$= \ln\left|\frac{x - 1}{x + 1}\right| + c.$$

In Example 10.39 you worked with the simplest type of partial fraction, in which there are two different linear factors in the denominator. This type will always result in two functions which can both be integrated to give logarithmic functions. You will now look at the other types of partial fraction.

A REPEATED FACTOR IN THE DENOMINATOR

EXAMPLE 10.40

Find $\int \frac{x + 4}{(2x - 1)(x + 1)^2} \, dx.$

Solution   First put the expression into partial fractions:

$$\frac{x + 4}{(2x - 1)(x + 1)^2} \equiv \frac{A}{(2x - 1)} + \frac{B}{(x + 1)} + \frac{C}{(x + 1)^2}$$

Where $x + 4 \equiv A(x + 1)^2 + B(2x - 1)(x + 1) + C(2x - 1).$

Let $x = -1$ $3 = -3C$ $\Rightarrow$ $C = -1$
Let $x = \frac{1}{2}$ $\frac{9}{2} = A(\frac{3}{2})^2$ $\Rightarrow$ $\frac{9}{2} = \frac{9}{4}A$ $\Rightarrow$ $A = 2$
Let $x = 0$ $4 = A - B - C$ $\Rightarrow$ $B = A - C - 4 = 2 + 1 - 4 = -1$

Substituting these values for A, B and C gives

$$\frac{x + 4}{(2x - 1)(x + 1)^2} \equiv \frac{2}{(2x - 1)} - \frac{1}{(x + 1)} - \frac{1}{(x + 1)^2}.$$

Now that the function is in partial fractions, each part can be integrated separately.

$$\int \frac{x + 4}{(2x - 1)(x + 1)^2} \, dx = \int \frac{2}{(2x - 1)} \, dx - \int \frac{1}{(x + 1)} \, dx - \int \frac{1}{(x + 1)^2} \, dx.$$

Using partial fractions in integration

The first two integrals give logarithmic functions as we saw in Example 10.39. The third, however, is of the form u^{-2} and therefore can be integrated by using the substitution $u = x + 1$, or by inspection (i.e. in your head). So

$$\int \frac{x + 4}{(2x - 1)(x + 1)^2}\, dx = \ln|2x - 1| - \ln|x + 1| + \frac{1}{x + 1} + c$$

$$= \ln\left|\frac{2x - 1}{x + 1}\right| + \frac{1}{x + 1} + c.$$

A quadratic factor in the denominator

EXAMPLE 10.41

Show that $\dfrac{x}{x^2 + 2} - \dfrac{1}{x + 1} \equiv \dfrac{x - 2}{(x^2 + 2)(x + 1)}$.

Hence find $\displaystyle\int \frac{x - 2}{(x^2 + 2)(x + 1)}\, dx$.

Solution

$$\frac{x}{x^2 + 2} - \frac{1}{x + 1} \equiv \frac{x(x + 1) - (x^2 + 2)}{(x^2 + 2)(x - 1)}$$

$$\equiv \frac{x^2 + x - x^2 - 2}{(x^2 + 2)(x - 1)}$$

$$\equiv \frac{x - 2}{(x^2 + 2)(x - 1)}$$

So $\displaystyle\int \frac{x - 2}{(x^2 + 2)(x + 1)}\, dx = \int \frac{x}{(x^2 + 2)}\, dx - \int \frac{1}{(x + 1)}\, dx$

$$= \tfrac{1}{2}\int \frac{2x}{x^2 + 2}\, dx - \int \frac{1}{x + 1}\, dx$$

$$= \tfrac{1}{2} \ln|x^2 + 2| - \ln|x + 1| + c$$

$$= \ln\left|\frac{\sqrt{x^2 + 2}}{x + 1}\right| + c.$$

> $\tfrac{1}{2} \ln|x^2 + 2| = \ln \sqrt{x^2 + 2}$
> Notice that $(x^2 + 2)$ is positive for all values of x

Note

Suppose the algebraic function could be decomposed to

$$\frac{x}{x^2 + 2} + \frac{1}{x^2 + 2}.$$

The first part of this can be integrated as in Example 10.41, but the second part cannot be integrated by any method you have met so far. If you go on to study Further Mathematics, you will meet integrals of this form then.

EXERCISE 10J

1 Express the functions in each of the following integrals in partial fractions, and hence perform the integration.

(a) $\int \dfrac{1}{(1-x)(3x-2)}\,dx$

(b) $\int \dfrac{7x-2}{(x-1)^2(2x+3)}\,dx$

(c) $\int \dfrac{x+1}{(2x+1)(x-1)}\,dx$

(d) $\int \dfrac{3x+3}{(x-1)(2x+1)}\,dx$

(e) $\int \dfrac{2}{x(1-x)(2-x)}\,dx$

(f) $\int \dfrac{1}{(x+1)(x+3)}\,dx$

(g) $\int \dfrac{2x^2+10x-3}{(x-1)(2x+1)}\,dx$

(h) $\int \dfrac{5x+1}{(x+2)(2x+1)^2}\,dx$

2 (a) (i) Express $\dfrac{3}{(1+x)(1-2x)}$ in partial fractions.

   (ii) Hence find
   $$\int_0^{0.1} \dfrac{3}{(1+x)(1-2x)}\,dx$$
   giving your answer to 5 decimal places.

(b) (i) Find the first three terms in the binomial expansion of $3(1+x)^{-1}(1-2x)^{-1}$.

   (ii) Use the first three terms of this expansion to find an approximation for
   $$\int_0^{0.1} \dfrac{3}{(1+x)(1-2x)}\,dx.$$

   (iii) What is the percentage error in your answer to part (ii)?

3 (a) Given that
$$\dfrac{x^2-x-24}{(x+2)(x-4)} \equiv A + \dfrac{B}{(x+2)} + \dfrac{C}{(x-4)},$$
find the values of the constants A, B and C.

(b) Find $\int_1^3 \dfrac{x^2-x-24}{(x+2)(x-4)}\,dx$.

[MEI]

4 The expression $\dfrac{x^2}{(x-4)^2(x-2)}$ is to be written in partial fractions of the form
$$\dfrac{A}{(x-4)^2} + \dfrac{B}{x-4} + \dfrac{C}{x-2}.$$

Show that $B = 0$ and find A and C.

Hence show that $\int_5^8 \dfrac{x^2}{(x-4)^2(x-2)}\,dx = 6 + \ln 2$.

[MEI, part]

EXERCISE 10J

5 **(a)** Express $\dfrac{3}{(1 + 2x)(2 + x)}$ in partial fractions.

(b) Show that the area of the region enclosed between $y = \dfrac{3}{(1 + 2x)(2 + x)}$, the x-axis and the lines $x = 2$ and $x = 3$ is $\ln\left(\tfrac{28}{25}\right)$.

6 The function f(x) is defined by

$$f(x) = \dfrac{4x^2 + 4x + 3}{(2x^2 + 1)(x - 2)}.$$

(a) Show that $\dfrac{3}{x - 2} - \dfrac{2x}{2x^2 + 1} = f(x)$.

(b) Evaluate $\int_3^5 f(x)\,dx$, giving your answer to 3 significant figures.

7 $f(x) \equiv \dfrac{5x^2 - 8x + 1}{2x(x - 1)^2} \equiv \dfrac{A}{x} + \dfrac{B}{x - 1} + \dfrac{C}{(x - 1)^2}$

(a) Find the value of the constants A, B and C.

(b) Hence find $\int f(x)\,dx$.

(c) Hence show that

$$\int_4^9 f(x)\,dx = \ln\left(\tfrac{32}{3}\right) - \tfrac{5}{24}.$$

[Edexcel]

8 $f(x) = \dfrac{25}{(3 + 2x)^2(1 - x)},\quad |x| < 1$

(a) Express f(x) as a sum of partial fractions.

(b) Hence find $\int f(x)\,dx$

[Edexcel (part)]

USING TRIGONOMETRIC IDENTITIES IN INTEGRATION

Sometimes, when it is not immediately obvious how to integrate a function involving trigonometric functions, it may help to rewrite the function using one of the trigonometric identities.

The following identities might be helpful.

Fundamental identities

$\sin^2\theta + \cos^2\theta \equiv 1 \qquad \tan^2\theta + 1 \equiv \sec^2\theta \qquad 1 + \cot^2\theta \equiv \text{cosec}^2\theta$

Double-angle formulae

$\sin 2\theta \equiv 2\sin\theta\cos\theta$

$\cos 2\theta \equiv \cos^2\theta - \sin^2\theta \equiv 2\cos^2\theta - 1 \equiv 1 - 2\sin^2\theta$

EXAMPLE 10.42

Find $\int \sin^2 x \, dx$.

Solution A substitution cannot be used in this case. However, in Chapter 2 you learnt the identity

$$\cos 2x \equiv 1 - 2\sin^2 x.$$

(Remember that this is just one of the three expressions for $\cos 2x$.)

This identity may be rewritten as

$$\sin^2 x \equiv \tfrac{1}{2}(1 - \cos 2x).$$

By writing $\sin^2 x$ in this form, you will be able to perform the integration:

$$\int \sin^2 x \, dx = \tfrac{1}{2} \int (1 - \cos 2x) \, dx$$
$$= \tfrac{1}{2}(x - \tfrac{1}{2}\sin 2x) + c$$
$$= \tfrac{1}{2}x - \tfrac{1}{4}\sin 2x + c.$$

You can integrate $\cos^2 x$ in the same way, by using $\cos^2 x = \tfrac{1}{2}(\cos 2x + 1)$. Other even powers of $\sin x$ or $\cos x$ can also be integrated in a similar way, but you have to use the identity twice or more.

EXAMPLE 10.43

Find $\int \cos^4 x \, dx$.

Solution First express $\cos^4 x$ as $(\cos^2 x)^2$:

$$\cos^4 x \equiv [\tfrac{1}{2}(\cos 2x + 1)]^2$$
$$\equiv \tfrac{1}{4}(\cos^2 2x + 2\cos 2x + 1).$$

Next, apply the same identity to $\cos^2 2x$:

$$\cos^2 2x \equiv \tfrac{1}{2}(\cos 4x + 1).$$

Hence $\cos^4 x \equiv \tfrac{1}{4}(\tfrac{1}{2}\cos 4x + \tfrac{1}{2} + 2\cos 2x + 1)$
$$\equiv \tfrac{1}{4}(\tfrac{1}{2}\cos 4x + 2\cos 2x + \tfrac{3}{2})$$
$$\equiv \tfrac{1}{8}\cos 4x + \tfrac{1}{2}\cos 2x + \tfrac{3}{8}.$$

This can now be integrated:

$$\int \cos^4 x \, dx = \int (\tfrac{1}{8}\cos 4x + \tfrac{1}{2}\cos 2x + \tfrac{3}{8}) \, dx$$
$$- \tfrac{1}{32}\sin 4x + \tfrac{1}{4}\sin 2x + \tfrac{3}{8}x + c.$$

USING TRIGONOMETRIC IDENTITIES IN INTEGRATION

For odd powers of $\sin x$ or $\cos x$, a different technique is used, as in the next example.

EXAMPLE 10.44

Find $\int \cos^3 x \, dx$.

Solution First write $\cos^3 x \equiv \cos x \cos^2 x$.

Now remember that

$$\cos^2 x + \sin^2 x \equiv 1 \quad \Rightarrow \quad \cos^2 x \equiv 1 - \sin^2 x.$$

This gives

$$\cos^3 x \equiv \cos x(1 - \sin^2 x) \equiv \cos x - \cos x \sin^2 x.$$

The first part of this expression, $\cos x$, is easily integrated to give $\sin x$.

The second part is more complicated, but it can be seen that it is of a type we have met already, as it is a product of two functions, one of which is a function of $\sin x$ and the other of which is the derivative of $\sin x$. This can be integrated either by making the substitution $u = \sin x$ or simply in your head (by inspection). So

$$\int \cos^3 x \, dx = \int (\cos x - \cos x \sin^2 x) \, dx$$
$$= \sin x - \tfrac{1}{3}\sin^3 x + c.$$

Any odd power of $\sin x$ or $\cos x$ can be integrated in this way, but again it may be necessary to use the identity more than once. For example:

$$\sin^5 x = \sin x \, (\sin^2 x)(\sin^2 x)$$
$$= \sin x(1 - \cos^2 x)^2$$
$$= \sin x(1 - 2\cos^2 x + \cos^4 x)$$
$$= \sin x - 2\sin x \cos^2 x + \sin x \cos^4 x.$$

This can now be integrated.

EXAMPLE 10.45

Using a suitable identity evaluate $\int_0^{\frac{\pi}{8}} \tan^2 2x \, dx$.

Solution $\tan^2 2x$ cannot be integrated directly, so use the identity

$$\tan^2 \theta + 1 = \sec^2 \theta$$
$$\therefore \tan^2 2x = \sec^2 2x - 1.$$

Hence

$$\int_0^{\pi/8} \tan^2 2x \, dx = \int_0^{\pi/8} (\sec^2 2x - 1) \, dx$$

$$= \left[\tfrac{1}{2} \tan 2x - x\right]_0^{\pi/8} \quad \text{since } \int \sec^2 x \, dx = \tan x + c$$

$$= \tfrac{1}{2} \tan \tfrac{\pi}{4} - \tfrac{\pi}{8} - 0$$

$$= \tfrac{1}{2} - \tfrac{\pi}{8}$$

$$= \tfrac{1}{8}(4 - \pi)$$

EXERCISE 10K

1 Integrate the following functions with respect to x.
   (a) $\sin x - 2\cos x$
   (b) $3\cos x + 2\sin x$
   (c) $5\sin x + 4\cos x$

2 Integrate the following functions by using the substitution given, or otherwise.
   (a) $\cos 3x$, $u = 3x$
   (b) $\sin(1 - x)$, $u = 1 - x$
   (c) $\sin x \cos^3 x$, $u = \cos x$
   (d) $\dfrac{\sin x}{2 - \cos x}$, $u = 2 - \cos x$
   (e) $\tan x$, $u = \cos x$ (write $\tan x$ as $\dfrac{\sin x}{\cos x}$)
   (f) $\sin 2x(1 + \cos 2x)^2$, $u = 1 + \cos 2x$

3 Use a suitable substitution to integrate the following functions.
   (a) $2x\sin(x^2)$
   (b) $\cos x \, e^{\sin x}$
   (c) $\sec^2 x \tan x$
   (d) $\dfrac{\cos x}{\sin^2 x}$

4 Evaluate the following definite integrals by using suitable substitutions.
   (a) $\int_0^{\pi/2} \cos\left(2x - \tfrac{\pi}{2}\right) dx$
   (b) $\int_0^{\pi/4} \cos x \sin^3 x \, dx$
   (c) $\int_0^{\sqrt{\pi}} x\sin(x^2) \, dx$
   (d) $\int_0^{\pi/4} \sec^2 x \, e^{\tan x} \, dx$
   (e) $\int_0^{\pi/4} \dfrac{\sec^2 x}{1 + \tan x} \, dx$

5 (a) Use a graphical calculator or computer to sketch the graph of the function $y = \sin x(\cos x - 1)^2$ for $0 \leqslant x \leqslant 4\pi$.
   (b) Use definite integration to find the area between the positive part of one cycle of the curve and the x-axis.

6 Use a suitable trigonometric identity to perform the following integrations.
   (a) $\int \cos^2 x \, dx$
   (b) $\int \sin^3 x \, dx$
   (c) $\int \sin^4 x \, dx$
   (d) $\int \cos^5 x \, dx$

Exercise 10K

7 (a) Express $\tan^2 x$ in terms of $\sec^2 x$.

(b) State $\int \sec^2 x \, dx$.

(c) Work out the exact value of $\int_0^{\frac{\pi}{16}} 3\tan^2 4x \, dx$.

8 (a) Prove that $\int \sin^2 x \, dx = \frac{1}{2}x - \frac{1}{4}\sin 2x + c$.

(b) Evaluate $\int_0^{\frac{\pi}{4}} x \sin^2 x \, dx$, leaving your answer in terms of π.

9 (a) Find $\int \tan 4x \, dx$.

(b) Using a suitable trigonometric identity find $\int \tan^2 4x \, dx$.

(c) By putting $\tan^3 4x = \tan 4x \times \tan^2 4x$ show that

$$\int \tan^3 4x \, dx = \tfrac{1}{8}\tan^2 4x - \tfrac{1}{4}\ln|\sec 4x| + c.$$

(d) Find $\int \tan^4 4x \, dx$.

10 (a) Show that $\int (\cos^4 x - \sin^4 x) \, dx \equiv \int \cos 2x \, dx$.

Hence find $\int_0^{\frac{\pi}{4}} (\cos^4 x - \sin^4 x) \, dx$.

(b) Show that $\int (\cos^4 x + \sin^4 x) \, dx \equiv \tfrac{1}{4}\int (3 + \cos 4x) \, dx$.

Hence find $\int_0^{\frac{\pi}{4}} (\cos^4 x + \sin^4 x) \, dx$.

Finding volumes of revolution by integration

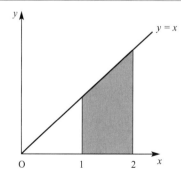

FIGURE 10.13

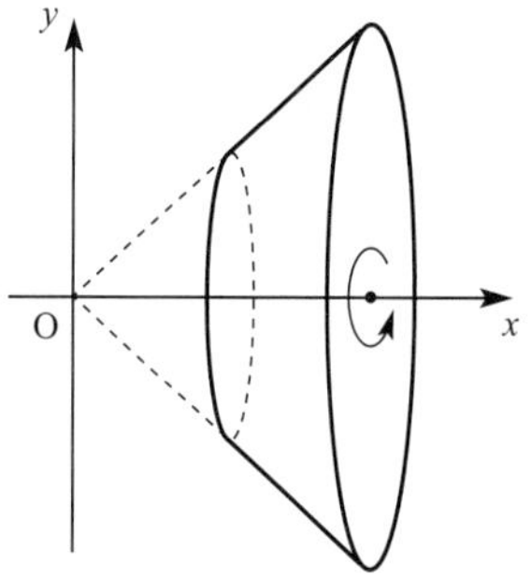

FIGURE 10.14

When the shaded region in figure 10.13 is rotated through 360° about the x-axis, the solid obtained, illustrated in figure 10.14, is called a *solid of revolution*. In this particular case, the volume of the solid could be calculated as the difference between the volumes of two cones, i.e. using $V = \tfrac{1}{3}\pi r^2 h$ gives

$$\text{volume of solid of revolution} = \tfrac{1}{3}\pi \times 2^2 \times 2 - \tfrac{1}{3}\pi \times 1^2 \times 1$$

$$= \tfrac{7}{3}\pi \text{ cubic units}.$$

If the line $y = x$ in figure 10.13 was replaced by a curve, such a simple calculation would no longer be possible.

Solids formed by rotation about the x-axis

The shaded region in figure 10.15 is bounded by $y = f(x)$, the x-axis and the lines $x = a$ and $x = b$.

Now look at the solid formed by rotating the shaded region through 360° about the x-axis.

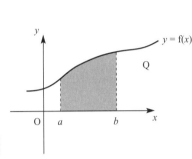

FIGURE 10.15

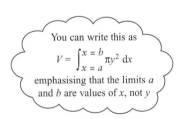

FIGURE 10.16

The volume of the solid of revolution (which is usually called the *volume of revolution*) can be found by imagining that the solid can be sliced into thin discs.

The disc shown in figure 10.16 is approximately cylindrical with radius y and thickness δx, so its volume V is given by

$$\delta V = \pi y^2 \delta x.$$

The volume of the solid is the limit of the sum of all these elementary discs as $\delta x \to 0$,

i.e. the limit as $\delta x \to 0$ of $\sum_{\text{over all discs}} \delta V = \sum_{x=a}^{x=b} \pi y^2 \delta x$.

The limiting values of sums such as these are integrals, so

$$V = \int_a^b \pi y^2 \, dx.$$

> You can write this as
> $$V = \int_{x=a}^{x=b} \pi y^2 \, dx$$
> emphasising that the limits a and b are values of x, not y

The limits are a and b because x takes values from a to b.

Taking π out of the integral the volume of revolution about the x-axis from $x = a$ to $x = b$ is

$$V = \pi \int_a^b y^2 \, dx.$$

Note: Since the integration is 'with respect to x', indicated by the dx and the fact that the limits a and b are values of x, it cannot be evaluated unless the function y is also written in terms of x.

FINDING VOLUMES OF REVOLUTON BY INTEGRATION

Pure Mathematics: Core 4

EXAMPLE 10.46

The region between the curve $y = x^2$, the x-axis and the lines $x = 1$ and $x = 3$ is rotated through $360°$ about the x-axis. Find the volume of revolution which is formed.

Solution The region is shaded in the diagram.

Using $V = \pi \int_a^b y^2 \, dx$

$$\begin{aligned}\text{volume} &= \pi \int_1^3 (x^2)^2 \, dx \\ &= \pi \int_1^3 x^4 \, dx \\ &= \pi \left[\frac{x^5}{5} \right]_1^3 \\ &= \frac{\pi}{5}(243 - 1) \\ &= \frac{242\pi}{5}. \end{aligned}$$

Since in this case
$y = x^2$
$y^2 = (x^2)^2 = x^4$

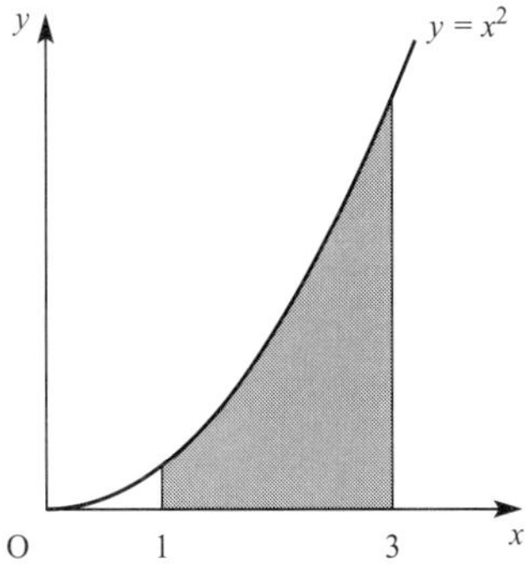

FIGURE 10.17

The volume is $\frac{242\pi}{5}$ cubic units or 152 cubic units (3 s.f.).

Note

Unless a decimal answer is required, it is usual to leave π in the answer, which is then exact.

EXAMPLE 10.47

(a) Find the volume of a spherical ball of radius 2 cm using integration.
(b) Verify your result using the formula for the volume of a sphere.

Solution (a) The volume is obtained by rotating the top half of the circle $x^2 + y^2 = 4$ through $360°$ about the x-axis.

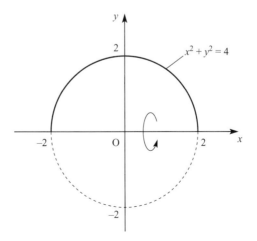

FIGURE 10.18

Using $V = \pi \int_a^b y^2 \, dx$ and $y^2 = 4 - x^2$ from the circle equation

volume $= \pi \int_{-2}^{2} (4 - x^2) \, dx$

$= \pi \left[4x - \dfrac{x^3}{3} \right]_{-2}^{2}$

$= \pi \left[\left(8 - \dfrac{8}{3} \right) - \left(-8 + \dfrac{8}{3} \right) \right]$

$= \dfrac{32\pi}{3} \text{ cm}^3.$

(b) Volume of a sphere

$= \dfrac{4}{3} \pi r^3$

$= \dfrac{4}{3} \pi \times 2^3$

$= \dfrac{32\pi}{3} \text{ cm}^3$

which verifies the result in part **(a)**.

EXERCISE 10L

1 In each part of this question a region is defined in terms of the lines which form its boundaries. Draw a sketch of the region and find the volume of the solid obtained by rotating it through 360° about the x-axis:
  (a) $y = 2x$, the x-axis and the lines $x = 1$ and $x = 3$;
  (b) $y = x + 2$, the x-axis, the y-axis and the line $x = 2$;
  (c) $y = x^2 + 1$, the x-axis and the lines $x = 0$ and $x = 1$;
  (d) $y = \sqrt{x}$, the x-axis and the line $x = 4$;
  (e) $y = \dfrac{1}{\sqrt{x}}$, the x-axis and the lines $x = 1$ and $x = 4$;
  (f) $y = e^{\frac{x}{2}}$, the x-axis and the lines $x = 0$ and $x = 3$.
  (g) $y = \dfrac{r}{h} x$ and the x-axis from $x = 0$ to the line $x = h$. Comment.
  (h) $y^2 = r^2 - x^2$ and the x-axis from $x = -r$ to $x = r$. Comment.

2 (a) Find the coordinates of A and B, the points of intersection of the circle $x^2 + y^2 = 25$ and the line $y = 4$.

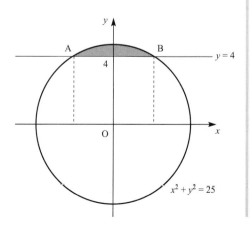

(b) A napkin ring is formed by rotating the shaded area through 360° about the x-axis. By considering the shaded area as the difference between two areas, and hence the volume of the napkin ring as the difference between two volumes, find the volume of the napkin ring.

3 (a) Sketch the line $4y = 3x$ for $x \geq 0$.
(b) Identify the area between this line and the x-axis which, when rotated through 360° about the x-axis, would give a cone of base radius 3 and height 4.
(c) Calculate the volume of the cone using:
  (i) integration;
  (ii) a formula.

4 (a) Sketch the graph of $y = (x-2)^2$ for values of x between $x = -1$ and $x = +5$. Shade in the region under the curve, between $x = 0$ and $x = 2$.
(b) Calculate the area you have shaded.
(c) Show that $(x-2)^4 = x^4 - 8x^3 + 24x^2 - 32x + 16$.
(d) The shaded region is rotated about the x-axis to form a volume of revolution. Calculate this volume, using your answer to part **(c)** or otherwise.

[MEI]

5 The diagram shows the graph of $y = x^3$.

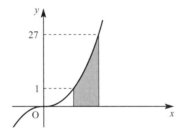

(a) Show that the volume generated when the shaded region shown in the diagram is rotated about the x-axis is $\frac{2186}{7}\pi$.

6 The diagram shows the curve $y = \frac{x+1}{\sqrt{x}}$.

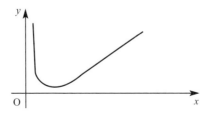

Show that the volume generated when the region enclosed between the curve, the x-axis and the lines $x = 2$ and $x = 4$ is rotated around the x-axis is $\pi(10 + \ln x)$.

7 The region R is enclosed between $y = \sin x + \cos x$, the x- and y-axes and the line $x = \frac{\pi}{6}$.

   (a) Show that when R is rotated about the x-axis the volume V of revolution is given by
   $$V = \pi \int_0^{\frac{\pi}{6}} (1 + \sin 2x) \, dx.$$

   (b) Evaluate V.

8 Sketch the graph of $y = \sin x$ from $x = 0$ to $x = \pi$ and on your diagram shade the region enclosed between the curve and the x-axis.

   Calculate the volume of revolution when the shaded region is rotated about the x-axis.

9 (a) Use the compound-angle formula to prove that $\cos 2\theta = 2\cos^2 \theta - 1$.

   Part of the graph of $y = x\cos x$ from $x = 0$ to $x = \frac{\pi}{2}$ is shown below.

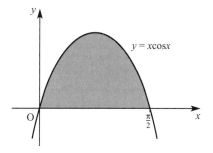

   The shaded region is rotated about the x-axis.

   (b) Show that the volume V generated by the rotation is given by
   $$V = \frac{\pi}{2} \left(\int_0^{\frac{\pi}{2}} x^2 \, dx + \int_0^{\frac{\pi}{2}} x^2 \cos 2x \, dx \right).$$

   (c) Evaluate V to 3 decimal places.

10 The diagram shows part of a curve C with equation $y = x^2 + 3$. The shaded region is bounded by C, the x-axis and the lines with equations $x = 1$ and $x = 3$. The shaded region is rotated $360°$ about the x-axis.

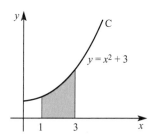

   Using calculus, calculate the volume of the solid generated. Give your answer as an exact multiple of π.

   [Edexcel]

VOLUMES OF REVOLUTION WITH EQUATIONS IN PARAMETRIC FORM

You now know that the formula for the volume of revolution about the x-axis from $x = a$ to $x = b$ is

$$V = \pi \int_a^b y^2 \, dx.$$

If x and y are given in terms of a parameter such as t or θ then y and dx have to be changed in the integral.

The following two examples illustrate the method.

EXAMPLE 10.48

A curve C is given parametrically by the equations

$$x = 2t, \quad y = t^2.$$

Find the volume of revolution obtained when C is rotated about the x-axis from $t = 1$ to $t = 2$.

Solution If $y = t^2$ then $y^2 = t^4$,

And if $x = 2t$ then $\dfrac{dx}{dt} = 2$.

As before (with integration by substitution) we treat this as a fraction such that $dx = 2 \, dt$.

The integral becomes:

$$V = \pi \int_1^2 t^4 \times 2 \, dt = \frac{2\pi}{5}[t^5]_1^2 = \frac{2\pi}{5}(32 - 1) = \frac{62}{5}\pi.$$

Note

You could have eliminated the parameter t to give $y = \frac{1}{4}x^2$ and, working out the x-limits to be from 2 to 4 (i.e. when $t = 1$, $x = 2$ and when $t = 2$, $x = 4$), the integral would have been

$$V = \frac{\pi}{16}\int_2^4 x^4 \, dx = \frac{\pi}{5 \times 16}[x^5]_2^4 = \frac{\pi}{5 \times 16}(1024 - 32) = \frac{62}{5}\pi.$$

EXAMPLE 10.49

A curve is defined by the parametric equations

$$x = 2\cos\theta, \; y = 3\sin\theta, \; 0 \leq \theta \leq \frac{\pi}{2}.$$

Find the magnitude of the volume generated when the curve is rotated about the x-axis from $\theta = 0$ to $\theta = \frac{\pi}{2}$.

Solution If $y = 3\sin\theta$ then $y^2 = 9\sin^2\theta$,

and if $x = 2\cos\theta$ then $\dfrac{dx}{d\theta} = -2\sin\theta$, so $dx = -2\sin\theta\, d\theta$

The integral becomes:

$$V = \pi \int_0^{\frac{\pi}{2}} 9\sin^2\theta \times -2\sin\theta\, d\theta = -18\pi \int_0^{\frac{\pi}{2}} \sin^3\theta\, d\theta.$$

$$= -18\pi \int_0^{\frac{\pi}{2}} (1 - \cos2\theta)\sin\theta\, d\theta = -18\pi \int_0^{\frac{\pi}{2}} (\sin\theta - \cos^2\theta \sin\theta)\, d\theta$$

$$= -18\pi \left[-\cos\theta + \tfrac{1}{3}\cos^3\theta\right]_0^{\frac{\pi}{2}}$$

$$= -12\pi.$$

So the volume of revolution has magnitude 12π cubic units.

EXERCISE 10M

1 A curve C is given parametrically by the equations

$$x = t^2, \quad y = 4t.$$

Find the volume of revolution obtained when C is rotated about the x-axis from $t = 1$ to $t = 2$.

2 A curve C is given parametrically by the equations

$$x = t, \quad y = \frac{1}{t}.$$

Find the volume of revolution obtained when C is rotated about the x-axis from $t = 2$ to $t = 4$.

3 A curve is defined by the parametric equations

$$x = 4\sin\theta, \quad y = 3\cos\theta, \quad 0 \leqslant \theta \leqslant 2\pi.$$

Find the magnitude of the volume generated when the curve is rotated about the x-axis from $\theta = 0$ to $\theta = \frac{\pi}{2}$.

4 A curve is defined by the parametric equations

$$x = 2 + \cos\theta, \quad y = 3 + \sin\theta, \quad 0 \leqslant \theta \leqslant 2\pi.$$

Find the magnitude of the volume generated when the curve is rotated about the x-axis from $\theta = 0$ to $\theta = \frac{\pi}{4}$.

5 You are given the parametric equations

$$x = t + \frac{1}{t}, \quad y = t - \frac{1}{t}, \quad t \neq 0.$$

Find the volume generated when the curve is rotated through 360° about the x-axis from $t = 1$ to $t = 3$.

EXERCISE 10M

6 A curve has parametric equations

$$x = \frac{t}{1+t}, \quad y = \frac{\sqrt{t}}{1-t}, \quad t \neq \pm 1$$

Find the volume generated when the curve is rotated through 360° about the x-axis from $t = 0$ to $t = \frac{1}{2}$.

7 A curve is defined by the parametric equations

$$x = 3 + \sin\theta, \quad y = 1 + 4\cos\theta, \quad 0 \leq \theta \leq 2\pi.$$

Find the volume generated when the curve is rotated about the x-axis from $\theta = \frac{\pi}{2}$ to $\theta = \pi$.

8 A curve is defined by the parametric equations

$$x = \sin^2\theta, \quad y = 1 + 2\cos\theta, \quad 0 \leq \theta \leq 2\pi.$$

Find the volume generated when the curve is rotated about the x-axis from $\theta = 0$ to $\theta = \pi$.

9 A curve is defined by the parametric equations

$$x = \tan\theta, \quad y = \tan^2\theta, \quad 0 \leq \theta \leq 2\pi.$$

Find the volume generated when the curve is rotated about the x-axis from $\theta = 0$ to $\theta = \frac{\pi}{4}$.

10 A curve has parametric equations $x = \sin t$ and $y = e^t$, where t is in radians. The curve is rotated about the x-axis from $t = 0$ to $t = \frac{\pi}{2}$. Show that the volume generated is

$$\tfrac{\pi}{5}(2e^\pi + 1).$$

SOLVING DIFFERENTIAL EQUATIONS

Finding an expression for $f(x)$ from a differential equation involving derivatives of $f(x)$ is called solving the equation.

Some differential equations may be solved simply by integration.

EXAMPLE 10.50

Solve the differential equation

$$\frac{dy}{dx} = 3x^2 - 2.$$

Solution Integrating gives

$$y = \int (3x^2 - 2)\, dx$$
$$y = x^3 - 2x + c.$$

The general solution of the differential equation

Notice that when you solve a differential equation, you get not just one solution, but a whole family of solutions, as c can take any value. This is called the *general solution* of the differential equation. The family of solutions for the differential equation in Example 10.50 would be translations in the y-direction of the curve $y = x^3 - 2x$.

THE METHOD OF SEPARATION OF VARIABLES

It is not difficult to solve a differential equation like the one in Example 10.50, because the right-hand side is a function of x only. So long as the function can be integrated, the equation can be solved.

Now consider the differential equation

$$\frac{dy}{dx} = xy.$$

This cannot be solved directly by integration, because the right-hand side is a function of both x and y. However, as you will see in the next example, you can solve this and similar differential equations where the right-hand side consists of a function of x and a function of y multiplied together.

EXAMPLE 10.51

Find, for $y > 0$, the general solution of the differential equation

$$\frac{dy}{dx} = xy.$$

Solution The equation may be rewritten as

$$\frac{1}{y}\frac{dy}{dx} = x$$

so that the right-hand side is now a function of x only.

Integrating both sides with respect to x gives

$$\int \frac{1}{y}\frac{dy}{dx}\,dx = \int x\,dx.$$

As $\frac{dy}{dx}\,dx$ can be written as dy

$$\int \frac{1}{y}\,dy = \int x\,dx.$$

Both sides may now be integrated separately:

$$\ln|y| = \tfrac{1}{2}x^2 + c.$$

> Since we have been told that $y > 0$, we may drop the modulus symbol. In this case, $|y| = y$

(There is no need to put a constant of integration on both sides of the equation.)

We now need to rearrange the solution above to give y in terms of x. Making both sides powers of e gives

$$e^{\ln y} = e^{\frac{1}{2}x^2 + c}$$
$$\Rightarrow y = e^{\frac{1}{2}x^2 + c}$$

> Notice that the right-hand side is $e^{\frac{1}{2}x^2+c}$ and not $e^{\frac{1}{2}x^2} + e^c$

which is rearranged to give

$$y = e^{\frac{1}{2}x^2} \times e^c.$$

This expression can be simplified by replacing e^c with a new constant A. So

$$y = Ae^{\frac{1}{2}x^2}.$$

Note

Usually the first part of this process is carried out in just one step

$$\frac{dy}{dx} = xy$$

and can immediately be rewritten as

$$\int \frac{1}{y} \, dy = \int x \, dx.$$

This method is called *separation of variables*. It can be helpful to do this by thinking of the differential equation as if $\frac{dy}{dx}$ were a fraction, and trying to rearrange the equation to obtain all the x terms on one side and all the y terms on the other. Then just insert an integration sign on each side. Remember that dy and dx must both end up on the top line (numerator).

EXAMPLE 10.52

Find the general solution of the differential equation

$$\frac{dy}{dx} = e^{-y}.$$

Solution Separating the variables gives

$$\int \frac{1}{e^{-y}} \, dy = \int dx$$

which is rearranged to give

$$\int e^y \, dy = \int dx.$$

The right-hand side can be thought of as integrating 1 with respect to x:

$$e^y = x + c.$$

Taking logarithms of both sides:

$$y = \ln(x + c).$$

Note $\ln(x + c)$ is not the same as $\ln x + c$.

EXERCISE 10N

1 Solve the following differential equations by integration.

(a) $\dfrac{dy}{dx} = x^2$ (b) $\dfrac{dy}{dx} = \cos x$

(c) $\dfrac{dy}{dx} = e^x$ (d) $\dfrac{dy}{dx} = \sqrt{x}$

2 Find the general solutions of the following differential equations by separating the variables.

(a) $\dfrac{dy}{dx} = xy^2$ (b) $\dfrac{dy}{dx} = \dfrac{x^2}{y}$

(c) $\dfrac{dy}{dx} = y$ (d) $\dfrac{dy}{dx} = e^{x-y}$

(e) $\dfrac{dy}{dx} = \dfrac{y}{x}$ (f) $\dfrac{dy}{dx} = x\sqrt{y}$

(g) $\dfrac{dy}{dx} = y^2 \cos x$ (h) $\dfrac{dy}{dx} = \dfrac{x(y^2 + 1)}{y(x^2 + 1)}$

(i) $\dfrac{dy}{dx} = xe^y$ (j) $\dfrac{dy}{dx} = \dfrac{x \ln x}{y^2}$

PARTICULAR SOLUTIONS

You have already seen that a differential equation has an infinite number of different solutions corresponding to different values of the constant of integration. In Example 10.50, we found that

$$\dfrac{dy}{dx} = 3x^2 - 2$$

had the general solution $y = x^3 - 2x + c$.

Figure 10.19 shows the curves of the solutions corresponding to some different values of c.

If you are given some more information, you can find out which of the possible solutions is the one that matches the situation in question. For example, you might be told that when $x = 1$, $y = 0$. This tells you that the correct solution is the one whose curve passes through the point $(1, 0)$. You can use this information to find out the value of c for this particular solution by substituting the values $x = 1$ and $y = 0$ into the general solution:

$$y = x^3 - 2x + c$$
$$0 = 1 - 2 + c$$
$$\Rightarrow c = 1.$$

Solving differential equations

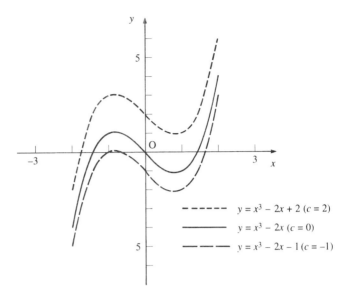

FIGURE 10.19

So the solution in this case is $y = x^3 - 2x + 1$.

This is called the *particular solution*.

EXAMPLE 10.53

(a) Find the general solution of the differential equation $\frac{dy}{dx} = y^2$.
(b) Find the particular solution for which $y = 1$ when $x = 0$.

Solution (a) Separating the variables gives $\int \frac{1}{y^2} \, dy = \int dx$

$$\therefore -\frac{1}{y} = x + c.$$

The general solution is $y = -\frac{1}{x + c}$.

(b) When $x = 0$, $y = 1$, which gives

$$1 = -\frac{1}{c} \quad \Rightarrow \quad c = -1.$$

So the particular solution is

$$y = -\frac{1}{x - 1} \quad \text{or} \quad y = \frac{1}{1 - x}.$$

EXAMPLE 10.54

The acceleration of an object is inversely proportional to its velocity at any given time and the direction of motion is taken to be positive. When the velocity is $1 \, \text{m s}^{-1}$, the acceleration is $3 \, \text{m s}^{-2}$.

(a) Find a differential equation to model this situation.
(b) Find the particular solution of this differential equation for which the initial velocity is 2 m s^{-1}.
(c) In this case, how long does the object take to reach a velocity of 8 m s^{-1}?

Solution (a) $\dfrac{dv}{dt} \propto \dfrac{1}{v} \Rightarrow \dfrac{dv}{dt} = \dfrac{k}{v}$

When $v = 1$, $\dfrac{dv}{dt} = 3$ so $k = 3$, which gives

$$\dfrac{dv}{dt} = \dfrac{3}{v}.$$

(b) Separating the variables:

$$\int v \, dv = \int 3 \, dt$$
$$\tfrac{1}{2}v^2 = 3t + c.$$

When $t = 0$, $v = 2$ so $c = 2$, which gives

$$\tfrac{1}{2}v^2 = 3t + 2$$
$$v^2 = 6t + 4.$$

Since the direction of motion is positive

$$v = \sqrt{6t + 4}.$$

(c) When $v = 8$ $64 = 6t + 4$

$60 = 6t \quad \Rightarrow \quad t = 10.$

The object takes 10 seconds.

EXERCISE 10O

1 Find the particular solution of each of the following differential equations.

(a) $\dfrac{dy}{dx} = x^2 - 1$, $y = 2$ when $x = 3$

(b) $\dfrac{dy}{dx} = x^2 y$, $y = 1$ when $x = 0$

(c) $\dfrac{dy}{dx} = xe^{-y}$, $y = 0$ when $x = 0$

(d) $\dfrac{dy}{dx} = y^2$, $y = 1$ when $x = 1$

(e) $\dfrac{dy}{dx} = x(y + 1)$, $y = 0$ when $x = 1$

(f) $\dfrac{dy}{dx} = y^2 \sin x$, $y = 1$ when $x = 0$

EXERCISE 10O

2 A cold liquid at temperature $\theta°C$, where $\theta < 20$, is standing in a warm room. The temperature of the liquid obeys the differential equation

$$\frac{d\theta}{dt} = 2(20 - \theta)$$

where the time t is measured in seconds.

(a) Find the general solution of this differential equation.
(b) Find the particular solution for which $\theta = 5$ when $t = 0$.
(c) In this case, how long does the liquid take to reach a temperature of 18°C?

3 A population of rabbits increases so that the number of rabbits N (in hundreds), after t years is modelled by the differential equation

$$\frac{dN}{dt} = N.$$

(a) Find the general solution for N in terms of t.
(b) Find the particular solution for which $N = 10$ when $t = 0$.
(c) What will happen to the number of rabbits when t becomes very large? Why is this not a realistic model for an actual population of rabbits?

4 An object is moving so that its velocity $v \left(= \frac{ds}{dt}\right)$ is inversely proportional to its displacements s from a fixed point. If its velocity is 1 m s^{-1} when its displacement is 2 m, find a differential equation to model the situation. Find the general solution of your differential equation.

5 (a) Write $\dfrac{1}{y(3-y)}$ in partial fractions.

(b) Find $\displaystyle\int \dfrac{1}{y(3-y)} \, dy$.

(c) Solve the differential equation

$$x\frac{dy}{dx} = y(3 - y)$$

where $x = 2$ when $y = 2$, giving y as a function of x.
[MEI]

6 The rate of change of the population of triffids T varies with time t and is given by

$$\frac{dT}{dt} = kT \sin kt$$

(a) If $T = T_0$ when $t = 0$ find an expression for T in terms of k, t and T_0.
(b) Show that the ratio of the maximum to minimum sizes of the population is $e^2 : 1$.

7 A colony of bacteria which is initially of size 1500 increases at a rate proportional to its size so that, after t hours, its population N satisfies the equation

$$\frac{dN}{dt} = kN.$$

(a) If the size of the colony increases to 3000 in 20 hours, solve the differential equation to find N in terms of t.

(b) What size is the colony when $t = 80$?

(c) How long did it take, to the nearest minute, for the population to increase from 2000 to 3000?

[MEI]

8 (a) Show that

$$\frac{x^2 + 1}{x^2 - 1} \equiv 1 + \frac{2}{x^2 - 1}.$$

(b) Find the partial fractions for

$$\frac{2}{(x - 1)(x + 1)}.$$

(c) Solve the differential equation

$$(x^2 - 1)\frac{dy}{dx} = -(x^2 + 1)y \quad \text{(where } x > 1\text{)}$$

given that $y = 1$ when $x = 3$. Express y as a function of x.

[MEI]

9 A hemispherical bowl of radius a has its axis vertical and is full of water. At time $t = 0$ water starts running out of a small hole in the bottom of the bowl so that the depth of water in the bowl at time t is x. The rate at which the volume of water is decreasing is proportional to x. Given that the volume of water in the bowl when the depth is x is $\pi(ax^2 - \frac{1}{3}x^3)$, show that there is a positive constant k such that

$$\pi(2ax - x^2)\frac{dx}{dt} = -kx.$$

Given that the bowl is empty after a time T, show that

$$k = \frac{3\pi a^2}{2T}.$$

[MEI]

10 The square horizontal cross-section of a container has side 2 m. Water is poured in at the constant rate of 0.08 m³s⁻¹ and, at the same time, leaks out of a hole in the base at the rate of $0.12x$ m³s⁻¹, where x m is the depth of the water in the container at time t s. So the volume, V m³, of the water in the container at time t is given by $V = 4x$ and the rate of change of volume is given by

$$\frac{dV}{dt} = 0.08 - 0.12x.$$

Use these results to find an equation for $\frac{dx}{dt}$ in terms of x and solve this to find x in terms of t if the container is empty initially.

Determine to the nearest 0.1 s the time taken for the depth to rise from 0.1 to 0.5 m.

[MEI]

11 (a) In a simple model of the growth of sprogletts in t days their number N is given by $\frac{dN}{dt} = kN$. When $t = 0$, $N = N_0$. Show that $N = N_0 e^{kt}$.

(b) One day there were 250 sprogletts and 7 days later there were 268 sprogletts. Show that $k \approx 0.01$ and that in another 7 days after the second count the model predicts that there will be 287 sprogletts.

As the number of sprogletts increases the growth rate is found to decrease and an improved model is

$$\frac{dN}{dt} = ke^{-at} N$$

Again, when $t = 0$, $N = N_0$.

(c) Show that $\ln \frac{N}{N_0} = \frac{k}{a}(1 - e^{-at})$.

(d) Give, in terms of k, a and N_0, an expression for the number of sprogletts after a long period of time.

12 (a) Express $\frac{1}{(3x - 1)x}$ in partial fractions.

A model for the way in which a population of animals in a closed environment varies with time is given, for $P > \frac{1}{3}$, by

$$\frac{dP}{dt} = \tfrac{1}{2}(3P^2 - P)\sin t$$

where P is the size of the population in thousands at time t.

(b) Given that $P = \frac{1}{2}$ when $t = 0$, use the method of separation of variables to show that

$$\ln\left(\frac{3P - 1}{P}\right) = \tfrac{1}{2}(1 - \cos t).$$

(c) Calculate the smallest positive value of t for which $P = 1$.

(d) Rearrange the equation at the end of part **(b)** to show that

$$P = \frac{1}{3 - e^{\frac{1}{2}(1 - \cos t)}}.$$

Hence find the two values between which the number of animals in the population oscillates.

[MEI]

CHOOSING AN APPROPRIATE METHOD OF INTEGRATION

You have now met the following standard integrals (neglecting c).

f(x)	$\int f(x)\,dx$	f(x)	$\int f(x)\,dx$
x^n ($n \neq -1$)	$\frac{1}{n+1}x^{n+1}$	$\sin x$	$-\cos x$
$\frac{1}{x}$	$\ln|x|$	$\cos x$	$\sin x$
e^x	e^x	$\sec^2 x$	$\tan x$
		$\tan x$	$\ln|\sec x|$

If you are asked to integrate any of these standard functions, you may simply write down the answer. For other integrations, the table below may help.

Type of function to be integrated	Examples	Method of integration
Simple variations of any of the standard functions	$\cos(2x+1)$ e^{3x} $(2x+3)^4$	Substitution may be used, but it should be possible to do these by inspection
Product of two functions of the form $f'(x) \times g[f(x)]$ Note that $f'(x)$ means $\frac{d}{dx}[f(x)]$	$2xe^{x^2}$ $x^2(x^3+1)^6$	Substitution $u = f(x)$ or possibly by inspection
Other products, particularly when one function is a small positive integral power of x or a polynomial in x	xe^x $x^2 \sin x$	Integration by parts
Quotients of the form $\frac{f'(x)}{f(x)}$ or functions which can easily be converted to this form.	$\frac{x}{x^2+1}$ $\frac{\sin x}{\cos x}$	Substitution $u = f(x)$, or better by inspection: $k\ln|f(x)| + c$, where k is known
Polynomial quotients which may be split into partial fractions and integrate term by term	$\frac{x+1}{x(x-1)}$ $\frac{x-4}{x^2-x-2}$	Split into partial fractions
Even powers of $\sin x$ or $\cos x$	$\sin^2 x$ $\cos^4 x$	Use the double-angle formulae to transform the function before integrating
Odd powers of $\sin x$ or $\cos x$	$\cos^3 x$	Use $\cos^2 x + \sin^2 x = 1$ and write in the form $f'(x) \times g[f(x)]$

It is impossible to give an exhaustive list of possible types of integration, but the tables on this page cover the most common situations that you will meet.

Exercise 10P

1. Choose an appropriate method and integrate the following functions. You may find it helpful first to discuss in class which method to use.

   (a) $\int \cos(3x - 1)\, dx$

   (b) $\int \dfrac{2x + 1}{(x^2 + x - 1)^2}\, dx$

   (c) $\int \sec^2 x \tan^2 x\, dx$

   (d) $\int e^{1-x}\, dx$

   (e) $\int x^2 \sin 2x\, dx$

   (f) $\int \cos^2 x\, dx$

   (g) $\int \ln 2x\, dx$

   (h) $\int \dfrac{x}{(x^2 + 1)^3}\, dx$

   (i) $\int \sqrt{2x - 3}\, dx$

   (j) $\int \dfrac{4x + 1}{(x - 1)^2(x + 2)}\, dx$

   (k) $\int \sin^3 2x\, dx$

   (l) $\int x^3 \ln x\, dx$

   (m) $\int \dfrac{5}{2x^2 - 7x + 3}\, dx$

   (n) $\int (x + 1)e^{2+2x}\, dx$

2. Evaluate the following definite integrals.

   (a) $\displaystyle\int_8^{24} \dfrac{dx}{\sqrt{3x - 8}}$

   (b) $\displaystyle\int_8^{24} \dfrac{dx}{3x - 8}$

   (c) $\displaystyle\int_8^{24} \dfrac{9x}{3x - 8}\, dx$

   (d) $\displaystyle\int_0^{\frac{\pi}{3}} \sin^3 x\, dx$

   (e) $\displaystyle\int_1^3 x^2 \ln x\, dx$

3. Evaluate $\displaystyle\int_0^2 \dfrac{x^2}{\sqrt{1 + x^3}}\, dx$, using the substitution $u = 1 + x^3$, or otherwise.

   [MEI]

4. Find $\displaystyle\int_0^{\frac{\pi}{4}} \dfrac{\sin\theta}{\cos^4\theta}\, d\theta$ in terms of $\sqrt{2}$.

   [MEI]

5. (a) Find $\displaystyle\int_0^{\frac{\pi}{2}} \sin^2\theta\, d\theta$, leaving your answer in terms of π.

   (b) Using the substitution $u = \ln x$, or otherwise, find $\displaystyle\int_1^2 \dfrac{\ln x}{x}\, dx$, giving your answer to 2 decimal places.

   [MEI]

6. Find $\displaystyle\int_0^{\frac{\pi}{4}} x\cos 2x\, dx$, expressing your answer in terms of π.

   [MEI]

7. (a) Find $\displaystyle\int xe^{-2x}\, dx$.

   (b) Evaluate $\displaystyle\int_0^1 \dfrac{x}{(4 + x^2)}\, dx$, giving your answer correct to 3 significant figures.

   [MEI]

8. (a) Find $\displaystyle\int \sin(2x - 3)\, dx$.

   (b) Use the method of integration by parts to evaluate $\displaystyle\int_0^2 xe^{2x}\, dx$.

   (c) Using the substitution $t = x^2 - 9$, or otherwise, find $\displaystyle\int \dfrac{x}{x^2 - 9}\, dx$.

   [MEI]

9 Evaluate

   (a) $\int_0^1 (2x^2 + 1)(2x^3 + 3x + 4)^{\frac{1}{2}} \, dx$

   (b) $\int_1^e \frac{\ln x}{x^3} \, dx$

[MEI]

10 Find $\int_0^{\frac{\pi}{2}} \sin x \cos^3 x \, dx$ and $\int_0^1 te^{-2t} \, dt$.

[MEI]

NUMERICAL INTEGRATION OF FUNCTIONS

In *Pure Mathematics: Core 2* you used the trapezium rule to find approximate values to integrals we could not do analytically. The formula was obtained by dividing the area required into n strips, each of width h. Each strip was approximated to be a trapezium in shape. Figure 10.20 illustrates the method.

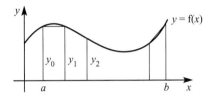

FIGURE 10.20

The trapezium rule is

$$\text{Area} = \int_a^b y \, dx \approx \tfrac{1}{2} h \{y_0 + y_n + 2(y_1 + y_2 + y_3 + \ldots + y_{n-1})\}$$

where $h = (b - a)/n$.

It is sometimes possible to tell from the shape of the graph whether the trapezium rule gives a value which is an overestimate or an underestimate. These are illustrated in figures 10.21 and 10.22.

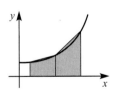

A concave graph gives an overestimate.

FIGURE 10.21

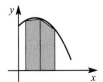

A convex graph gives an underestimate.

FIGURE 10.22

In some cases it is not possible to determine whether the trapezium rule approximation to the area is an overestimate or an underestimate. Such a case is shown in figure 10.23.

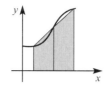

Not possible to tell whether it is an overestimate or underestimate.

FIGURE 10.23

EXAMPLE 10.55

You are given that $f(x) = \sqrt{1 + x}$. Sketch the graph of $y = f(x)$. Estimate $\int_0^3 f(x)\,dx$ using the trapezium rule with strip widths **(a)** 1.0 and **(b)** 0.5. State whether your answers are an overestimate or an underestimate.

Solution Working out points on the curve gives

x	−1	0	1	2	3
y	0	1	$\sqrt{2}$	$\sqrt{3}$	2

Sketching gives

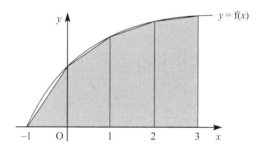

FIGURE 10.24

(a) With $h = 1.0$ then

$$\int_0^3 f(x)\,dx \approx \tfrac{1}{2} \times 1.0 \times \{1 + 2 + 2(\sqrt{2} + \sqrt{3})\} \approx 4.6463 \text{ (5 s.f.)}$$

(b) With $h = 0.5$ then

$$\int_0^3 f(x)\,dx \approx \tfrac{1}{2} \times 0.5 \times \{1 + 2 + 2(\sqrt{1.5} + \sqrt{2} + \sqrt{2.5} + \sqrt{3} + \sqrt{3.5})\}$$
$$\approx 4.6615 \text{ (5 s.f.)}$$

From the diagram both approximations to the integral are underestimates.

Consider example 10.55 in more detail.

Clearly the more trapezia you divide the area into the better they will fit the area under the curve. In the example you expect the answer with $h = 0.5$ to be a better approximation to the integral than when $h = 1.0$, but by how much?

The exact value of the integral is

$$\int_0^3 f(x)\, dx = \int_0^3 (1+x)^{\frac{1}{2}}\, dx = \tfrac{2}{3}\left[(1+x)^{\frac{3}{2}}\right]_0^3 = \tfrac{2}{3}(8-1) = 4.\dot{6}.$$

So, when $h = 1.0$ the error was $4.6666 - 4.6463 \approx 0.0203$ and when $h = 0.5$ the error was $4.6666 - 4.6615 \approx 0.0051$.

You can see that 0.0051 is approximately $\tfrac{1}{4}$ of 0.0203. That is, dividing h by 2 you have divided the error by 4.

If you were to divide h by 3 you would find that the error would be divided by 9.

These observations indicate the general rule that

> the error in the trapezium rule is approximately proportional to h^2
> i.e. error $\approx kh^2$.

The proof of this rule requires the use of the Taylor series which is in the Further Pure 3 specification.

EXAMPLE 10.56

Find the error between the exact value of $\int_0^3 \sqrt{1+x}\, dx$ and the trapezium rule approximation when $h = 1.0$.

Solution Let $\int_0^3 \sqrt{1+x}\, dx = I$.

From example 10.55 you know that the trapezium rule approximation with $h = 1.0$ gives

$$4.6463 + k \times 1^2 = I. \qquad \qquad ①$$

Putting $h = 0.5$ gives

$$4.6615 + k \times 0.5^2 = I. \qquad \qquad ②$$

Solving (1) and (2) to find k gives

$$k = 0.203.$$

So, when $h = 1.0$ the error is approximately 0.203 (as we found above). Further, if $h = 0.5$ the error is approximately 0.0051 (also as before).

EXERCISE 10Q

1 You are given that $f(x) = e^{\sin x}$.
   (a) Find $f(x)$ to 5 significant figures, when x takes values $0, 0.2, 0.4, 0.6, 0.8, 1.0$ and 1.2. Make sure your calculator is set to radians.
   (b) Sketch the graph of $y = f(x)$ for $0 \leqslant x \leqslant 1.2$.

EXERCISE 10Q

(c) Use the trapezium rule to find approximations, correct to 4 significant figures, to the integral $\int_0^{1.2} f(x)\, dx$ with (i) 3 strips and (ii) 6 strips.
(d) State whether your answers are overestimates or underestimates.
(e) Using your answers to part (c) find an approximate value for the error between the trapezium rule answer with 6 strips and the exact value of the integral.

2 (a) Use the trapezium rule to find approximate values, to 5 significant figures, of $\int_1^3 (1 + \ln x)\, dx$ with (i) 2 strips, (ii) 4 strips and (iii) 8 strips.
   (b) Which of your answers is the closest to the exact value?
   (c) Find an approximate value for the error between the exact value and your best approximation to the exact value.

3 Find an approximate answer, to 3 significant figures, of the error between the exact value of $\int_1^2 \sqrt{1 + x^2}\, dx$ and an approximation found using the trapezium rule with 4 strips.

4 You are given that $\int \dfrac{1}{1 + x^2}\, dx = \tan^{-1} x + c$.
   (a) Find the exact value of $\int_0^1 \dfrac{1}{1 + x^2}\, dx$.
   (b) Using the trapezium rule with $h = 0.25$ find an approximate value for $\int_0^1 \dfrac{1}{1 + x^2}\, dx$ to 4 decimal places.
   (c) Find an approximate value for the error between your answer in part (b) and the exact value of the integral.
   (d) Find an approximate value for π and state the approximate size of the error between this value and the exact value.

5 Find an approximate value for π using the trapezium rule with 5 strips given that $\int_0^{\frac{1}{2}} \dfrac{1}{\sqrt{1 - x^2}}\, dx = \dfrac{\pi}{6}$. Using the trapezium rule with one strip obtain an approximate value for the error between your value of π and the exact value of π.

6 You are given that $I = \int_2^3 \dfrac{e^x}{1 + e^x}\, dx$
   (a) Using the trapezium rule with 1, 2 and 4 strips find approximate values of I to 4 significant figures.
   (b) Which of the values you have found is the closest to the exact value of I?
   (c) Find the error between your best approximation and the exact value of I.
   (d) Using the substitution $u = e^x$, or otherwise, show that the exact value of I is $\ln\left(\dfrac{1 + e^3}{1 + e^2}\right)$. By evaluating this to 4 significant figures show that your answer to (c) is correct to 2 significant figures.

7 You are given that $f(x) = \ln(1 + x^3)$.

  (a) Copy and complete this table, giving values to 5 significant figures:

x	1.0	1.2	1.4	1.5	1.6	1.8	2.0
$\ln(1 + x^3)$							

  (b) Sketch the graph of $y = \ln(1 + x^3)$ for $1 \leq x \leq 2$.

  (c) Using the trapezium rule with 2 strips and then with 5 strips obtain, to 4 significant figures, approximate values for $\int_1^2 \ln(1 + x^3)\,dx$.

  (d) Find an approximate value for the error between the trapezium rule approximation obtained with 5 strips and the exact value of the integral.

8 Obtain approximate values for $\int_0^{\pi} \dfrac{\sin x}{1 + \sin x}\,dx$ using the trapezium rule with 2, 3 and 6 strips. Which value is the most accurate? For this value find the approximate error from the exact value of the integral.

9 Obtain approximate values of $\int_{0.5}^{1} \sin(e^x)\,dx$ using the trapezium rule with 1, 2 and then 4 trapezia. How would you obtain a better approximation?

10 You are given that $I = \int_0^{\frac{\pi}{4}} \tan x\,dx$.

  (a) Show that $I = \frac{1}{2}\ln 2$.

  (b) Using the trapezium rule with 2 strips and then with 4 strips find approximate values for I.

  (c) Using your results state the best approximation for $\ln 2$ and calculate the error between this value and the exact value.

Exercise 10R   Examination-style questions

1 Find $\int x^2 e^x\,dx$.

  Given that $x = 0$ when $y = 0$, solve the differential equation

  $$\frac{dy}{dx} = x^2 e^{2y+x}.$$

2 (a) Express $\dfrac{13 - 2x}{(2x - 3)(x + 1)}$ in partial fractions.

  (b) Given that $y = 4$ at $x = 2$, use your answer to part **(a)** to find the solution of the differential equation

  $$\frac{dy}{dx} = \frac{y(13 - 2x)}{(2x - 3)(x + 1)}, \quad x > 1.5.$$

  Express your answer in the form $y = f(x)$.

  [Edexcel]

EXERCISE 10R

3 (a) Use the substitution $t = 2x + 1$ to find $\int \dfrac{x+1}{\sqrt{2x+1}}\,dx$.

(b) Hence evaluate $\int_0^4 \dfrac{x+1}{\sqrt{2x+1}}\,dx$.

4 (a) Given that $2y = x - \sin x\cos x$, show that $\dfrac{dy}{dx} = \sin^2 x$.

(b) Hence find $\int \sin^2 x\,dx$.

(c) Hence, using integration by parts, find $\int x\sin^2 x\,dx$.

[Edexcel]

5

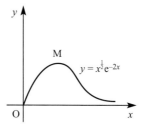

The diagram shows the curve with equation $y = x^{\frac{1}{2}}e^{-2x}$.

(a) Find the x-coordinate of M, the maximum point of the curve.

The finite region enclosed by the curve, the x-axis and the line $x = 1$ is rotated through 2π about the x-axis.

(b) Find, in terms of π and e, the volume of the solid generated.

6 Find

(a) $\int (1 - 2\sin^2 y)\,dy$.

(b) $\int x\cos x\,dx$.

(c) Hence find the general solution of the differential equation
$$\dfrac{dy}{dx} = \dfrac{x\cos x}{1 - 2\sin^2 y}, \quad -\dfrac{\pi}{4} \leqslant y \leqslant \dfrac{\pi}{4}.$$

7 (a) Find $\int x(x^2 + 3)^5\,dx$.

(b) Show that $\int_1^e \dfrac{1}{x^2}\ln x\,dx = 1 - \dfrac{2}{e}$.

(c) Given that $p > 1$, show that $\int_1^p \dfrac{1}{(x+1)(2x-1)}\,dx = \dfrac{1}{3}\ln\dfrac{4p-2}{p+1}$.

[Edexcel]

8 A laboratory is at a constant temperature of 20 °C. A cool liquid of temperature θ °C (where $\theta < 20$) is warming up at a rate that is proportional to the difference between the laboratory temperature and the temperature of the liquid.

Given that when $\theta = 10\ °C$ the rate of change of temperature is 2 °C per minute, show that

$$\frac{d\theta}{dt} = 0.2(20 - \theta)$$

where t is the time in minutes.

(a) Find θ in terms of t given that when $t = 0$, $\theta = 5\ °C$.
(b) How long does it take for the liquid to reach 18 °C?

9 By using the substitution $t = 1 + 2x$ show that

$$\int_0^1 \frac{x}{(1 + 2x)^2}\ dx = \tfrac{1}{4}\ln 3 - \tfrac{1}{6}.$$

The diagram shows part of the curve $y = \dfrac{4\sqrt{x}}{1 + 2x}$.

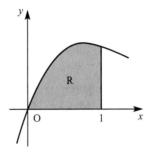

The region R (shown shaded) is bounded by the curve, the x-axis and the line $x = 1$. Find the volume generated when R is rotated about the x-axis.

10 Use integration by parts to find the exact values of:

(a) $\displaystyle\int_1^2 \ln x\ dx$

(b) $\displaystyle\int_1^2 x\ln x\ dx$

(c) $\displaystyle\int_1^2 x^2\ln x\ dx$.

11 $f(x) \equiv \dfrac{5x^2 - 8x + 1}{2x(x - 1)^2} \equiv \dfrac{A}{x} + \dfrac{B}{x - 1} + \dfrac{C}{(x - 1)^2}.$

(a) Find the values of constants A, B and C.
(b) Hence find $\int f(x)\ dx$.
(c) Hence show that

$$\int_4^9 f(x)\ dx = \ln\left(\tfrac{32}{3}\right) - \tfrac{5}{24}.$$

[Edexcel]

EXERCISE 10R

12 (a) Show that $\dfrac{3}{x-4} - \dfrac{x}{x^2+1} = \dfrac{2x^2+4x+3}{(x^2+1)(x-4)}$.

   (b) Find $\displaystyle\int \dfrac{2x^2+4x+3}{(x^2+1)(x-4)}\,dx$.

   (c) Solve $\dfrac{dy}{dx} = \dfrac{(2x^2+4x+3)y}{(x^2+1)(x-4)}$ given that $y = 128$ when $x = 0$, expressing your answer for y in terms of x.

13 (a) By first writing $\dfrac{6x}{1+3x}$ in the form $A + \dfrac{B}{1+3x}$, find $\displaystyle\int \dfrac{6x}{1+3x}\,dx$.

   (b) Using the substitution $u = 1 + 3x$, find $\displaystyle\int \dfrac{6x}{(1+3x)^2}\,dx$.

   (c) Show, using integration by parts, that
   $$\int \dfrac{6x}{(1+3x)^2}\,dx = -\dfrac{2x}{1+3x} + \tfrac{2}{3}\ln(1+3x) + c.$$

   (d) Prove that the answer to part (c) is equivalent to that found in part (b).

14 (a) Use the identities for $\cos(A+B)$ and $\cos(A-B)$ to prove that
   (i) $2\cos A \cos B \equiv \cos(A+B) + \cos(A-B)$,
   (ii) $\cos^2 A \equiv \tfrac{1}{2}(1 + \cos 2A)$.

   (b) Find $\displaystyle\int \cos 3x \cos x\,dx$.

   (c) Use the substitution $x = \cos t$ to evaluate
   $$\int_0^{\frac{1}{2}} \dfrac{x^2}{(1-x^2)^{\frac{1}{2}}}\,dx.$$
   [Edexcel]

15 The diagram shows part of the curve with equation $y = 1 + \dfrac{c}{x}$, where c is a positive constant.

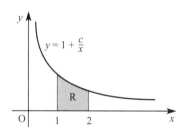

The point P with x-coordinate p lies on the curve. Given that the gradient of the curve at P is -4,

(a) show that $c = 4p^2$.

Given also that the y-coordinate of P is 5,

(b) prove that $c = 4$.

The region R is bounded by the curve, the x-axis and the lines $x = 1$ and $x = 2$, as shown in the diagram. The region R is rotated through $360°$ about the x-axis.

(c) Show that the volume of the solid generated can be written in the form $\pi(k + q\ln 2)$, where k and q are constants to be found.

16 A curve C is given parametrically by the equations

$$x = 5\cos\theta, \quad y = 2\sin\theta, \quad 0 \leq \theta \leq 2\pi.$$

(a) The curve, from $\theta = 0$ to $\theta = \pi$, is rotated 2π radians about the x-axis. Show that the volume of revolution is given by:

$$V = 20\pi \int_0^\pi (\cos^2\theta \sin\theta - \sin\theta) \, d\theta.$$

Hence find this volume leaving your answer in terms of π.

(b) By eliminating θ from the parametric equations find an expression for y^2 in terms of x. Hence show that the volume is

$$V = \frac{4\pi}{25} \int_a^b (25 - x^2) \, dx$$

where the values of a and b are to be stated.

Evaluate this integral and show that the answer is the same as that found in part **(a)**.

17 A curve C has parametric equations $x = 2at$, $y = at^2$, where a is a constant.

(a) Show that the volume of revolution obtained when the curve C, from $t = 0$ to $t = 1$, is rotated $360°$ about the x-axis is

$$2\pi a^3 \int_0^1 x^4 \, dx$$

and find this volume in terms of a and π.

(b) By eliminating t from the equations find an expression for y in terms of x. Hence obtain the following expression for the volume of revolution:

$$\frac{\pi}{16a^2} \int_0^{2a} x^4 \, dx$$

Show that this gives the same answer as that in part **(a)**.

18 The speed, v m s^{-1}, of a lorry at time t seconds is modelled by

$$v = 5(e^{0.1t} - 1) \sin(0.1t), \quad 0 \leq t \leq 30.$$

(a) Copy and complete the following table, showing the speed of the lorry at 5-second intervals. Use radian measure for $0.1t$ and give your values of v to 2 decimal places where appropriate.

t	0	5	10	15	20	25
v		1.56	7.23	17.36		

(b) Verify that, according to this model, the lorry is moving more slowly at $t = 25$ than at $t = 24.5$.

The distance, s metres, travelled by the lorry during the first 25 seconds is given by $s = \int_0^{25} v \, dt$.

(c) Estimate s by using the trapezium rule with all the values from your table.

EXERCISE 10R

19 The diagram shows the cross-section of a road tunnel and its concrete surround.

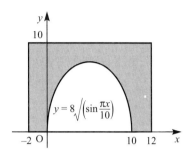

The curved section of the tunnel is modelled by the curve with equation $y = 8\sqrt{\left(\sin\frac{\pi x}{10}\right)}$, in the interval $0 \leqslant x \leqslant 10$. The concrete surround is represented by the shaded area bounded by the curve, the x-axis and the lines $x = -2$, $x = 12$ and $y = 10$. The units on both axes are metres.

(a) Using this model, copy and complete the table below, giving the values of y to 2 decimal places.

x	0	2	4	6	8	10
y	0	6.13				0

The area of the cross-section of the tunnel is given by $\int_0^{10} y \, dx$.

(b) Estimate this area, using the trapezium rule with all the values from your table.
(c) Deduce an estimate of the cross-sectional area of the concrete surround.
(d) State, with a reason, whether your answer in part (c) overestimates or underestimates the true value.
(e) How would you improve the accuracy of your estimation of the cross-sectional area?

[Edexcel (adapted)]

20 A student tests the accuracy of the trapezium rule by evaluating I, where

$$I = \int_{0.5}^{1.5} \left(\frac{3}{x} + x^4\right) dx.$$

(a) Complete the student's table, giving values to 2 decimal places where appropriate.

x	0.5	0.75	1	1.25	1.5
$\frac{3}{x} + x^4$	6.06	4.32			

(b) Use the trapezium rule, with all the values from your table, to calculate an estimate for the value of I.
(c) Use integration to calculate the exact value of I.
(d) Verify that the answer obtained by the trapezium rule is within 3% of the exact value.

[Edexcel]

KEY POINTS

1. **Integrals by inspection**

    $\int x^n \, dx = \dfrac{1}{n+1} x^{n+1} + c$ provided $n \neq -1$

    $\int (ax+b)^n \, dx = \dfrac{1}{a} \times \dfrac{1}{n+1} (ax+b)^{n+1} + c$

    $\int e^x \, dx = e^x + c, \quad \int e^{ax+b} \, dx = \dfrac{1}{a} e^{ax+b} + c$

    $\int \dfrac{1}{x} \, dx = \ln|x| + c, \quad \int \dfrac{1}{ax+b} \, dx = \dfrac{1}{a} \ln|ax+b| + c$

    $\int \dfrac{f'(x)}{f(x)} \, dx = \ln|f(x)| + c$

2. **Trigonometric functions**

    $\int \cos x \, dx = \sin x + c \qquad \int \sec^2 x \, dx = \tan x + c$

    $\int \sin x \, dx = -\cos x + c \qquad \int \cot x \, dx = \ln|\sin x| + c$

    $\int \tan x \, dx = \ln|\sec x| + c$

3. **Integration by substitution**

    To find $\int_a^b f(g(x)) \, dx$ put $u = g(x)$, replace dx in the integral and change the limits.

    Note: the new integral must be in u only (i.e. have no x terms).
    If there are no limits you must replace u by $g(x)$ for your final answer.

4. **Integration by parts**

    $\int u \dfrac{dv}{dx} \, dx = uv - \int v \dfrac{du}{dx} \, dx$

5. **Partial fractions**

    Some functions may be integrated by first splitting them into partial fractions.

6. **Volumes of revolution**

    About the x-axis:

    Cartesian: $V = \pi \int_a^b y^2 \, dx$

    Parametric: $V = \pi \int_{t=t_1}^{t=t_2} (g(t))^2 \times f'(t) \, dt$ where $x = f(t), y = g(t)$.

7 **Solving differential equations by separating the variables**

Treat the equation as a fraction and rewrite it to give equations of the type

$$\int g(y)\, dy = \int f(x)\, dx.$$

The general solution has a constant of integration (c).

The particular solution is found by evaluating the constant of integration using the given information.

8 **Numerical integration**

Trapezium rule:

$$\text{Area} = \int_a^b y\, dx \approx \tfrac{1}{2}h\{y_0 + y_n + 2(y_1 + y_2 + \ldots + y_{n-1})\}$$

where $h = \dfrac{b - a}{n}$

Error $\approx kh^2$ where k is a constant.

Chapter eleven

VECTORS

The true genius is a mind of large general powers, accidentally determined to some particular direction.

Samuel Johnson

A quantity which has both size and direction is called a *vector*. The velocity of an aircraft through the sky is an example of a vector, having size (e.g. 600 mph) and direction (on a course of 254°). By contrast the mass of the aircraft (100 tonnes) is completely described by its size and no direction is associated with it; such a quantity is called a *scalar*.

Vectors are used extensively in mechanics to represent quantities such as force, velocity and momentum, and in geometry to represent displacements. They are an essential tool in three-dimensional coordinate geometry and it is this application of vectors which is the subject of this chapter. However before coming on to this, you need to be familiar with the associated vocabulary and notation, in two and three dimensions.

VECTORS IN TWO DIMENSIONS

TERMINOLOGY

In two dimensions, it is common to represent a vector by a drawing of a straight line with an arrowhead. The length represents the size, or magnitude, of the vector and the direction is indicated by the line and the arrowhead. Direction is usually given as the angle the vector makes with the positive *x*-axis, with the anticlockwise direction taken to be positive.

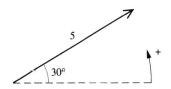

FIGURE 11.1

The vector in figure 11.1 has magnitude 5, direction +30°. This is written (5, 30°) and said to be in *magnitude–direction form* or in *polar form*. The general form of a vector written in this way is (r, θ) where r is its magnitude and θ its direction.

Note

In the special case when the vector is representing real travel, as in the case of the velocity of an aircraft, the direction may be described by a compass bearing with the angle measured from north, clockwise. However, this is not done in this chapter; all directions are taken to be measured anticlockwise from the positive *x*-direction.

An alternative way of describing a vector is in terms of *components* in given directions. The vector in figure 11.2 is 4 units in the *x*-direction, and 2 in the *y*-direction, and this is denoted by $\binom{4}{2}$.

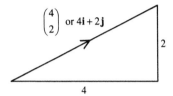

FIGURE 11.2

This may also be written as 4**i** + 2**j**, where **i** is a vector of magnitude 1, a unit vector, in the *x*-direction and **j** is a unit vector in the *y*-direction (figure 11.3).

FIGURE 11.3

In a book, a vector may be printed in bold, for example **a** or **OA**, or as a line between two points with an arrow above it to indicate its direction, such as $\overrightarrow{OA}$. When you write a vector by hand, it is usual to put a wavy line under it, for example $\underset{\sim}{a}$, or to put an arrow above it, as in $\overrightarrow{OA}$.

To convert a vector from component form to magnitude–direction form, or vice-versa, is just a matter of applying trigonometry to a right-angled triangle.

In general, the vector **a** with components *x* and *y* is denoted by $\binom{x}{y}$ and it has length or magnitude |**a**| and direction θ with the positive *x*-axis (figure 11.4), where

> magnitude $|\mathbf{a}| = \sqrt{x^2 + y^2}$
>
> direction $\theta = \tan^{-1}\left(\frac{y}{x}\right)$.

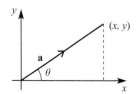

FIGURE 11.4

EXAMPLE 11.1

Write the vector **a** = 4**i** + 2**j** in magnitude–direction form.

Solution

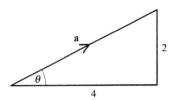

FIGURE 11.5

The magnitude of **a** is given by the length |**a**| in figure 11.5:

$$|\mathbf{a}| = \sqrt{4^2 + 2^2} \quad \text{(using Pythagoras' theorem)}$$

$$= 4.47 \quad \text{(to 3 significant figures)}$$

The direction is given by the angle θ:

$$\tan \theta = \tfrac{2}{4} = 0.5$$
$$\theta = 26.6° \quad \text{(to 3 significant figures)}$$

The vector **a** is (4.47, 26.6°).

The magnitude of a vector is also called its modulus and is denoted by the symbols | |. In the example **a** = 4**i** + 2**j**, the modulus of **a**, written |**a**|, is 4.47. Another convention for writing the magnitude of a vector is to use the same letter, but in italics and not bold type; thus the magnitude of **a** may be written *a*.

EXAMPLE 11.2

Write the vector (5, 60°) in component form.

Solution In the right-angled triangle OPX

OX = 5cos60° = 2.5
XP = 5sin60° = 4.33   (to 2 decimal places).

$\overrightarrow{OP}$ is $\begin{pmatrix} 2.5 \\ 4.33 \end{pmatrix}$ or 2.5**i** + 4.33**j**.

This technique can be written as a general rule, for all values of θ:

$$(r, \theta) \Rightarrow \begin{pmatrix} r\cos\theta \\ r\sin\theta \end{pmatrix} \text{ or } (r\cos\theta)\mathbf{i} + (r\sin\theta)\mathbf{j}$$

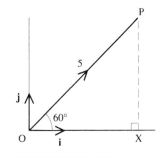

FIGURE 11.6

EXAMPLE 11.3

Write the vector (10, 290°) in component form.

Solution   In this case $r = 10$ and $\theta = 290°$.

$$(10, 290°) \Rightarrow \begin{pmatrix} 10\cos 290° \\ 10\sin 290° \end{pmatrix} = \begin{pmatrix} 3.42 \\ -9.40 \end{pmatrix} \text{ to 2 decimal places}$$

This may also be written $3.42\mathbf{i} - 9.40\mathbf{j}$.

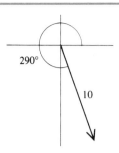

FIGURE 11.7

Note

In Example 11.3 the signs looked after themselves. The component in the **i** direction came out positive and that in the **j** direction negative, as must be the case for a direction in the fourth quadrant (270° < θ < 360°). This will always be the case when the conversion is from magnitude–direction form into component from.

The situation is not quite so straightforward when the conversion is carried out the other way, from component form to magnitude–direction form. In that case, it is best to draw a diagram and use it to see the approximate size of the angle required. This is shown in the next example.

EXAMPLE 11.4

Write $-5\mathbf{i} + 4\mathbf{j}$ in magnitude–direction form.

Solution

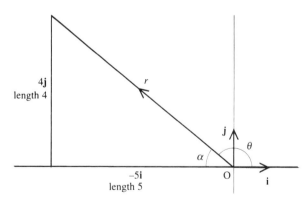

FIGURE 11.8

In this case, the magnitude $r = \sqrt{5^2 + 4^2} = \sqrt{41}$
$\phantom{\text{In this case, the magnitude } r } = 6.40$ (to 2 decimal places).

The direction is given by the angle θ in the diagram, but first find the angle α:

$\tan \alpha = \frac{4}{5} \Rightarrow \alpha = 38.7°$ (to nearest 0.1°)
so $\theta = 180 - \alpha = 141.3°$

The vector is (6.40, 141.3°) in magnitude–direction form.

Using your calculator

Most graphical calculators include the facility to convert from polar coordinates (r, θ) to rectangular coordinates (x, y), and vice versa. This is the same as converting one form of a vector into the other. Once you are clear what is involved, you will probably prefer to do such conversions on your calculator.

EQUAL VECTORS

The statement that two vectors **a** and **b** are equal means two things:

1. The direction of **a** is the same as the direction of **b**.
2. The magnitude of **a** is the same as the magnitude of **b**.

If the vectors are given in component form, each component of **a** equals the corresponding component of **b**.

POSITION VECTORS

Saying the vector **a** is given by $4\mathbf{i} + 2\mathbf{j}$ tells you the components of the vector, or equivalently its magnitude and direction. It does not tell you where the vector is situated; indeed it could be anywhere.

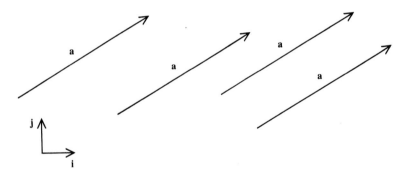

FIGURE 11.9

All of the lines in figure 11.9 represent the vector **a**.

There is, however, one special case which is an exception to the rule, that of a vector which starts at the origin. This is called a *position vector*. Thus the line joining the origin to the point (3, 5) is the position vector $\binom{3}{5}$ or $3\mathbf{i} + 5\mathbf{j}$. Another way of expressing this is to say that the point (3, 5) has the position vector $\binom{3}{5}$.

EXAMPLE 11.5

Points L, M and N have coordinates (4, 3), (−2, −1) and (2, 2).

(a) Write down, in component form, the position vector of L and the vector $\overrightarrow{MN}$.

(b) What do your answers to part (a) tell you about the lines OL and MN?

EXERCISE 11A   Pure Mathematics: Core 4

Solution **(a)** The position vector of L is $\vec{OL} = \begin{pmatrix} 4 \\ 3 \end{pmatrix}$.

The vector $\vec{MN}$ is also $\begin{pmatrix} 4 \\ 3 \end{pmatrix}$ (see figure 11.10).

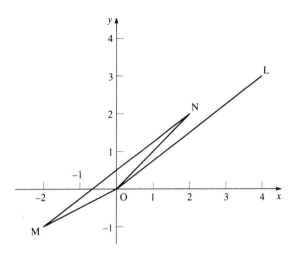

FIGURE 11.10

(b) Since $\vec{OL} = \vec{MN}$, lines OL and MN are parallel and equal in length.

Note A line joining two points, like MN in figure 11.10, is often called a *line segment*, meaning that it is just that particular part of the infinite straight line that passes through those two points.

EXERCISE 11A **1** Express the following vectors in component form (they are all drawn to the same scale).

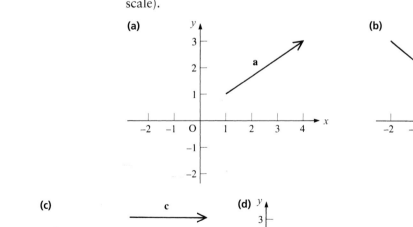

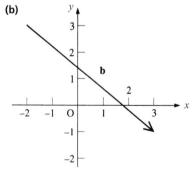

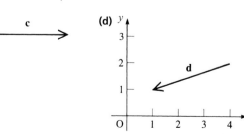

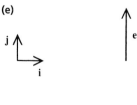

2 Draw diagrams to show these vectors and then write them in magnitude–direction form. You may find it helpful to use your calculator to check your answers.
   (a) $2\mathbf{i} + 3\mathbf{j}$
   (b) $\begin{pmatrix} 3 \\ -2 \end{pmatrix}$
   (c) $\begin{pmatrix} -4 \\ -4 \end{pmatrix}$
   (d) $-\mathbf{i} + 2\mathbf{j}$
   (e) $3\mathbf{i} - 4\mathbf{j}$

3 Draw diagrams to show these vectors and then write them in component form. You may find it helpful to use your calculator to check your answers.
   (a) $(5, 45°)$
   (b) $(10, 210°)$
   (c) $\left(4, \frac{\pi}{2}\right)$
   (d) $(8, 2\pi)$
   (e) $\left(4, \frac{5\pi}{4}\right)$

4 Write, in component form, the vectors represented by the line segments joining:
   (a) $(2, 3)$ to $(4, 1)$
   (b) $(4, 0)$ to $(6, 0)$
   (c) $(0, 0)$ to $(0, -4)$
   (d) $(0, -4)$ to $(0, 0)$
   (e) $(-3, -4)$ to $(-4, -3)$
   (f) $(-4, -3)$ to $(-3, -4)$
   (g) $(0, 0)$ to $(8, 0)$
   (h) $(8, 0)$ to $(0, 0)$
   (i) $(3, 1)$ to $(5, -3)$
   (j) $(3, -1)$ to $(7, 3)$

5 The points A, B and C have coordinates $(2, 3)$, $(0, 4)$ and $(-2, 1)$.
   (a) Write down the position vectors of A and C.
   (b) Write down the vectors of the line segments joining AB and CB.
   (c) What do your answers to parts (a) and (b) tell you about
      (i) AB and OC?
      (ii) CB and OA?
   (d) Describe the quadrilateral OABC.

MULTIPLYING A VECTOR BY A SCALAR

When a vector is multiplied by a number (a scalar) its length is altered but its direction remains the same if the scalar is positive, and reversed if the scalar is negative.

FIGURE 11.11

The vector **2a** in figure 11.11 is twice as long as the vector **a** but in the same direction.

When the vector is in component form, each component is multiplied by the number. For example:

Vectors in two dimensions

$$2 \times (3i - 5j) = 6i - 10j$$

$$2 \times \begin{pmatrix} 3 \\ -5 \end{pmatrix} = \begin{pmatrix} 6 \\ -10 \end{pmatrix}$$

The negative of a vector

In figure 11.12 the vector −**a** has the same length as the vector **a** but in the opposite direction.

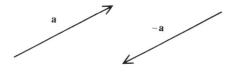

FIGURE 11.12

When **a** is given in component form, the components of −**a** are the same as those for **a** but with their signs reversed. So:

$$-\begin{pmatrix} 23 \\ -11 \end{pmatrix} = \begin{pmatrix} -23 \\ +11 \end{pmatrix}$$

Adding vectors

When vectors are given in component form, they can be added component by component. This process can be seen geometrically by drawing them on graph paper, as in the example below.

EXAMPLE 11.6

Add the vectors 2**i** − 3**j** and 3**i** + 5**j**.

Solution   2**i** − 3**j** + 3**i** + 5**j** = 5**i** + 2**j**

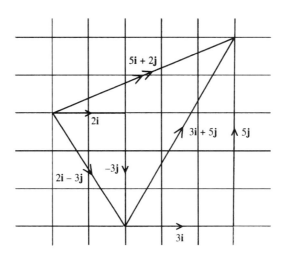

FIGURE 11.13

The sum of two (or more vectors) is called the *resultant* and is usually indicated by being marked with two arrowheads.

Adding vectors is like adding the legs of a journey to find its overall outcome (figure 11.14).

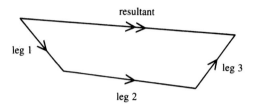

FIGURE 11.14

When vectors are given in magnitude–direction form, you can find their resultant by making a scale drawing, as in figure 11.14. If, however, you need to calculate their resultant, it is usually easiest to convert the vectors into component form, add component by component, and then convert the answer back to magnitude–direction form.

SUBTRACTING VECTORS

Subtracting one vector from another is the same as adding the negative of the vector.

EXAMPLE 11.7

Two vectors **a** and **b** are given by

$$\mathbf{a} = 2\mathbf{i} + 3\mathbf{j} \quad \mathbf{b} = -\mathbf{i} + 2\mathbf{j}.$$

(a) Find **a** − **b**.
(b) Draw diagrams showing **a**, **b**, **a** − **b**.

Solution (a) **a** − **b** = $(2\mathbf{i} + 3\mathbf{j}) - (-\mathbf{i} + 2\mathbf{j})$
            = $3\mathbf{i} + \mathbf{j}$

(b)

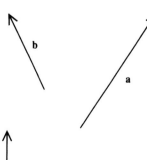
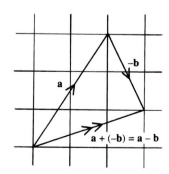

FIGURE 11.15

When you find the vector represented by the line segment joining two points, you are in effect subtracting their position vectors. If, for example, A is the point (2, 1) and B is the point (3, 5), $\overrightarrow{AB}$ is $\begin{pmatrix} 1 \\ 4 \end{pmatrix}$, as figure 11.16 shows.

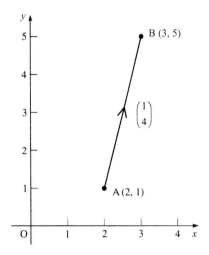

FIGURE 11.16

You find this by saying:

$$\overrightarrow{AB} = \overrightarrow{AO} + \overrightarrow{OB}$$
$$= \overrightarrow{OB} - \overrightarrow{OA}$$
$$= \mathbf{b} - \mathbf{a}$$

In this case, this gives:

$$\overrightarrow{AB} = \begin{pmatrix} 3 \\ 5 \end{pmatrix} - \begin{pmatrix} 2 \\ 1 \end{pmatrix}$$
$$= \begin{pmatrix} 1 \\ 4 \end{pmatrix}$$

as expected.

This is an important result, that

$$\overrightarrow{AB} = \overrightarrow{OB} - \overrightarrow{OA} = \mathbf{b} - \mathbf{a}$$

where **a** and **b** are the position vectors of A and B.

Geometrical figures

It is often useful to be able to express lines in a geometrical figure in terms of given vectors, as in the next example.

EXAMPLE 11.8

Figure 11.17 shows a hexagon ABCDEF. The hexagon is regular and consequently AD = 2BC. $\overrightarrow{AB} = \mathbf{p}$ and $\overrightarrow{BC} = \mathbf{q}$. Express, in terms of $\mathbf{p}$ and $\mathbf{q}$,

(a) $\overrightarrow{AC}$
(b) $\overrightarrow{AD}$
(c) $\overrightarrow{CD}$
(d) $\overrightarrow{DE}$
(e) $\overrightarrow{EF}$
(f) $\overrightarrow{BE}$

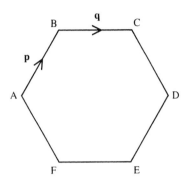

FIGURE 11.17

Solution

(a) $\overrightarrow{AC} = \overrightarrow{AB} + \overrightarrow{BC}$
    $= \mathbf{p} + \mathbf{q}$

(b) $\overrightarrow{AD} = 2\overrightarrow{BC}$
    $= 2\mathbf{q}$

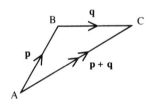

(c) Since $\overrightarrow{AC} + \overrightarrow{CD} = \overrightarrow{AD}$
    $\mathbf{p} + \mathbf{q} + \overrightarrow{CD} = 2\mathbf{q}$
    and so $\overrightarrow{CD} = \mathbf{q} - \mathbf{p}$

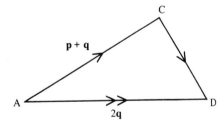

(d) $\overrightarrow{DE} = -\overrightarrow{BA}$
    $= -\mathbf{p}$

(e) $\overrightarrow{EF} = -\overrightarrow{BC}$
    $= -\mathbf{q}$

(f) $\overrightarrow{BE} = \overrightarrow{BC} + \overrightarrow{CD} + \overrightarrow{DE}$
    $= \mathbf{q} + (\mathbf{q} - \mathbf{p}) + -\mathbf{p}$
    $= 2\mathbf{q} - 2\mathbf{p}$

Notice that $\overrightarrow{BE} = 2\overrightarrow{CD}$.

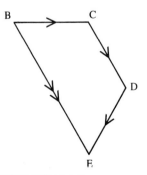

FIGURE 11.18

Unit vectors

A unit vector is a vector with a magnitude of 1, like **i** and **j**. To find the unit vector in the same direction as a given vector, divide that vector by its magnitude.

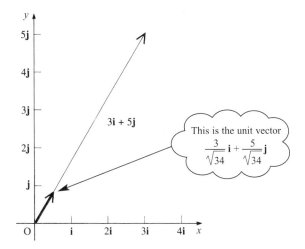

FIGURE 11.19

Thus the vector $3\mathbf{i} + 5\mathbf{j}$ (figure 11.19) has magnitude $\sqrt{3^2 + 5^2} = \sqrt{34}$, and so the vector $\frac{3}{\sqrt{34}}\mathbf{i} + \frac{5}{\sqrt{34}}\mathbf{j}$ is a unit vector. It has magnitude 1.

The unit vector in the direction of vector **a** is written as **â** and read as 'a hat'.

In general,

$$\text{if } \mathbf{a} = \begin{pmatrix} x \\ y \end{pmatrix} \quad \text{then} \quad \mathbf{\hat{a}} = \frac{1}{|\mathbf{a}|}\begin{pmatrix} x \\ y \end{pmatrix}.$$

EXERCISE 11B

1 Simplify:

(a) $\begin{pmatrix} 2 \\ 3 \end{pmatrix} + \begin{pmatrix} 4 \\ 5 \end{pmatrix}$

(b) $\begin{pmatrix} 2 \\ -1 \end{pmatrix} + \begin{pmatrix} -1 \\ 2 \end{pmatrix}$

(c) $\begin{pmatrix} 3 \\ 4 \end{pmatrix} + \begin{pmatrix} -3 \\ -4 \end{pmatrix}$

(d) $3\begin{pmatrix} 2 \\ 1 \end{pmatrix} + 2\begin{pmatrix} 1 \\ -2 \end{pmatrix}$

(e) $6(3\mathbf{i} - 2\mathbf{j}) - 9(2\mathbf{i} - \mathbf{j})$

2 The vectors **p**, **q** and **r** are given by

$$\mathbf{p} = 3\mathbf{i} + 2\mathbf{j} \quad \mathbf{q} = 2\mathbf{i} + 2\mathbf{j} \quad \mathbf{r} = -3\mathbf{i} - \mathbf{j}.$$

Find, in component form, the following vectors.

(a) $\mathbf{p} + \mathbf{q} + \mathbf{r}$

(b) $\mathbf{p} - \mathbf{q}$

(c) $\mathbf{p} + \mathbf{r}$

(d) $3(\mathbf{p} - \mathbf{q}) + 2(\mathbf{p} + \mathbf{r})$

(e) $4\mathbf{p} - 3\mathbf{q} + 2\mathbf{r}$

3 In the diagram, PQRS is a parallelogram and $\vec{PQ}$ = a, $\vec{PS}$ = b.

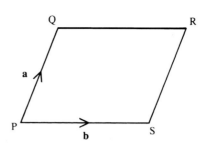

(a) Write, in terms of **a** and **b**, the following vectors.

(i) $\vec{QR}$ (ii) $\vec{PR}$ (iii) $\vec{QS}$

(b) The mid-point of PR is M. Find (i) $\vec{PM}$, (ii) $\vec{QM}$.

(c) Explain why this shows you that the diagonals of a parallelogram bisect each other.

4 In the diagram, ABCD is a kite. AC and BD meet at M.

$\vec{AB} = i + j$ and $\vec{AD} = i - 2j$.

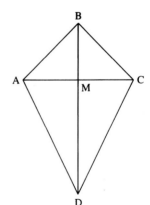

(a) Use the facts that the diagonals of a kite meet at right angles and that M is the mid-point of AC, to find the following, in terms of **i** and **j**.

(i) $\vec{AM}$ (ii) $\vec{AC}$

(iii) $\vec{BC}$ (iv) $\vec{CD}$

(b) Verify that $|\vec{AB}| = |\vec{BC}|$ and $|\vec{AD}| = |\vec{CD}|$.

EXERCISE 11B

5 In the diagram, ABC is a triangle. L, M and N are the mid-points of the sides BC, CA and AB.

$\overrightarrow{AB} = \mathbf{p}$ and $\overrightarrow{AC} = \mathbf{q}$.

(a) Find, in terms of **p** and **q**, $\overrightarrow{BC}, \overrightarrow{MN}, \overrightarrow{LM}$ and $\overrightarrow{LN}$.

(b) Explain how your results from part (a) show you that the sides of triangle LMN are parallel to those of triangle ABC, and half their lengths.

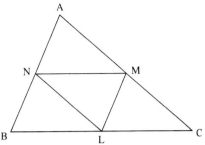

6 Find unit vectors in the same directions as the following vectors.

(a) $\begin{pmatrix} 2 \\ 3 \end{pmatrix}$

(b) $3\mathbf{i} + 4\mathbf{j}$

(c) $\begin{pmatrix} -2 \\ -2 \end{pmatrix}$

(d) $5\mathbf{i} - 12\mathbf{j}$

(e) $6\mathbf{i}$

(f) $\begin{pmatrix} -2 \\ 4 \end{pmatrix}$

(g) $\begin{pmatrix} -1 \\ 2 \end{pmatrix}$

(h) $\begin{pmatrix} 3 \\ 6 \end{pmatrix}$

(i) $\begin{pmatrix} r\cos\alpha \\ r\sin\alpha \end{pmatrix}$

(j) $\begin{pmatrix} 1 \\ \tan\beta \end{pmatrix}$

VECTORS IN THREE DIMENSIONS

POINTS

In three dimensions, a point has three coordinates, usually called x, y and z.

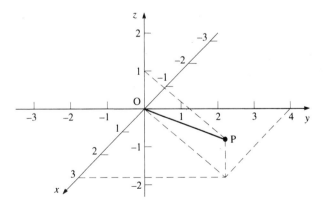

FIGURE 11.20

The axes are conventionally arranged as shown in figure 11.20, where the point P is (3, 4, 1). Even on correctly drawn three-dimensional grids, it is often hard to see the relationship between the points, lines and planes, so it is seldom worth your while trying to plot points accurately.

If the vector **a** has components x, y and z then it is written in the form $\begin{pmatrix} x \\ y \\ z \end{pmatrix}$.

Alternatively we can use the orthogonal unit vectors **i**, **j** and **k**. You saw that **i** and **j** were vectors of length one unit in the x- and y-directions. Extending these, we may include the vector **k** which is of length one unit in the z-direction. These follow the standard axes in that they are orthogonal (meaning mutually perpendicular) and are shown in figure 11.21.

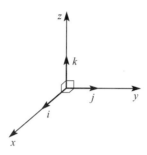

FIGURE 11.21

So the vector **a** may be written

$$\mathbf{a} = \begin{pmatrix} x \\ y \\ z \end{pmatrix} = x\mathbf{i} + y\mathbf{j} + z\mathbf{k}.$$

MAGNITUDE

The vector $\mathbf{a} = \begin{pmatrix} x \\ y \\ z \end{pmatrix}$ has, from Pythagoras' theorem, length, or magnitude

$$|\mathbf{a}| = \sqrt{x^2 + y^2 + z^2}.$$

EXAMPLE 11.9

Find the length of the vector $\begin{pmatrix} 3 \\ -4 \\ 12 \end{pmatrix}$.

Solution From Pythagoras' theorem the length of the vector is

$$\sqrt{3^2 + (-4)^2 + 12^2} = \sqrt{169}$$
$$= 13.$$

UNIT VECTORS

As in two dimensions, the vector **a** has unit vector **â** given by $\hat{\mathbf{a}} = \dfrac{\mathbf{a}}{|\mathbf{a}|}$.

EXAMPLE 11.10

Find the unit vector in the direction of $\begin{pmatrix} 2 \\ 1 \\ -2 \end{pmatrix}$.

Solution The length of the vector is

$$\sqrt{2^2 + 1^2 + (-2)^2} = \sqrt{9} = 3$$

so the unit vector is given by

$$\tfrac{1}{3}\begin{pmatrix} 2 \\ 1 \\ -2 \end{pmatrix}.$$

EXERCISE 11C

1 Simplify:

(a) $\begin{pmatrix} 1 \\ 3 \\ 5 \end{pmatrix} + \begin{pmatrix} 3 \\ -2 \\ 4 \end{pmatrix}$

(b) $\begin{pmatrix} 1 \\ 3 \\ 5 \end{pmatrix} - \begin{pmatrix} 3 \\ -2 \\ 4 \end{pmatrix}$

(c) $2\begin{pmatrix} 2 \\ -1 \\ 4 \end{pmatrix} + 3\begin{pmatrix} -1 \\ 4 \\ 0 \end{pmatrix}$

(d) $2\begin{pmatrix} 2 \\ -3 \\ -1 \end{pmatrix} - \tfrac{1}{2}\begin{pmatrix} 6 \\ -2 \\ 8 \end{pmatrix}$

(e) $5\begin{pmatrix} -3 \\ 4 \\ -1 \end{pmatrix} - 2\begin{pmatrix} -8 \\ 9 \\ -3 \end{pmatrix}$

(f) $x\begin{pmatrix} 1 \\ 0 \\ 0 \end{pmatrix} + y\begin{pmatrix} 0 \\ 1 \\ 0 \end{pmatrix} + z\begin{pmatrix} 0 \\ 0 \\ 1 \end{pmatrix}$

2 The vectors **p**, **q** and **r** are given by **p** = 4**i** − 2**j** + 3**k**, **q** = −**i** + 2**j** + 4**k** and **r** = 3**i** + 4**j** + 5**k**. Find:

(a) **p** + **q** + **r**
(b) **p** − (**q** − **r**)
(c) 2**p** + 3**q** + 4**r**
(d) 3(**q** + **r**) − **p**
(e) 4(**p** − **q**) − 3(**p** − **r**)
(f) *a***p** + *b***q** + *c***r**

3 For each of the following vectors find **(i)** their length and **(ii)** the unit vector in the same direction.

(a) $\begin{pmatrix} 2 \\ -1 \\ 2 \end{pmatrix}$

(b) $\begin{pmatrix} 4 \\ 12 \\ -3 \end{pmatrix}$

(c) $\begin{pmatrix} -4 \\ 4 \\ -2 \end{pmatrix}$

(d) 6**i** + 3**j** + 2**k**

(e) 6**i** − 7**j** + 6**k**

(f) **i** + **j** + **k**

Coordinate geometry using vectors

Two-dimensional coordinate geometry involves the study of points, given as coordinates, and lines, given as Cartesian equations. The same work may also be treated using vectors.

Since most two-dimensional problems are readily solved using the methods of Cartesian coordinate geometry, as introduced in Chapter 2 of *Pure Mathematics: Core 1 and Core 2*, why go to the trouble of relearning it all in vectors? The answer is that vector methods are very much easier than Cartesian methods to use in many three-dimensional situations.

The vector joining two points

In figure 11.22, start by looking at two points A (2, –1) and B (4, 3). The points have position vectors $\overrightarrow{OA} = \begin{pmatrix} 2 \\ -1 \end{pmatrix}$ and $\overrightarrow{OB} = \begin{pmatrix} 4 \\ 3 \end{pmatrix}$, or alternatively $2\mathbf{i} - \mathbf{j}$ and $4\mathbf{i} + 3\mathbf{j}$.

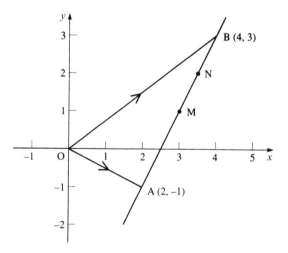

FIGURE 11.22

The vector joining A to B is $\overrightarrow{AB}$ and this is given by

$$\overrightarrow{AB} = \overrightarrow{OB} - \overrightarrow{OA}$$

$$= \begin{pmatrix} 4 \\ 3 \end{pmatrix} - \begin{pmatrix} 2 \\ -1 \end{pmatrix} = \begin{pmatrix} 2 \\ 4 \end{pmatrix}.$$

Since $\overrightarrow{AB} = \begin{pmatrix} 2 \\ 4 \end{pmatrix}$, then it follows that the length of AB is given by

$$|\overrightarrow{AB}| = \sqrt{2^2 + 4^2}$$

$$= \sqrt{20}.$$

You can find the position vectors of points on AB as follows.

The mid-point, M, has position vector $\overrightarrow{OM}$, given by

$$\overrightarrow{OM} = \overrightarrow{OA} + \tfrac{1}{2}\overrightarrow{AB}$$
$$= \begin{pmatrix} 2 \\ -1 \end{pmatrix} + \tfrac{1}{2}\begin{pmatrix} 2 \\ 4 \end{pmatrix}$$
$$= \begin{pmatrix} 3 \\ 1 \end{pmatrix}.$$

In the same way, the position vector of the point N, three-quarters of the distance from A to B, is given by

$$\overrightarrow{ON} = \begin{pmatrix} 2 \\ -1 \end{pmatrix} + \tfrac{3}{4}\begin{pmatrix} 2 \\ 4 \end{pmatrix}$$
$$= \begin{pmatrix} 3\tfrac{1}{2} \\ 2 \end{pmatrix}$$

and it is possible to find the position vector of any other point of subdivision of the line AB in the same way.

The same methods apply in three dimensions.

EXAMPLE 11.11

Given that $\overrightarrow{OA} = \begin{pmatrix} 1 \\ 3 \\ 2 \end{pmatrix}$ and $\overrightarrow{OB} = \begin{pmatrix} 3 \\ -1 \\ 8 \end{pmatrix}$, find

(a) the vector $\overrightarrow{AB}$;

(b) the position vector of the mid-point of AB.

Solution **(a)** $\overrightarrow{AB} = \overrightarrow{OB} - \overrightarrow{OA}$

So $\overrightarrow{AB} = \begin{pmatrix} 3 \\ -1 \\ 8 \end{pmatrix} - \begin{pmatrix} 1 \\ 3 \\ 2 \end{pmatrix} = \begin{pmatrix} 2 \\ -4 \\ 6 \end{pmatrix}$

(b) The position vector of the mid-point of AB is

$$\overrightarrow{OA} + \tfrac{1}{2}\overrightarrow{AB} = \begin{pmatrix} 1 \\ 3 \\ 2 \end{pmatrix} + \tfrac{1}{2}\begin{pmatrix} 2 \\ -4 \\ 6 \end{pmatrix} = \begin{pmatrix} 2 \\ 1 \\ 5 \end{pmatrix}.$$

THE DISTANCE BETWEEN TWO POINTS

From *Pure Mathematics: Core 1*, Chapter 2 you know that, in two dimensions, the distance d between the points (x_1, y_1) and (x_2, y_2) is given by

$$d^2 = (x_1 - x_2)^2 + (y_1 - y_2)^2.$$

In three dimensions the distance between (x_1, y_1, z_1) and (x_2, y_2, z_2) is

$$d^2 = (x_1 - x_2)^2 + (y_1 - y_2)^2 + (z_1 - z_2)^2.$$

Proof

Let A have the position vector $\begin{pmatrix} x_2 \\ y_2 \\ z_2 \end{pmatrix}$ and B have the position vector $\begin{pmatrix} x_1 \\ y_1 \\ z_1 \end{pmatrix}$.

Then the vector $\overrightarrow{AB}$ is

$$\overrightarrow{OB} - \overrightarrow{OA} = \begin{pmatrix} x_1 \\ y_1 \\ z_1 \end{pmatrix} - \begin{pmatrix} x_2 \\ y_2 \\ z_2 \end{pmatrix} = \begin{pmatrix} x_1 - x_2 \\ y_1 - y_2 \\ z_1 - z_2 \end{pmatrix}$$

and the magnitude of this vector is the distance d between the points A and B, so

$$d^2 = (x_1 - x_2)^2 + (y_1 - y_2)^2 + (z_1 - z_2)^2.$$

EXAMPLE 11.12

The distance between the vectors $3\mathbf{i} - 2\mathbf{j} + 5\mathbf{k}$ and $a\mathbf{i} + \mathbf{k}$ is 6 units. Find the possible values of a.

Solution

Using $\quad d^2 = (x_1 - x_2)^2 + (y_1 - y_2)^2 + (z_1 - z_2)^2$

then $\quad 36 = (3 - a)^2 + (-2 - 0)^2 + (5 - 1)^2$

$\quad 36 = (3 - a)^2 + 4 + 16$

$(3 - a)^2 = 16$

$3 - a = \pm 4$

$a = -1 \text{ or } 7$

EXERCISE 11D

1 For each of these pairs of points, A and B, write down
   (i) the vector $\overrightarrow{AB}$;
   (ii) $|\overrightarrow{AB}|$;
   (iii) the position vector of the mid-point of AB.

   (a) A is (2, 3), B is (4, 11)   (b) A is (4, 3), B is (0, 0)
   (c) A is (−2, −1), B is (4, 7)   (d) A is (−3, 4), B is (3, −4)
   (e) A is (−10, −8), B is (−5, 4)   (f) A is (1, 4, 2), B is (3, −2, 0)
   (g) A is (2, −3, 1), B is (6, 1, 3)   (h) A is (0, 5, −3), B is (4, 3, −5)
   (i) A is (6, 2, −3), B is (2, −2, 5) (j) A is (1, 2, 3), B is (3, 1, 2)

Exercise 11D

2 Find the distance between the following pairs of points.
   (a) (3, 1) and (6, 5)
   (b) (4, 6) and (9, −6)
   (c) (20, −1) and (−4, 6)
   (d) (1, 1) and (5, 5)
   (e) (3, 2) and (3, −2)
   (f) (1, 4, 3) and (3, 6, 4)
   (g) (−3, 1, 5) and (3, 4, 3)
   (h) (4, −4, 0) and (−2, 2, 7)
   (i) (1, 2, 3) and (3, 4, 1)
   (j) (1, 0, 1) and (0, 1, 0)

3 A quadrilateral has vertices A (3, 2, −4), B (5, −3, 2), C (0, 3, 4) and D (2, −1, 3). Find the length of each side of ABCD.

4 The distance between (5, −3, 6) and (a, 4, 2) is 9. Calculate the possible values of a.

5 The distance between (8, p, 5) and (−4, −2, p) is 13. Find p.

THE VECTOR EQUATION OF A STRAIGHT LINE

Figure 11.23 shows a straight line l. The point A with position vector **a** lies on l. The vector **b** is in the direction of the line.

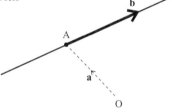

FIGURE 11.23

To get to any point on the line from the origin you go to the line along the vector **a** and then move along the line by some multiple, say t, of **b**.

Thus for any point $\mathbf{r} = \begin{pmatrix} x \\ y \\ z \end{pmatrix}$ on the line:

$$\mathbf{r} = \mathbf{a} + t\mathbf{b}$$

This is the *vector equation of a straight line* and **b** is the *direction vector* of the line.

EXAMPLE 11.13

Find the vector equation of a line passing through the point $2\mathbf{i} - 3\mathbf{j} + 4\mathbf{k}$ with direction vector $\mathbf{i} + 4\mathbf{j} - 3\mathbf{k}$.

Solution Using $\mathbf{r} = \mathbf{a} + t\mathbf{b}$, the vector equation is

$$\mathbf{r} = \begin{pmatrix} 2 \\ -3 \\ 4 \end{pmatrix} + t \begin{pmatrix} 1 \\ 4 \\ -3 \end{pmatrix}.$$

EXAMPLE 11.14

Find the vector equation of the line passing through the points A (1, 2, 3) and B (5, 4, 3).

Solution Figure 11.24 illustrates the question.

The direction vector is $\vec{AB} = \begin{pmatrix} 5 \\ 4 \\ 3 \end{pmatrix} - \begin{pmatrix} 1 \\ 2 \\ 3 \end{pmatrix} = \begin{pmatrix} 4 \\ 2 \\ 0 \end{pmatrix}$.

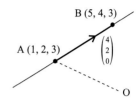

FIGURE 11.24

Hence the vector equation of the line is

$$\mathbf{r} = \begin{pmatrix} 1 \\ 2 \\ 3 \end{pmatrix} + t\begin{pmatrix} 4 \\ 2 \\ 0 \end{pmatrix}.$$

However, the direction vector $\begin{pmatrix} 4 \\ 2 \\ 0 \end{pmatrix}$ is parallel to $\begin{pmatrix} 2 \\ 1 \\ 0 \end{pmatrix}$ so we can simplify the vector equation of the straight line through A and B to

$$\mathbf{r} = \begin{pmatrix} 1 \\ 2 \\ 3 \end{pmatrix} + t\begin{pmatrix} 2 \\ 1 \\ 0 \end{pmatrix}.$$

Further, we could have used the position vector of B rather than A at the start of the right-hand side of the equation. It is possible, therefore, to have many different forms of the equation of a line.

Note

In Example 11.14 you saw how to find the vector equation of a line which passes through two given points. If, in general, these points have position vectors **c** and **d** the equation of the line may be written in the form

$\mathbf{r} = \mathbf{c} + t(\mathbf{d} - \mathbf{c})$.

EXAMPLE 11.15

Determine whether or not the point (7, 5, 2) lies on the line $\mathbf{r} = \begin{pmatrix} 1 \\ 2 \\ 3 \end{pmatrix} + t\begin{pmatrix} 2 \\ 1 \\ 0 \end{pmatrix}$.

Solution Remember that $\mathbf{r} = \begin{pmatrix} x \\ y \\ z \end{pmatrix}$. The method is to find t from the x-equation, say, and then check both the y- and z-equations.

Put $x = 7$ then $7 = 1 + 2t \Rightarrow t = 3$.

Then, substituting for t in the y and z equations:

$y = 2 + 3 \times 1 = 5$ correct

$z = 3 + 0 = 3$ incorrect

Hence $(7, 5, 2)$ does not lie on the line.

This section is not examined in the Core 4 examination. It may prove useful for further study.

CARTESIAN FORM OF THE EQUATION OF A STRAIGHT LINE

Recalling that the vector equation of a straight line is $\mathbf{r} = \mathbf{a} + t\mathbf{b}$ where $\mathbf{r} = \begin{pmatrix} x \\ y \\ z \end{pmatrix}$ and putting

$\mathbf{a} = \begin{pmatrix} a_1 \\ a_2 \\ a_3 \end{pmatrix}$ and $\mathbf{b} = \begin{pmatrix} b_1 \\ b_2 \\ b_3 \end{pmatrix}$ gives

$$\begin{pmatrix} x \\ y \\ z \end{pmatrix} = \begin{pmatrix} a_1 \\ a_2 \\ a_3 \end{pmatrix} + t \begin{pmatrix} b_1 \\ b_2 \\ b_3 \end{pmatrix}.$$

The x-equation is $x = a_1 + tb_1$ from which $t = \frac{x - a_1}{b_1}$. Making t the subject of the y- and z-equations as well gives the Cartesian equations of the straight line as:

$$\frac{x - a_1}{b_1} = \frac{y - a_2}{b_2} = \frac{z - a_3}{b_3}$$

EXAMPLE 11.16

Express $\mathbf{r} = \begin{pmatrix} 1 \\ 2 \\ 3 \end{pmatrix} + t \begin{pmatrix} 2 \\ 1 \\ -3 \end{pmatrix}$ in Cartesian form.

Solution Making t the subject of each line gives:

$$\frac{x - 1}{2} = y - 2 = \frac{z - 3}{-3}$$

It is common practice to avoid negative values in the denominator, so rewriting the equation gives:

$$\frac{x - 1}{2} = y - 2 = \frac{3 - z}{3}$$

Note If any of b_1, b_2, b_3 is zero the equation is written in two parts.

EXAMPLE 11.17

Write $\mathbf{r} = \begin{pmatrix} 1 \\ 2 \\ 3 \end{pmatrix} + t \begin{pmatrix} 2 \\ 1 \\ 0 \end{pmatrix}$ in Cartesian form.

Solution The first two lines give t in terms of x and y but the third does not involve t. So the Cartesian equations are written

$$\frac{x-1}{2} = y - 2 \quad \text{and} \quad z = 3.$$

EXAMPLE 11.18

Find the Cartesian form of the equation of the line through $(4, 2, 3)$ in the direction $\begin{pmatrix} 1 \\ 0 \\ 0 \end{pmatrix}$.

Solution The equation of the line is

$$r = \begin{pmatrix} 4 \\ 2 \\ 3 \end{pmatrix} + t \begin{pmatrix} 1 \\ 0 \\ 0 \end{pmatrix}$$

giving $x = 4 + t$, $y = 2$ and $z = 3$.

The first part $\frac{x-4}{1}$ does not really give any further information: x may take any value, and this is understood when the equation of the line is written as

$$y = 2 \quad \text{and} \quad z = 3.$$

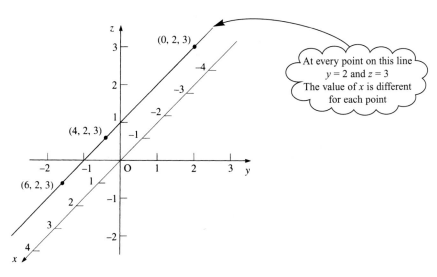

FIGURE 11.25

COORDINATE GEOMETRY USING VECTORS

The vector forms of the equations of the lines given in the last two examples are

$$\mathbf{r} = \begin{pmatrix} x \\ y \\ z \end{pmatrix} = \begin{pmatrix} 1 \\ 2 \\ 3 \end{pmatrix} + t\begin{pmatrix} 2 \\ 1 \\ 0 \end{pmatrix} \quad \text{and} \quad \mathbf{r} = \begin{pmatrix} x \\ y \\ z \end{pmatrix} = \begin{pmatrix} 4 \\ 2 \\ 3 \end{pmatrix} + t\begin{pmatrix} 1 \\ 0 \\ 0 \end{pmatrix}.$$

These are considerably simpler than the equivalent Cartesian forms. You will usually find it much easier to work with the equation of the line in vector form.

To convert from the Cartesian form to the vector form, you can reverse the procedure as in Example 11.19. Usually, however, you would just write down the answer by looking at the numbers in the three numerators and denominators.

EXAMPLE 11.19

Write the equation of this line in vector form.

$$\frac{x-5}{2} = \frac{y+1}{1} = \frac{z+3}{6}$$

Solution

$$\frac{x-5}{2} = \frac{y+1}{1} = \frac{z+3}{6} = t$$

$$\frac{x-5}{2} = t \implies x = 5 + 2t$$

$$\frac{y+1}{1} = t \implies y = -1 + t$$

$$\frac{z+3}{6} = t \implies z = -3 + 6t$$

So $\quad \mathbf{r} = \begin{pmatrix} x \\ y \\ z \end{pmatrix} = \begin{pmatrix} 5 + 2t \\ -1 + t \\ -3 + 6t \end{pmatrix}$

which is written

$$\mathbf{r} = \begin{pmatrix} 5 \\ -1 \\ -3 \end{pmatrix} + t\begin{pmatrix} 2 \\ 1 \\ 6 \end{pmatrix}.$$

This line passes through $(5, -1, -3)$ in the direction $\begin{pmatrix} 2 \\ 1 \\ 6 \end{pmatrix}$.

STRAIGHT LINE IN TWO DIMENSIONS

The vector form is the same as for three dimensions, that is

$$\mathbf{r} = \mathbf{a} + t\mathbf{b},$$

but in this case $\mathbf{r} = \begin{pmatrix} x \\ y \end{pmatrix}$, $\mathbf{a} = \begin{pmatrix} a_1 \\ a_2 \end{pmatrix}$ and $\mathbf{b} = \begin{pmatrix} b_1 \\ b_2 \end{pmatrix}$.

The Cartesian form is found by making t the subject of the x- and y-equations, giving

$$\frac{x - a_1}{b_1} = \frac{y - a_2}{b_2}$$

This can be rearranged to make y the subject:

$$y = \frac{b_2(x - a_1)}{b_1} + a_2 = \frac{b_2}{b_1}x + \frac{a_2 b_1 - a_1 b_2}{b_1}$$

which is equivalent to the more familiar $y = mx + c$.

EXAMPLE 11.20

Write $\mathbf{r} = \begin{pmatrix} 2 \\ -1 \end{pmatrix} + t\begin{pmatrix} 2 \\ 4 \end{pmatrix}$ in Cartesian form.

Solution Write $\mathbf{r}$ as $\begin{pmatrix} x \\ y \end{pmatrix}$, so the equation of the line becomes:

$$\begin{pmatrix} x \\ y \end{pmatrix} = \begin{pmatrix} 2 \\ -1 \end{pmatrix} + t\begin{pmatrix} 2 \\ 4 \end{pmatrix}$$

or
$$x = 2 + 2t$$
$$y = -1 + 4t$$

The last two equations can be rewritten as:

$$\frac{x - 2}{2} = t \quad \text{and} \quad \frac{y + 1}{4} = t$$

$$\Rightarrow \quad \frac{x - 2}{2} = \frac{y + 1}{4}$$

The equation is now in Cartesian form and may be tidied up to give $y = 2x - 5$.

When converting from Cartesian form to vector form, you need first to find any point on the line, and then to convert the gradient into a vector with the same direction, as shown in the following example.

EXERCISE 11E

EXAMPLE 11.21

Write $y = \tfrac{1}{3}x + 2$ in vector form.

Solution First find any point on the line. For example, when $x = 0$, $y = 2$ and so the point $(0, 2)$ with position vector $\binom{0}{2}$ is on the line. Then convert the gradient into a vector with the same direction. The equation of the line is of the form $y = mx + c$ and so its gradient m is $\tfrac{1}{3}$.

The vector $\binom{3}{1}$ has gradient $\tfrac{1}{3}$.

So a vector equation of the line is

$$\mathbf{r} = \binom{0}{2} + t\binom{3}{1}.$$

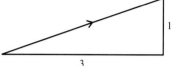

FIGURE 11.26

Note Remember, vector equations can be written in many ways.

EXERCISE 11E

1. Find the equation of each of these lines in vector form.
   (a) joining $(2, 1)$ to $(4, 5)$
   (b) joining $(3, 5)$ to $(0, 8)$
   (c) joining $(-6, -6)$ to $(4, 4)$
   (d) through $(5, 3)$ in the same direction as $\mathbf{i} + \mathbf{j}$
   (e) through $(2, 1)$ parallel to $6\mathbf{i} + 3\mathbf{j}$
   (f) through $(0, 0)$ parallel to $\binom{-1}{4}$
   (g) joining $(0, 0)$ to $(-2, 8)$
   (h) joining $(3, -12)$ to $(-1, 4)$

2. Write these lines in Cartesian form.
   (a) $\mathbf{r} = \binom{1}{2} + t\binom{1}{3}$
   (b) $\mathbf{r} = \binom{-2}{0} + t\binom{-2}{-1}$
   (c) $\mathbf{r} = \binom{1}{0} + t\binom{4}{4}$
   (d) $\mathbf{r} = \binom{4}{3} + t\binom{1}{1}$
   (e) $\mathbf{r} = \binom{2}{5} + t\binom{4}{0}$

3. Write these lines in vector form.
   (a) $y = 2x + 3$
   (b) $y = x - 4$
   (c) $y = \tfrac{1}{2}x - 1$
   (d) $y = -\tfrac{1}{4}x$
   (e) $x + 2y = 8$

4 Find the equations of the following lines in vector form.

(a) through (2, 4, −1) in the direction $\begin{pmatrix} 3 \\ 6 \\ 4 \end{pmatrix}$

(b) through (1, 0, −1) in the direction $\begin{pmatrix} 1 \\ 0 \\ 0 \end{pmatrix}$

(c) through (1, 0, 4) and (6, 3, −2)

(d) through (0, 0, 1) and (2, 1, 4)

(e) through (1, 2, 3) and (−2, −4, −6)

5 Write the equations of the following lines in Cartesian form.

(a) $\mathbf{r} = \begin{pmatrix} 2 \\ 4 \\ -1 \end{pmatrix} + t \begin{pmatrix} 3 \\ 6 \\ 4 \end{pmatrix}$

(b) $\mathbf{r} = \begin{pmatrix} 1 \\ 0 \\ -1 \end{pmatrix} + t \begin{pmatrix} 1 \\ 3 \\ 4 \end{pmatrix}$

(c) $\mathbf{r} = \begin{pmatrix} 3 \\ 0 \\ 4 \end{pmatrix} + t \begin{pmatrix} 1 \\ 0 \\ 2 \end{pmatrix}$

(d) $\mathbf{r} = \begin{pmatrix} 0 \\ 4 \\ 1 \end{pmatrix} + t \begin{pmatrix} 2 \\ 0 \\ 4 \end{pmatrix}$

(e) $\mathbf{r} = \begin{pmatrix} -2 \\ -7 \\ 3 \end{pmatrix} + t \begin{pmatrix} 0 \\ 1 \\ 0 \end{pmatrix}$

6 Write the equations of the following lines in vector form.

(a) $\dfrac{x-3}{5} = \dfrac{y+2}{3} = \dfrac{z-1}{4}$

(b) $\dfrac{x+6}{6} = \dfrac{y}{2} = \dfrac{z+4}{3}$

(c) $x = \dfrac{y}{2} = \dfrac{z+1}{3}$

(d) $x = y = z$

(e) $x = 2$ and $y = z$

7 The quadrilateral PQRS has vertices P (1, 4, −2), Q (8, 0, 2), R (11, −6, 4) and S (4, −2, 0).

(a) Find the vectors $\overrightarrow{PQ}, \overrightarrow{QR}, \overrightarrow{RS}, \overrightarrow{PS}$.

(b) Find $|\overrightarrow{PQ}|, |\overrightarrow{QR}|, |\overrightarrow{RS}|, |\overrightarrow{PS}|$.

(c) Find the vector equations of the lines along each side of the quadrilateral.

(d) What can you say about the quadrilateral?

8 Points A and B have position vectors given by

$$\overrightarrow{OA} = 4\mathbf{i} - \mathbf{j} + 6\mathbf{k} \quad \text{and} \quad \overrightarrow{OB} = 2\mathbf{i} + 3\mathbf{j} - 2\mathbf{k}.$$

(a) Find the vector equation of the line through A and B.

(b) Find the position vector of the point where this line crosses the x, y plane.

(c) Does the point (1, 5, 6) lie on the line?

(d) Find p and q if $\begin{pmatrix} 6 \\ p \\ q \end{pmatrix}$ lies on the line.

Coordinate Geometry Using Vectors

Intersection of Two Lines in Two Dimensions

EXAMPLE 11.22

Find the position vector of the point where the following lines intersect:

$$\mathbf{r} = \begin{pmatrix} 2 \\ 3 \end{pmatrix} + \lambda \begin{pmatrix} 1 \\ 2 \end{pmatrix} \quad \text{and} \quad \mathbf{r} = \begin{pmatrix} 6 \\ 1 \end{pmatrix} + \mu \begin{pmatrix} 1 \\ -3 \end{pmatrix}$$

Notice that different letters are used for the parameters in the two equations to avoid confusion.

Solution When the lines intersect, the position vector is the same for each of them:

$$\mathbf{r} = \begin{pmatrix} x \\ y \end{pmatrix} = \begin{pmatrix} 2 \\ 3 \end{pmatrix} + \lambda \begin{pmatrix} 1 \\ 2 \end{pmatrix} = \begin{pmatrix} 6 \\ 1 \end{pmatrix} + \mu \begin{pmatrix} 1 \\ -3 \end{pmatrix}$$

This gives two simultaneous equations for λ and μ:

$$x: \quad 2 + \lambda = 6 + \mu \quad \Rightarrow \quad \lambda - \mu = 4$$
$$y: \quad 3 + 2\lambda = 1 - 3\mu \quad \Rightarrow \quad 2\lambda + 3\mu = -2$$

Solving these gives $\lambda = 2$ and $\mu = -2$. Substituting in either equation gives:

$$\mathbf{r} = \begin{pmatrix} 4 \\ 7 \end{pmatrix}$$

which is the position vector of the point of intersection.

EXAMPLE 11.23

Find the coordinates of the point of intersection of the lines joining A (1, 6) to B (4, 0), and C (1, 1) to D (5, 3).

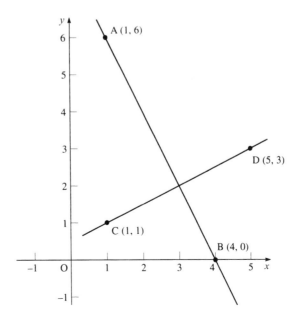

FIGURE 11.27

Solution $\overrightarrow{AB} = \begin{pmatrix} 4 \\ 0 \end{pmatrix} - \begin{pmatrix} 1 \\ 6 \end{pmatrix} = \begin{pmatrix} 3 \\ -6 \end{pmatrix}$

and so the vector equation of line AB is:

$$\mathbf{r} = \overrightarrow{OA} + \lambda \overrightarrow{AB}$$
$$\mathbf{r} = \begin{pmatrix} 1 \\ 6 \end{pmatrix} + \lambda \begin{pmatrix} 3 \\ -6 \end{pmatrix}$$

$\overrightarrow{CD} = \begin{pmatrix} 5 \\ 3 \end{pmatrix} - \begin{pmatrix} 1 \\ 1 \end{pmatrix} = \begin{pmatrix} 4 \\ 2 \end{pmatrix}$

and so the vector equation of line CD is:

$$\mathbf{r} = \overrightarrow{OC} + \mu \overrightarrow{CD}$$
$$\mathbf{r} = \begin{pmatrix} 1 \\ 1 \end{pmatrix} + \mu \begin{pmatrix} 4 \\ 2 \end{pmatrix}$$

The intersection of these lines is at

$$\mathbf{r} = \begin{pmatrix} 1 \\ 6 \end{pmatrix} + \lambda \begin{pmatrix} 3 \\ -6 \end{pmatrix} = \begin{pmatrix} 1 \\ 1 \end{pmatrix} + \mu \begin{pmatrix} 4 \\ 2 \end{pmatrix}.$$

x: $1 + 3\lambda = 1 + 4\mu \implies 3\lambda - 4\mu = 0$ ①
y: $6 - 6\lambda = 1 + 2\mu \implies 6\lambda + 2\mu = 5$ ②

Solve ① and ② simultaneously:

① : $3\lambda - 4\mu = 0$
② × 2: $12\lambda + 4\mu = 10$
Add $15\lambda = 10$

$\implies \lambda = \tfrac{2}{3}$

Substitute $\lambda = \tfrac{2}{3}$ in the equation for AB:

$\implies \mathbf{r} = \begin{pmatrix} 1 \\ 6 \end{pmatrix} + \tfrac{2}{3} \begin{pmatrix} 3 \\ -6 \end{pmatrix}$

$\implies \mathbf{r} = \begin{pmatrix} 3 \\ 2 \end{pmatrix}$

The point of intersection has coordinates (3, 2).

Note Alternatively, you could have found $\mu = \tfrac{1}{2}$ and substituted in the equation for $\overrightarrow{CD}$.

INTERSECTION OF TWO LINES IN THREE DIMENSIONS

In three-dimensional space two lines will do one of the following:

- be parallel (either separate lines or the same line, i.e. collinear);
- intersect at a point;
- be skew (i.e. they do not intersect and are not parallel).

COORDINATE GEOMETRY USING VECTORS

The method to determine which of these three situations is the case is:

1. Check whether the lines are parallel by looking at their direction vectors – if these are equal or are multiples of each other the lines will be parallel.
2. If they are parallel, check whether they are collinear by seeing whether any point on one line lies on the second line.
3. If they are not parallel solve the simultaneous equations obtained from two of the x-, y- and z-equations.
4. Having solved for two of the equations, find the value of the third variable (x, y or z). If the values are the same in both equations the lines intersect. If they are different the lines are skew.

The following examples illustrate three different situations.

EXAMPLE 11.24

The line l_1 has equation $\mathbf{r} = \begin{pmatrix} 1 \\ 2 \\ 3 \end{pmatrix} + \lambda \begin{pmatrix} 3 \\ 2 \\ 4 \end{pmatrix}$ and the line l_2 has equation $\mathbf{r} = \begin{pmatrix} 4 \\ 3 \\ 2 \end{pmatrix} + \mu \begin{pmatrix} 6 \\ 4 \\ 8 \end{pmatrix}$. Determine whether they are parallel, intersect or are skew.

Solution By inspection, the direction vectors $\begin{pmatrix} 3 \\ 2 \\ 4 \end{pmatrix}$ and $\begin{pmatrix} 6 \\ 4 \\ 8 \end{pmatrix}$ are parallel.

Pick a point on l_1, for example when $\lambda = 1$, $\mathbf{r} = \begin{pmatrix} 4 \\ 4 \\ 7 \end{pmatrix}$. This does not lie on l_2, since putting $\mu = 0$ satisfies the x-value but not the y- or z-values.

Hence l_1 and l_2 are parallel.

EXAMPLE 11.25

The line l_1 has equation $\mathbf{r} = \begin{pmatrix} 1 \\ 2 \\ 3 \end{pmatrix} + \lambda \begin{pmatrix} 3 \\ 2 \\ 4 \end{pmatrix}$ and the line l_2 has equation $\mathbf{r} = \begin{pmatrix} 5 \\ 0 \\ 1 \end{pmatrix} + \mu \begin{pmatrix} 1 \\ -4 \\ -6 \end{pmatrix}$. Determine whether they are parallel, intersect or are skew.
If they intersect find the point of intersection.

Solution By inspection, the direction vectors $\begin{pmatrix} 3 \\ 2 \\ 4 \end{pmatrix}$ and $\begin{pmatrix} 1 \\ -4 \\ -6 \end{pmatrix}$ are not parallel.

Solve the simultaneous equations in x and y:

x: $1 + 3\lambda = 5 + \mu \quad \Rightarrow \quad 3\lambda = 4 + \mu$
y: $2 + 2\lambda = -4\mu \quad \Rightarrow \quad 2\lambda = -2 - 4\mu$
$\quad \Rightarrow \quad \lambda = 1 \quad \text{and} \quad \mu = -1$

In l_1 $z = 3 + 4\lambda = 7$
and in l_2 $z = 1 - 6\mu = 7$.

These are equal so the lines intersect.

The point of intersection is found by substituting $\lambda = 1$ in l_1 or $\mu = -1$ in l_2 (it is best to do one and then check with the other), giving $(4, 4, 7)$.

EXAMPLE 11.26

The line l_1 has equation $\mathbf{r} = \begin{pmatrix} 1 \\ 2 \\ 3 \end{pmatrix} + \lambda \begin{pmatrix} 3 \\ 2 \\ 4 \end{pmatrix}$ and the line l_2 has equation $\mathbf{r} = \begin{pmatrix} 5 \\ 0 \\ 1 \end{pmatrix} + \mu \begin{pmatrix} 1 \\ -4 \\ 6 \end{pmatrix}$. Determine whether they are parallel, intersect or are skew. If they intersect, find the point of intersection.

Solution By inspection, the direction vectors $\begin{pmatrix} 3 \\ 2 \\ 4 \end{pmatrix}$ and $\begin{pmatrix} 1 \\ -4 \\ 6 \end{pmatrix}$ are not parallel.

Solve the simultaneous equations in x and y:

x: $1 + 3\lambda = 5 + \mu$ $\Rightarrow$ $3\lambda = 4 + \mu$
y: $2 + 2\lambda = -4\mu$ $\Rightarrow$ $2\lambda = -2 - 4\mu$
    $\Rightarrow$ $\lambda = 1$ and $\mu = -1$

In l_1 $z = 3 + 4\lambda = 7$
and in l_2 $z = 1 + 6\mu = -5$.

These are not equal so the lines do not intersect. That is, l_1 and l_2 are skew.

EXERCISE 11F

1 Find the position vector of the point of intersection of each of these pairs of lines.

(a) $\mathbf{r} = \begin{pmatrix} 2 \\ 1 \end{pmatrix} + \lambda \begin{pmatrix} 1 \\ 0 \end{pmatrix}$ and $\mathbf{r} = \begin{pmatrix} 3 \\ 0 \end{pmatrix} + \mu \begin{pmatrix} 1 \\ 1 \end{pmatrix}$

(b) $\mathbf{r} = \begin{pmatrix} 2 \\ -1 \end{pmatrix} + \lambda \begin{pmatrix} 1 \\ 2 \end{pmatrix}$ and $\mathbf{r} = \mu \begin{pmatrix} 1 \\ 1 \end{pmatrix}$

(c) $\mathbf{r} = \begin{pmatrix} 0 \\ 5 \end{pmatrix} + \lambda \begin{pmatrix} -2 \\ -2 \end{pmatrix}$ and $\mathbf{r} = \begin{pmatrix} 0 \\ -7 \end{pmatrix} + \mu \begin{pmatrix} 1 \\ 2 \end{pmatrix}$

(d) $\mathbf{r} = \begin{pmatrix} -2 \\ -3 \end{pmatrix} + \lambda \begin{pmatrix} -1 \\ 3 \end{pmatrix}$ and $\mathbf{r} = \begin{pmatrix} 1 \\ 3 \end{pmatrix} + \mu \begin{pmatrix} 2 \\ -1 \end{pmatrix}$

(e) $\mathbf{r} = \begin{pmatrix} 2 \\ 7 \end{pmatrix} + \lambda \begin{pmatrix} 1 \\ -1 \end{pmatrix}$ and $\mathbf{r} = \begin{pmatrix} 5 \\ 1 \end{pmatrix} + \mu \begin{pmatrix} 1 \\ 2 \end{pmatrix}$

2 In this question the origin is taken to be at a harbour and the unit vectors **i** and **j** to have lengths of 1 km in the directions E and N.

EXERCISE 11F

A cargo vessel leaves the harbour and its position vector t hours later is given by

$$\mathbf{r}_1 = 12t\mathbf{i} + 16t\mathbf{j}.$$

A fishing boat is trawling nearby and its position at time t is given by

$$\mathbf{r}_2 = (10 - 3t)\mathbf{i} + (8 + 4t)\mathbf{j}.$$

(a) How far apart are the two boats when the cargo vessel leaves harbour?
(b) How fast is each boat travelling?
(c) What happens?

3 The points A (1, 0), B (7, 2) and C (13, 7) are the vertices of a triangle. The mid-points of the sides BC, CA and AB are L, M and N.
(a) Write down the position vectors of L, M and N.
(b) Find the vector equations of the lines AL, BM and CN.
(c) Find the intersections of these pairs of lines:
  (i) AL and BM
  (ii) BM and CN
(d) What do you notice?

4 State whether or not the following pairs of lines are parallel.

(a) $\mathbf{r} = \begin{pmatrix} 1 \\ -2 \\ 0 \end{pmatrix} + \lambda \begin{pmatrix} -6 \\ 2 \\ -4 \end{pmatrix}$ and $\mathbf{r} = \begin{pmatrix} 4 \\ 2 \\ 1 \end{pmatrix} + \mu \begin{pmatrix} -3 \\ 1 \\ -2 \end{pmatrix}$

(b) $\mathbf{r} = \begin{pmatrix} 5 \\ 3 \\ 1 \end{pmatrix} + \lambda \begin{pmatrix} 4 \\ 6 \\ 2 \end{pmatrix}$ and $\mathbf{r} = \begin{pmatrix} 5 \\ 3 \\ 1 \end{pmatrix} + \mu \begin{pmatrix} 2 \\ 3 \\ -1 \end{pmatrix}$

(c) $\mathbf{r} = \begin{pmatrix} -3 \\ 5 \\ 1 \end{pmatrix} + \lambda \begin{pmatrix} 5 \\ -10 \\ 5 \end{pmatrix}$ and $\mathbf{r} = \begin{pmatrix} 3 \\ -1 \\ 4 \end{pmatrix} + \mu \begin{pmatrix} -3 \\ 6 \\ -3 \end{pmatrix}$

(d) $\mathbf{r} = \begin{pmatrix} 0 \\ 1 \\ 0 \end{pmatrix} + \lambda \begin{pmatrix} 5 \\ -3 \\ 6 \end{pmatrix}$ and $\mathbf{r} = \begin{pmatrix} 1 \\ 2 \\ 3 \end{pmatrix} + \mu \begin{pmatrix} 4 \\ -5 \\ 5 \end{pmatrix}$

(e) $\mathbf{r} = 3\mathbf{i} + 5\mathbf{j} - 2\mathbf{k} + \lambda(2\mathbf{i} - \mathbf{j} + 3\mathbf{k})$ and $\mathbf{r} = \mathbf{i} + 4\mathbf{j} - 5\mathbf{k} + \mu(2\mathbf{i} + \mathbf{j} + 3\mathbf{k})$

(f) $\mathbf{r} = -4\mathbf{i} + 3\mathbf{j} + 6\mathbf{k} + \lambda(6\mathbf{i} - 3\mathbf{j} + 15\mathbf{k})$ and $\mathbf{r} = -8\mathbf{i} + 2\mathbf{j} + 5\mathbf{k} + \mu(8\mathbf{i} - 4\mathbf{j} + 20\mathbf{k})$

5 Determine whether the following pairs of lines intersect or are skew. If they intersect find the coordinate of the point of intersection.

(a) $\mathbf{r} = \begin{pmatrix} 1 \\ 3 \\ 2 \end{pmatrix} + \lambda \begin{pmatrix} 1 \\ 2 \\ -1 \end{pmatrix}$ and $\mathbf{r} = \begin{pmatrix} 6 \\ 1 \\ -3 \end{pmatrix} + \mu \begin{pmatrix} -1 \\ 2 \\ 1 \end{pmatrix}$

(b) $\mathbf{r} = \begin{pmatrix} 2 \\ -1 \\ 5 \end{pmatrix} + s \begin{pmatrix} 3 \\ -2 \\ 4 \end{pmatrix}$ and $\mathbf{r} = \begin{pmatrix} 5 \\ -1 \\ 5 \end{pmatrix} + t \begin{pmatrix} -3 \\ 1 \\ -2 \end{pmatrix}$

(c) $\mathbf{r} = \begin{pmatrix} 1 \\ 4 \\ 6 \end{pmatrix} + \lambda \begin{pmatrix} 2 \\ 1 \\ 0 \end{pmatrix}$ and $\mathbf{r} = \begin{pmatrix} 2 \\ -3 \\ 9 \end{pmatrix} + \mu \begin{pmatrix} 1 \\ 2 \\ -3 \end{pmatrix}$

(d) $\mathbf{r} = -3\mathbf{i} + 4\mathbf{j} + \mathbf{k} + \lambda(2\mathbf{i} - 3\mathbf{j} + \mathbf{k})$ and $\mathbf{r} = 5\mathbf{i} + 2\mathbf{j} + \mathbf{k} + \mu(2\mathbf{i} + 2\mathbf{j} + \mathbf{k})$

(e) $\mathbf{r} = 2\mathbf{i} - \mathbf{k} + \lambda(3\mathbf{i} - \mathbf{j} - 4\mathbf{k})$ and $\mathbf{r} = \mathbf{i} - 5\mathbf{j} + 3\mathbf{k} + \mu(-\mathbf{i} + \mathbf{j} + \mathbf{k})$

(f) $\mathbf{r} = 6\mathbf{i} - 3\mathbf{j} + 5\mathbf{k} + \lambda(2\mathbf{i} + 3\mathbf{j} + \mathbf{k})$ and $\mathbf{r} = 11\mathbf{i} + \mathbf{j} + 2\mathbf{k} + \mu(8\mathbf{i} + 5\mathbf{j} - 7\mathbf{k})$

THE SCALAR PRODUCT

The scalar product (or dot product) of **a** and **b** is defined as

$$\mathbf{a}.\mathbf{b} = |\mathbf{a}||\mathbf{b}|\cos\theta$$

where **a** and **b** are two vectors and θ is the angle between them, as shown in figure 11.28.

FIGURE 11.28

Further, if $\mathbf{a} = \begin{pmatrix} a_1 \\ a_2 \\ a_3 \end{pmatrix}$ and $\mathbf{b} = \begin{pmatrix} b_1 \\ b_2 \\ b_3 \end{pmatrix}$ then

$$\mathbf{a}.\mathbf{b} = a_1 b_1 + a_2 b_2 + a_3 b_3 = |\mathbf{a}||\mathbf{b}|\cos\theta.$$

This result can be obtained as follows:

$$\mathbf{a}.\mathbf{b} = (a_1\mathbf{i} + a_2\mathbf{j} + a_3\mathbf{k}).(b_1\mathbf{i} + b_2\mathbf{j} + b_3\mathbf{k})$$

$$= a_1 b_1 \mathbf{i}.\mathbf{i} + a_1 b_2 \mathbf{i}.\mathbf{j} + a_1 b_3 \mathbf{i}.\mathbf{k} + a_2 b_1 \mathbf{j}.\mathbf{i} + a_2 b_2 \mathbf{j}.\mathbf{j} + a_2 b_3 \mathbf{j}.\mathbf{k} + a_3 b_1 \mathbf{k}.\mathbf{i} + a_3 b_2 \mathbf{k}.\mathbf{j} + a_3 b_3 \mathbf{k}.\mathbf{k}$$

Since **i**, **j** and **k** are mutually perpendicular the angle between any pair is 90°. But $\cos 90° = 0$ so the scalar product of any pair of **i**, **j** or **k** is zero, for example $\mathbf{i}.\mathbf{j} = \mathbf{j}.\mathbf{k} = \mathbf{k}.\mathbf{i} = 1 \times 1 \times \cos 90° = 0$.

Also $\mathbf{i}.\mathbf{i} = \mathbf{j}.\mathbf{j} = \mathbf{k}.\mathbf{k} = 1 \times 1 \times \cos 0° = 1$.

Hence:

$$\mathbf{a}.\mathbf{b} = a_1 b_1 + a_2 b_2 + a_3 b_3$$

THE SCALAR PRODUCT

PROPERTIES OF THE SCALAR PRODUCT

1. It is commutative, i.e.   **a.b** = **b.a**
2. It is distributive, i.e.   **a.**(**b** + **c**) = **a.b** + **a.c**
   This result was used to show that **a.b** = $a_1 b_1 + a_2 b_2 + a_3 b_3$.
3. If **a.b** = 0 then $\cos\theta = 0$ so $\theta = 90°$, i.e. **a** is perpendicular to **b**, provided that neither **a** nor **b** are zero vectors $0\mathbf{i} + 0\mathbf{j} + 0\mathbf{k}$.
4. If **a** is perpendicular to **b** then **a.b** = 0.
5. $\mathbf{a.a} = |\mathbf{a}||\mathbf{a}|\cos 0° = |\mathbf{a}|^2$

Since **a.b** = $a_1 b_1 + a_2 b_2 + a_3 b_3$, the scalar product is an ordinary number or *scalar*. There is another way of multiplying vectors which gives a vector as the answer; it is called the *vector product*. This appears in the Further Pure 3 unit.

DERIVATION OF THE SCALAR PRODUCT EQUATION IN THREE DIMENSIONS

To find the angle θ between the two vectors

$$\overrightarrow{OA} = \mathbf{a} = \begin{pmatrix} a_1 \\ a_2 \\ a_3 \end{pmatrix} \text{ and } \overrightarrow{OB} = \mathbf{b} = \begin{pmatrix} b_1 \\ b_2 \\ b_3 \end{pmatrix}$$

you can illustrate them on a diagram, as in figure 11.29, and then use the cosine rule.

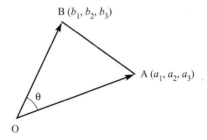

FIGURE 11.29

The cosine rule on this triangle gives:

$$AB^2 = OA^2 + OB^2 - 2\, OA\, OB \cos\theta$$

where, for example, OA is the length of the vector $\overrightarrow{OA}$, which is also $|\mathbf{a}|$.

So:

$$(b_1 - a_1)^2 + (b_2 - a_2)^2 + (b_3 - a_3)^2 = a_1^2 + a_2^2 + a_3^2 + b_1^2 + b_2^2 + b_3^2 - 2|\mathbf{a}||\mathbf{b}|\cos\theta$$

$$b_1^2 - 2a_1 b_1 + a_1^2 + b_2^2 - 2a_2 b_2 + a_2^2 + b_3^2 - 2a_3 b_3 + a_3^2$$
$$= a_1^2 + a_2^2 + a_3^2 + b_1^2 + b_2^2 + b_3^2 - 2|\mathbf{a}||\mathbf{b}|\cos\theta$$

$$-2a_1 b_1 - 2a_2 b_2 - 2a_3 b_3 = -2|\mathbf{a}||\mathbf{b}|\cos\theta$$

$$a_1 b_1 + a_2 b_2 + a_3 b_3 = |\mathbf{a}||\mathbf{b}|\cos\theta$$

But you have defined $a_1b_1 + a_2b_2 + a_3b_3$ as **a.b** so you therefore have a proof of the scalar product, that is of

$$\mathbf{a.b} = |\mathbf{a}||\mathbf{b}|\cos\theta.$$

EXAMPLE 11.27

Find the angle between the vectors $\begin{pmatrix}3\\4\end{pmatrix}$ and $\begin{pmatrix}5\\-12\end{pmatrix}$.

Solution Let $\mathbf{a} = \begin{pmatrix}3\\4\end{pmatrix} \Rightarrow |\mathbf{a}| = \sqrt{3^2 + 4^2} = 5$

and $\mathbf{b} = \begin{pmatrix}5\\-12\end{pmatrix} \Rightarrow |\mathbf{b}| = \sqrt{5^2 + (-12)^2} = 13.$

The scalar product:

$$\begin{pmatrix}3\\4\end{pmatrix} \cdot \begin{pmatrix}5\\-12\end{pmatrix} = 3 \times 5 + 4 \times (-12)$$
$$= 15 - 48$$
$$= -33$$

Substituting in $\mathbf{a.b} = |\mathbf{a}||\mathbf{b}|\cos\theta$ gives:

$$-33 = 5 \times 13 \times \cos\theta$$

$$\cos\theta = \frac{-33}{65}$$

$$\theta = 120.5°$$

EXAMPLE 11.28

The points P, Q and R are (1, 0, −1), (2, 4, 1), and (3, 5, 6). Find ∠QPR.

Solution First sketch the points P, Q and R as shown in figure 11.30.

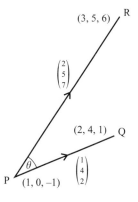

FIGURE 11.30

THE SCALAR PRODUCT

In this:

$$\vec{PQ} = \begin{pmatrix} 2 \\ 4 \\ 1 \end{pmatrix} - \begin{pmatrix} 1 \\ 0 \\ -1 \end{pmatrix} = \begin{pmatrix} 1 \\ 4 \\ 2 \end{pmatrix} \quad |\vec{PQ}| = \sqrt{1^2 + 4^2 + 2^2} = \sqrt{21}$$

Similarly:

$$\vec{PR} = \begin{pmatrix} 3 \\ 5 \\ 6 \end{pmatrix} - \begin{pmatrix} 1 \\ 0 \\ -1 \end{pmatrix} = \begin{pmatrix} 2 \\ 5 \\ 7 \end{pmatrix} \quad |\vec{PR}| = \sqrt{2^2 + 5^2 + 7^2} = \sqrt{78}$$

The scalar product is:

$$\vec{PQ}.\vec{PR} = \begin{pmatrix} 1 \\ 4 \\ 2 \end{pmatrix}.\begin{pmatrix} 2 \\ 5 \\ 7 \end{pmatrix}$$
$$= 1 \times 2 + 4 \times 5 + 2 \times 7$$
$$= 36$$

Substituting into $\vec{PQ}.\vec{PR} = |\vec{PQ}||\vec{PR}|\cos\theta$ gives:

$$36 = \sqrt{21}\ \sqrt{78}\ \cos\theta$$

$$\Rightarrow \quad \cos\theta = \frac{36}{\sqrt{21}\ \sqrt{78}}$$

$$\theta = 27.2°$$

Note

You must be careful to find the correct angle. To find ∠QPR (see figure 11.31), you need the scalar product $\vec{PQ}.\vec{PR}$. If you take $\vec{QP}.\vec{PR}$, you will obtain ∠Q'PR, which is 180° − ∠QPR. Hence the need to make a sketch to begin with.

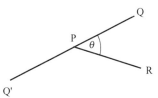

FIGURE 11.31

EXAMPLE 11.29

Show that $\begin{pmatrix} 1 \\ 3 \\ -2 \end{pmatrix}$ is perpendicular to $\begin{pmatrix} 4 \\ 2 \\ 5 \end{pmatrix}$.

Solution

$$\begin{pmatrix} 1 \\ 3 \\ -2 \end{pmatrix}.\begin{pmatrix} 4 \\ 2 \\ 5 \end{pmatrix} = 1 \times 4 + 3 \times 2 + -2 \times 5 = 4 + 6 - 10 = 0$$

Since their scalar product is zero the vectors are perpendicular.

EXAMPLE 11.30

Find a vector perpendicular to $2\mathbf{i} + 3\mathbf{j} + 4\mathbf{k}$ and $\mathbf{i} + 2\mathbf{j} + 3\mathbf{k}$.

Solution Let the perpendicular vector be $\begin{pmatrix} a \\ b \\ 1 \end{pmatrix}$.

Note that the perpendicular vector is not unique; there is an infinite set of them. You can therefore choose any number for any one of the components. It is often convenient to let the z-component be 1. If you cannot solve for a and b then $z = 0$.

Using the scalar product:

$$\begin{pmatrix} a \\ b \\ 1 \end{pmatrix} \cdot \begin{pmatrix} 2 \\ 3 \\ 4 \end{pmatrix} = 2a + 3b + 4 = 0$$

$$\begin{pmatrix} a \\ b \\ 1 \end{pmatrix} \cdot \begin{pmatrix} 1 \\ 2 \\ 3 \end{pmatrix} = a + 2b + 3 = 0$$

Solving these two simultaneous equations gives $a = 1$ and $b = -2$, and hence a vector perpendicular to $2\mathbf{i} + 3\mathbf{j} + 4\mathbf{k}$ and $\mathbf{i} + 2\mathbf{j} + 3\mathbf{k}$ is $\mathbf{i} - 2\mathbf{j} + \mathbf{k}$.

Even though the question is defined in terms of $\mathbf{i}$, $\mathbf{j}$ and $\mathbf{k}$ it is probably easier to work in column form. It is good practice to give your final answer in $\mathbf{i}$, $\mathbf{j}$, $\mathbf{k}$ form.

EXAMPLE 11.31

Find the component of $\begin{pmatrix} 1 \\ 2 \\ 2 \end{pmatrix}$ in the direction of $\begin{pmatrix} 2 \\ 3 \\ 6 \end{pmatrix}$.

Solution The component of $\begin{pmatrix} 1 \\ 2 \\ 2 \end{pmatrix}$ is the length d when $\begin{pmatrix} 1 \\ 2 \\ 2 \end{pmatrix}$ is projected on to $\begin{pmatrix} 2 \\ 3 \\ 6 \end{pmatrix}$ as shown in figure 11.32.

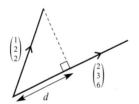

FIGURE 11.32

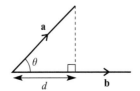

FIGURE 11.33

Figure 11.33 shows the general case for the vectors $\mathbf{a}$ and $\mathbf{b}$ where d is the component of $\mathbf{a}$ in the direction of $\mathbf{b}$. From the right-angled triangle:

$$d = |\mathbf{a}| \cos\theta$$

The scalar product $\mathbf{a}.\mathbf{b} = |\mathbf{a}||\mathbf{b}| \cos\theta$ means that:

$$d = \frac{\mathbf{a}.\mathbf{b}}{|\mathbf{b}|}$$

So, in this particular example:

$$d = \frac{1}{\sqrt{49}} \begin{pmatrix} 1 \\ 2 \\ 2 \end{pmatrix} \cdot \begin{pmatrix} 2 \\ 3 \\ 6 \end{pmatrix} = \frac{20}{7} = 2\frac{6}{7}$$

THE ANGLE BETWEEN TWO LINES

In two dimensions two distinct lines $\mathbf{r} = \mathbf{a} + \lambda\mathbf{b}$ and $\mathbf{r} = \mathbf{c} + \mu\mathbf{d}$ will meet at a point. The angle θ between the lines is the angle between their direction vectors, as shown in figure 11.34.

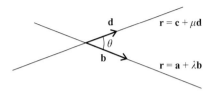

FIGURE 11.34

To find the value of θ use the scalar product.

$$\mathbf{b}.\mathbf{d} = |\mathbf{b}||\mathbf{d}|\cos\theta$$

EXAMPLE 11.32

Find the acute angle between the line l_1 with equation $\mathbf{r} = \begin{pmatrix} 1 \\ 2 \end{pmatrix} + s\begin{pmatrix} 3 \\ 1 \end{pmatrix}$ and the line l_2 with equation $\mathbf{r} = \begin{pmatrix} 3 \\ -4 \end{pmatrix} + t\begin{pmatrix} 2 \\ 4 \end{pmatrix}$.

Solution The angle between the lines is the angle between the direction vectors, so using the scalar product:

$$\begin{pmatrix} 3 \\ 1 \end{pmatrix}.\begin{pmatrix} 2 \\ 4 \end{pmatrix} = 6 + 4 = 10 = \sqrt{10} \times \sqrt{20}\cos\theta$$

$$\Rightarrow \quad \cos\theta = \frac{10}{\sqrt{200}} = \frac{1}{\sqrt{2}}$$

$$\theta = 45°$$

In three dimensions you can still specify the angle between two lines. Clearly, if the lines intersect the angle between them is, as before, equal to the angle between the direction vectors of the lines.

If the lines are skew we still define the angle between them as the angle between their direction vectors. Look at the lines l and m in figure 11.35, which do not meet. To justify taking the angle as that between their direction vectors translate m to m', a position where it intersects l. The angle between l and m' is θ, which is the angle between the direction vectors of l and m.

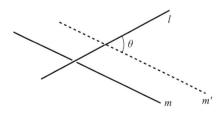

FIGURE 11.35

EXAMPLE 11.33

Find the angle between the lines

$$\mathbf{r} = \begin{pmatrix} 1 \\ 0 \\ 4 \end{pmatrix} + \lambda \begin{pmatrix} 2 \\ -1 \\ -1 \end{pmatrix} \quad \text{and} \quad \mathbf{r} = \begin{pmatrix} 2 \\ -1 \\ 3 \end{pmatrix} + \mu \begin{pmatrix} 3 \\ 0 \\ 1 \end{pmatrix}.$$

Solution The angle between the lines is the angle between their directions, $\begin{pmatrix} 2 \\ -1 \\ -1 \end{pmatrix}$ and $\begin{pmatrix} 3 \\ 0 \\ 1 \end{pmatrix}$.

Using $\quad \cos\theta = \dfrac{\mathbf{a}.\mathbf{b}}{|\mathbf{a}||\mathbf{b}|}$

$\Rightarrow \quad \cos\theta = \dfrac{2 \times 3 + (-1) \times 0 + (-1) \times 1}{\sqrt{2^2 + (-1)^2 + (-1)^2} \sqrt{3^2 + 0^2 + 1^2}}$

$\Rightarrow \quad \cos\theta = \dfrac{5}{\sqrt{6}\sqrt{10}}$

$\theta = 49.8°$

EXERCISE 11G

1. Find the angles between these vectors.
   (a) $2\mathbf{i} + 3\mathbf{j}$ and $4\mathbf{i} + \mathbf{j}$
   (b) $2\mathbf{i} - \mathbf{j}$ and $\mathbf{i} + 2\mathbf{j}$
   (c) $\begin{pmatrix} -1 \\ -1 \end{pmatrix}$ and $\begin{pmatrix} -1 \\ -2 \end{pmatrix}$
   (d) $4\mathbf{i} + \mathbf{j}$ and $\mathbf{i} + \mathbf{j}$
   (e) $\begin{pmatrix} 2 \\ 3 \end{pmatrix}$ and $\begin{pmatrix} -6 \\ 4 \end{pmatrix}$
   (f) $\begin{pmatrix} 3 \\ -1 \end{pmatrix}$ and $\begin{pmatrix} -6 \\ 2 \end{pmatrix}$

2. Points A, B, C and D are $(1, 0)$, $(9, 4)$, $(6, 1)$ and $(9, 7)$, respectively.
   (a) Write down the vector equation of line AB.
   (b) Write down the vector equation of line CD.
   (c) Find the position vector of the point of intersection.
   (d) Find the angle between the lines AB and CD.

3. The equations of the four sides AB, BC, CD, DA of a quadrilateral are:

   AB: $\mathbf{r} = \begin{pmatrix} 1 \\ 1 \end{pmatrix} + \lambda_1 \begin{pmatrix} 4 \\ 1 \end{pmatrix}$ \quad BC: $\mathbf{r} = \begin{pmatrix} 1 \\ 1 \end{pmatrix} + \lambda_2 \begin{pmatrix} 1 \\ 3 \end{pmatrix}$

   CD: $\mathbf{r} = \begin{pmatrix} 6 \\ 5 \end{pmatrix} + \lambda_3 \begin{pmatrix} 4 \\ 1 \end{pmatrix}$ \quad DA: $\mathbf{r} = \begin{pmatrix} 6 \\ 5 \end{pmatrix} + \lambda_4 \begin{pmatrix} 1 \\ 3 \end{pmatrix}$

   (a) Look carefully at the equations of the four lines and state, with reasons, what sort of quadrilateral it is.
   (b) Find the coordinates of the four vertices of the quadrilateral.
   (c) Find the internal angles of the quadrilateral.

EXERCISE 11G

4 The points A, B and C have coordinates (3, 2), (6, 3) and (5, 6), respectively.
  (a) Write down the vectors $\vec{AB}$ and $\vec{BC}$.
  (b) Show that the angle ABC is 90°.
  (c) Show that $|\vec{AB}| = |\vec{BC}|$.

  The figure ABCD is a square.

  (d) Find the coordinates of the point D.

5 Find the angles between these pairs of vectors.
  (a) $\begin{pmatrix} 2 \\ 1 \\ 3 \end{pmatrix}$ and $\begin{pmatrix} 2 \\ -1 \\ 4 \end{pmatrix}$
  (b) $\begin{pmatrix} 1 \\ -1 \\ 0 \end{pmatrix}$ and $\begin{pmatrix} 3 \\ 1 \\ 5 \end{pmatrix}$
  (c) $3\mathbf{i} + 2\mathbf{j} - 2\mathbf{k}$ and $-4\mathbf{i} - \mathbf{j} + 3\mathbf{k}$

6 Find the angles between these pairs of lines.
  (a) $\mathbf{r} = \begin{pmatrix} 2 \\ 1 \\ 3 \end{pmatrix} + \lambda \begin{pmatrix} 1 \\ 4 \\ 0 \end{pmatrix}$ and $\mathbf{r} = \begin{pmatrix} 6 \\ 10 \\ 4 \end{pmatrix} + \lambda \begin{pmatrix} 2 \\ 1 \\ 1 \end{pmatrix}$
  (b) $\mathbf{r} = \lambda \begin{pmatrix} 4 \\ 1 \\ 4 \end{pmatrix}$ and $\mathbf{r} = \begin{pmatrix} 7 \\ 0 \\ -3 \end{pmatrix} + \lambda \begin{pmatrix} 1 \\ 2 \\ -1 \end{pmatrix}$

7 The room illustrated in the diagram has rectangular walls, floor and ceiling. A string has been stretched in a straight line between the corners A and G.

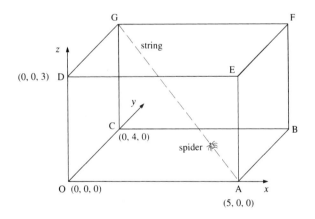

The corner O is taken as the origin. A is (5, 0, 0), C is (0, 4, 0) and D is (0, 0, 3), where the lengths are in metres.

  (a) Write down the coordinates of G.
  (b) Find the vector $\vec{AG}$ and the length of the string $|\vec{AG}|$.
  (c) Write down the equation of the line AG in vector form.

A spider walks up the string, starting from A.

(d) Find the position vector of the spider when it is at Q, one quarter of the way from A to G, and find the angle OQG.

(e) Show that when the spider is 1.5 m above the floor it is at its closest point to O, and find how far it is then from O.

[MEI]

8 The diagram shows an extension to a house. Its base and walls are rectangular and the end of its roof, EPF, is sloping, as illustrated.

(a) Write down the coordinates of A and F.
(b) Find, using vector methods, the angles FPQ and EPF.

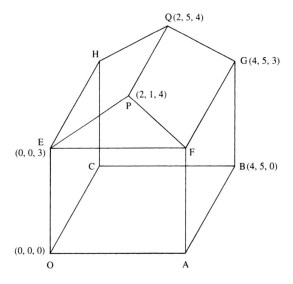

The owner decorates the room with two streamers which are pulled taut. One goes from O to G, the other from A to H. She says that they touch each other and that they are perpendicular to each other.

(c) Is she right?

9 Find a vector which is perpendicular to each of the following pairs of lines.

(a) $\mathbf{r} = \begin{pmatrix} 5 \\ 3 \\ 1 \end{pmatrix} + \lambda \begin{pmatrix} 1 \\ -2 \\ 4 \end{pmatrix}$ and $\mathbf{r} = \begin{pmatrix} 5 \\ 3 \\ 1 \end{pmatrix} + \mu \begin{pmatrix} 2 \\ 3 \\ -1 \end{pmatrix}$

(b) $\mathbf{r} = -4\mathbf{i} + 3\mathbf{j} + 6\mathbf{k} + \lambda(2\mathbf{i} - \mathbf{j} + 4\mathbf{k})$ and $\mathbf{r} = -8\mathbf{i} + 2\mathbf{j} + 5\mathbf{k} + \mu(\mathbf{i} + 3\mathbf{j} + 2\mathbf{k})$

10 A triangle has vertices A (1, 4, −2), B (8, 0, 2) and C (11, −6, 4).

(a) Find the vector equations of the lines along each side of the triangle.
(b) Find the internal angles of the triangle.

EXERCISE 11H Examination-style questions

1. The points P, Q, R have position vectors $\mathbf{p} = 3\mathbf{i} - 2\mathbf{j} + 3\mathbf{k}$, $\mathbf{q} = \mathbf{i} - 3\mathbf{j} + 5\mathbf{k}$ and $\mathbf{r} = 4\mathbf{i} + 9\mathbf{j} + \mathbf{k}$. Find:
   (a) the angle PQR;
   (b) the area of the triangle PQR.

2. The lines l_1 and l_2 have equations
   $$\mathbf{r} = \begin{pmatrix} 1 \\ 3 \\ 5 \end{pmatrix} + \lambda \begin{pmatrix} 1 \\ 2 \\ 1 \end{pmatrix} \quad \text{and} \quad \mathbf{r} = \begin{pmatrix} -3 \\ 4 \\ 2 \end{pmatrix} + \mu \begin{pmatrix} 2 \\ 1 \\ 1 \end{pmatrix}.$$
   (a) Show that $(7, 9, 7)$ lies on l_2.
   (b) Show that l_1 and l_2 are skew.
   (c) A is the point on l_1 where $\lambda = 3$. B is the point on l_2 where $\mu = 3$. Find the angle between AB and the line l_1.

3. Two lines l_1 and l_2 have equations
   $$\mathbf{r} = \mathbf{i} + 2\mathbf{k} + s(2\mathbf{i} - \mathbf{j} + 3\mathbf{k}) \text{ and } \mathbf{r} = 4\mathbf{i} + \mathbf{j} + 4\mathbf{k} + t(\mathbf{i} - \mathbf{j} + 2\mathbf{k}).$$
   (a) Show that l_1 and l_2 intersect.
   (b) Find the coordinates of the point P of intersection.
   (c) Find a vector perpendicular to both l_1 and l_2.
   (d) Find the vector equation of the line that passes through P and is perpendicular to both l_1 and l_2.

4. With respect to a fixed origin O the position vectors of the points K and L are $2\mathbf{i} - 4\mathbf{j} + 5\mathbf{k}$ and $6\mathbf{i} + 3\mathbf{k}$.
   (a) Find the position vector of the point M which is the mid-point of K and L.
   (b) Write down a vector equation of the line through O and M.
   (c) Find the angle between OM and KL.
   (d) Show that the area of the triangle OML is 9 square units.

5. The points A, B and C have position vectors
   $$\mathbf{a} = \mathbf{i} + 4\mathbf{j} + 2\mathbf{k},\ \mathbf{b} = 4\mathbf{i} + 6\mathbf{j} + \mathbf{k} \text{ and } \mathbf{c} = 8\mathbf{i} + \mathbf{j} + 3\mathbf{k}.$$
   (a) Find the vectors $\overrightarrow{AB}$ and $\overrightarrow{BC}$ and show that $\overrightarrow{AB}$ is perpendicular to $\overrightarrow{BC}$.
   (b) Find the lengths of AB and BC.
   (c) Find the angle BAC.
   (d) Hence find the area of the triangle ABC.

6 With respect to an origin O the position vectors of P and Q are

$$p = 6i - 3j + 2k \text{ and } q = -6i + 9j + 8k.$$

(a) Verify that the point S with coordinates (2, 1, 4) lies on the line through P and Q.
(b) Find the ratio of PS : PQ.
(c) Find the lengths of OP and PQ.
(d) State the length of PS.
(e) Find the area of the triangle OPS.
(f) State the area of the triangle OPQ.

7 The line l_1 has equation $r = (i + 2j + 3k) + \lambda(4i - 6j + 2k)$.
The line l_2 passes through the points (1, −2, 6) and (7, 1, −3).
(a) Find a vector equation of l_2.
(b) Show that l_1 and l_2 are skew.
(c) Find the angle between l_1 and l_2.
(d) State a translation that would move l_2 so that it would intersect with l_1.

8 Given that $a.b = |a||b|\cos\theta$ prove that

$$\begin{pmatrix}a_1\\a_2\\a_3\end{pmatrix} \cdot \begin{pmatrix}b_1\\b_2\\b_3\end{pmatrix} = a_1 b_1 + a_2 b_2 + a_3 b_3.$$

The points A (5, −3, 4) and B (7, 1, 0) are referred to Cartesian axes, origin O.
(a) Find the length of AB.
(b) Find the angle OAB.
(c) Find the angle between the line through AB and the x-axis.
(d) If the vector $\begin{pmatrix}2\\1\\p\end{pmatrix}$ is perpendicular to AB find the value of p.

9 The points A (24, 6, 0), B (30, 12, 12) and C (18, 6, 36) are referred to Cartesian axes, origin O.
(a) Find a vector equation for the line passing through the points A and B.

The point P lies on the line passing through A and B.

(b) Show that $\overrightarrow{CP}$ can be expressed as

$$(6 + t)i + tj + (2t - 36)k, \quad \text{where } t \text{ is a parameter.}$$

(c) Given that $\overrightarrow{CP}$ is perpendicular to $\overrightarrow{AB}$, find the coordinates of P.

(d) Hence, or otherwise, find the area of the triangle ABC, giving your answer to 3 significant figures.

[Edexcel]

EXERCISE 11H

10 With respect to a fixed origin O, the lines l_1 and l_2 are given by the equations

l_1: $\mathbf{r} = (2\mathbf{i} + 3\mathbf{j} - 2\mathbf{k}) + \lambda(-2\mathbf{i} + 4\mathbf{j} + \mathbf{k})$
l_2: $\mathbf{r} = (-6\mathbf{i} - 3\mathbf{j} + \mathbf{k}) + \mu(5\mathbf{i} + \mathbf{j} - 2\mathbf{k})$

where λ and μ are scalar parameters.

(a) Show that l_1 and l_2 meet and find the position vector of their point of intersection.
(b) Find, to the nearest $0.1°$, the acute angle between l_1 and l_2.

[Edexcel]

11 Referred to a fixed origin O, the points P, Q and R have position vectors $(2\mathbf{i} + \mathbf{j} + \mathbf{k})$, $(5\mathbf{j} + 3\mathbf{k})$ and $(5\mathbf{i} - 4\mathbf{j} + 2\mathbf{k})$ respectively.

(a) Find, in the form $\mathbf{r} = \mathbf{a} + t\mathbf{b}$, an equation of the line PQ.
(b) Show that the point S with position vector $(4\mathbf{i} - 3\mathbf{j} - \mathbf{k})$ lies on PQ.
(c) Show that the lines PQ and RS are perpendicular.
(d) Find the size of $\angle PQR$, giving your answer to $0.1°$.

[Edexcel]

12 With respect to an origin O, the position vectors of the points L and M are $\mathbf{i} - \mathbf{j} + 3\mathbf{k}$ and $2\mathbf{i} - 4\mathbf{j} + 2\mathbf{k}$ respectively.

(a) Write down the vector $\overrightarrow{LM}$.
(b) Show that $|\overrightarrow{OL}| = |\overrightarrow{LM}|$.
(c) Find $\angle OLM$, giving your answer to the nearest tenth of a degree.

[Edexcel]

KEY POINTS

1. A vector quantity has magnitude and direction.

2. A scalar quantity has magnitude only.

3. Vectors are typeset in bold, **a** or **OA**, or in the form $\overrightarrow{OA}$.
   They are handwritten either in the underlined form a, or as $\overrightarrow{OA}$.

4. Unit vectors in the x-, y- and z-directions are denoted by **i**, **j** and **k**, respectively.

5. A vector may be specified in
   - magnitude–direction form: (r, θ) (in two dimensions)
   - component form: $x\mathbf{i} + y\mathbf{j}$ or $\begin{pmatrix} x \\ y \end{pmatrix}$ (in two dimensions)

     $x\mathbf{i} + y\mathbf{j} + z\mathbf{k}$ or $\begin{pmatrix} x \\ y \\ z \end{pmatrix}$ (in three dimensions)

6. The length (or modulus or magnitude) of the vector **a** is written as a or as $|\mathbf{a}|$.
   - In two dimensions: if $\mathbf{a} = x\mathbf{i} + y\mathbf{j}$, $|\mathbf{a}| = \sqrt{x^2 + y^2}$.
   - In three dimensions: if $\mathbf{a} = x\mathbf{i} + y\mathbf{j} + z\mathbf{k}$, $|\mathbf{a}| = \sqrt{x^2 + y^2 + z^2}$.

7. Unit vector $\hat{\mathbf{a}} = \dfrac{\mathbf{a}}{|\mathbf{a}|} = \dfrac{1}{\sqrt{x^2 + y^2 + z^2}}(x\mathbf{i} + y\mathbf{j} + z\mathbf{k})$ and is a vector in the direction of **a** of length one unit.

8. The position vector $\overrightarrow{OP}$ of a point P is the vector joining the origin to P.

9. The vector $\overrightarrow{AB}$ is $\mathbf{b} - \mathbf{a}$, where **a** and **b** are the position vectors of A and B.

10. The vector **r** often denotes the position vector of a general point:

    $\mathbf{r} = x\mathbf{i} + y\mathbf{j} + z\mathbf{k}$

11. A vector equation of the line through a point with position vector **a** and with direction vector **b** is:

    $\mathbf{r} = \mathbf{a} + t\mathbf{b}$

12. A vector equation of the line through C and D with position vectors **c** and **d** is:

    $\mathbf{r} = \mathbf{c} + t(\mathbf{d} - \mathbf{c})$

13. A vector equation of the line through points A and B is given by:

    $\mathbf{r} = \overrightarrow{OA} + t\overrightarrow{AB}$
    $= \mathbf{a} + t(\mathbf{b} - \mathbf{a})$
    $= (1 - t)\mathbf{a} + t\mathbf{b}$

Key points

14 The equation of the line through (a_1, a_2, a_3) in the direction $\begin{pmatrix} b_1 \\ b_2 \\ b_3 \end{pmatrix}$ is given by:

$$\mathbf{r} = \begin{pmatrix} a_1 \\ a_2 \\ a_3 \end{pmatrix} + t \begin{pmatrix} b_1 \\ b_2 \\ b_3 \end{pmatrix} \qquad \text{in vector form}$$

15 The angle between two vectors, **a** and **b**, is θ where $\mathbf{a}.\mathbf{b} = |\mathbf{a}||\mathbf{b}|\cos\theta$

and $\mathbf{a}.\mathbf{b} = a_1 b_1 + a_2 b_2 \qquad$ (in two dimensions)
$\qquad \quad = a_1 b_1 + a_2 b_2 + a_3 b_3 \quad$ (in three dimensions).

16 If **a** and **b** are non-zero vectors and **a** is perpendicular to **b**:

$\mathbf{a}.\mathbf{b} = 0$

Answers

C3 C4

Answers (Core 3)

Chapter 1

Exercise 1A (Page 6)

1. $\dfrac{2a^2}{3b^3}$
2. $\dfrac{1}{9y}$
3. $\dfrac{x+3}{x-6}$
4. $\dfrac{x+3}{x+1}$
5. $\dfrac{2x-5}{2x+5}$
6. $\dfrac{3(a+4)}{20}$
7. $\dfrac{x(2x+3)}{(x+1)}$
8. $\dfrac{2}{5(p-2)}$
9. $\dfrac{a-b}{2a-b}$
10. $\dfrac{(x+4)(x-1)}{x(x+3)}$
11. $\dfrac{9}{20x}$
12. $\dfrac{x-3}{12}$
13. $\dfrac{a^2+1}{a^2-1}$
14. $\dfrac{5x-13}{(x-3)(x-2)}$
15. $\dfrac{2}{(x+2)(x-2)}$
16. $\dfrac{2p^2}{(p^2-1)(p^2+1)}$
17. $\dfrac{a^2-a+2}{(a+1)(a^2+1)}$
18. $-\dfrac{2(y^2+4y+8)}{(y+2)^2(y+4)}$
19. $\dfrac{x^2+x+1}{x+1}$
20. $-\dfrac{(3b+1)}{(b+1)^2}$
21. $\dfrac{13x-5}{6(x-1)(x+1)}$
22. $\dfrac{4(3-x)}{5(x+2)^2}$
23. $\dfrac{3a-4}{(a+2)(2a-3)}$
24. $\dfrac{3x^2-4}{x(x-2)(x+2)}$

Exercise 1B (Page 9)

1. (a) 84
   (b) 4
   (c) −2
   (d) 5.24 or 0.76
   (e) 3 or $\tfrac{1}{3}$
   (f) 0 or 3
   (g) 1.71 or 0.29

2. (a) $\dfrac{600}{x}$
   (b) $\dfrac{600}{x-1}$
   (c) $x^2-x-600=0$, $x=25$

ANSWERS

3 (a) $\frac{270}{x}, \frac{270}{x-10}$
(b) $x^2 - 10x - 9000 = 0$, $x = 100$
(c) Arrive 1 pm

4 Cost = £16, 16 staff left

5 12 thick slices

6 (a) 1.714 ohms
(b) 4 ohms
(c) Equivalent to half

7 $(x+1)(x^2 - x + 1)$
$\frac{x^2 - x + 1}{(x+2)}$

8 $-2, 3$

9 $\frac{1}{(x-1)(x-2)}$

10 $\frac{x}{x+4}$

EXERCISE 1C (Page 13)

1 (a) $x - 1$; 1
(b) $2x^2 - 3x + 5$; -9
(c) $x^2 + 3x - 5$; -4
(d) $2x^2 - 4x + 1$; 4
(e) $x^2 + 2x + 1$; -2
(f) $2x^2 - 3x - 4$; 2
(g) $x^3 - 3x^2 + 5x - 2$; 8
(h) $2x^3 + 5x^2 - 4x + 1$; 6
(i) $2x^2 - 3x + 8$; $-2x - 10$
(j) $3x^2 + 4x - 9$; $-23x + 26$

2 (a) $2x + 3 + \frac{8}{x-3}$
(b) $f'(x) = 2 - \frac{8}{(x-3)^2}$

4 (a) $2x^2 - 7x + 20 + \frac{20}{2x+5}$
(b) $p(x) = (2x^2 - 7x + 20)(2x + 5) + 20$
(c) 20

5 (a) $2x - 3$
(c) $\frac{3}{2}$

EXERCISE 1D (Page 18)

1 (a) one-to-one   function
(b) many-to-one   function
(c) one-to-one   function
(d) many-to-one   function
(e) many-to-many
(f) one-to-one   function
(g) one-to-one   function
(h) many-to-one   function

2 (a) one-to-one $x \in \mathbb{R}$ $y \in \mathbb{R}$
(b) many-to-one $x \neq 0$ $y > 0$
(c) one-to-many
(d) one-to-one $x \in \mathbb{R}$ $y \in \mathbb{R}$
(e) many-to-many
(f) one-to-one $x \neq 0$ $y \neq 0$
(g) many-to-one $x \in \mathbb{R}$ $y \geq -4$
(h) many-to-one $x \in \mathbb{R}$ $y \in \mathbb{R}$
(i) one-to-one $x \in \mathbb{R}$ $y > 0$
(j) many-to-many

3 (a) (i) -5
   (ii) 9
   (iii) -11
(b) (i) 3
   (ii) 5
   (iii) 10
(c) (i) 32
   (ii) 82.4
   (iii) 14
   (iv) -40

4 (a) $f(x) \leq 2$
(b) $0 \leq f(\theta) \leq 1$
(c) $y \in \{2, 3, 6, 11, 18\}$
(d) $y \in \mathbb{R}^+$
(e) $\mathbb{R}$
(f) $\{\frac{1}{2}, 1, 2, 4\}$
(g) $0 \leq y \leq 1$
(h) $f(\theta) \geq 1$ or $f(\theta) \leq -1$
(i) $0 < f(x) \leq 1$
(j) $f(x) \geq 3$

5 For f, every value of x (including $x = 3$) gives a unique output, whereas g(2) can equal either 4 or 6.

EXERCISE 1E (Page 23)

1 (a) $8x^3$
(b) $2x^3$
(c) $(x+2)^3$
(d) $x^3 + 2$
(e) $8(x+2)^3$
(f) $2(x^3 + 2)$
(g) $4x$
(h) $\{(x+2)^3 + 2\}^3$
(i) $x + 4$

Pure Mathematics: Core 3 — Answers

2. (a) $x^2 + 8x + 14$
   domain $x \in \mathbb{R}$
   range $y \geq -2$
   (b) $x = \pm 5$

3. (a) $\sqrt{x + 1}$
   domain $x \geq -1$
   range $y \geq 0$
   (b) $x = 8$
   (c) $x = 4$

4. (a) $gf(x) = \dfrac{1}{2(x + 1)}$
   domain $x \neq -1$
   (b) $0.229, -0.729$

5. (a) $gh = \dfrac{3}{x - 3}$
   (b) $fgh = \dfrac{6}{x - 3} + 3$
   (c) f and h

6. (a) fg
   (b) g^2
   (c) fg^2
   (d) gf

7. (a) $gf(x) = x^2 - 4x + 7$
   (c) $y \geq 3$
   (d) many-to-one
   (e) Yes

8. (b) $3p + q = 7$
   (c) $p = 3$ with $q = -2$ or $p = 5$ with $q = -8$

9. (a) 3
   (b) 2
   (c) 1
   (d) 61

10. (a) $fg(x) = 2x^2 - 2x + 3$
    (b) $fg(x) = 2\left(x - \tfrac{1}{2}\right)^2 + \tfrac{5}{2}$
    $p = 2, q = -\tfrac{1}{2}, r = \tfrac{5}{2}$
    (c) $fg(x) \geq \tfrac{5}{2}$

Exercise 1F (Page 30)

1. (a) $f^{-1}(x) = \dfrac{x - 7}{2}$
   (b) $f^{-1}(x) = 4 - x$
   (c) $f^{-1}(x) = \dfrac{2x - 4}{x}$
   (d) $f^{-1}(x) = \sqrt{x + 3}, x \geq -3$

2. (a), (b)

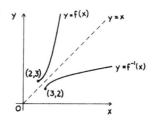

3. (a) $f(x)$ not defined for $x = 4$;
   $g(x)$ is defined for all x;
   $h(x)$ not defined for $x > 2$
   (b) $f^{-1}(x) = \dfrac{4x + 3}{x}$;
   $h^{-1}(x) = 2 - x^2, x \geq 0$
   (c) $g(x)$ is not one-to-one.
   (d) Suitable domain: $x \geq 0$
   (e) No: $fg(x) = \dfrac{3}{x^2 - 4}$, not defined for $x = \pm 2$;
   $gf(x) = \left(\dfrac{3}{x - 4}\right)^2$, not defined for $x = 4$.

4. (a) x (b) $\dfrac{1}{x}$ (c) $\dfrac{1}{x}$ (d) $\dfrac{1}{x}$

5. (a) $a = 3$
   (b)

   (c) $f(x) \geq 3$
   (d) many-to-one function; possible domain is $x \geq -2$

6. (a) $f^{-1}: x \mapsto \sqrt[3]{\dfrac{x - 3}{4}}, x \in \mathbb{R}$
   The graphs are reflections of each other in the line $y = x$.

7. (a) $f^{-1}: x \mapsto \dfrac{x + 2}{4}$
   $g^{-1}: x \mapsto \sqrt{x}$
   (b)

   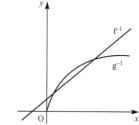

   (c) 2
   (d) $6 \pm 4\sqrt{2}$

Answers

8 (a) $a = -3, b = 1$
  (b) $(3, 1)$
  (c)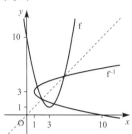
  (d) $x \geq 3, y \geq 1$
  (e) see graph; reflection in $y = x$

9 (b) $y \leq 4$
  (c) $f^{-1}: x \mapsto 4 - x, x \leq 4$
  (d) $x = -5$

10 (b) $x > 1$
   (c) $g^{-1}: x \mapsto x - 1$
   (d)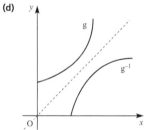

Exercise 1G (Page 34)

1 (a)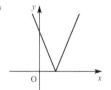

  (b)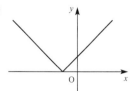

  (c)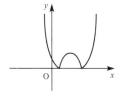

  (d)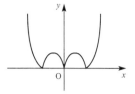

  (e)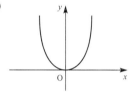

  (f)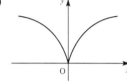

2 (a)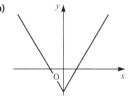

  (b)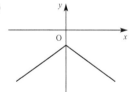

  (c)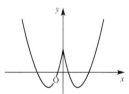

  (d)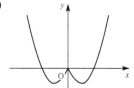

(e)

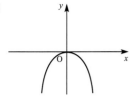

(f)

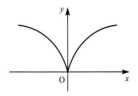

3 (a) $y = |x + 2|$

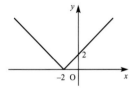

(b) $y = |3x - 2|$

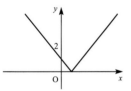

(c) $y = |x| + 2$

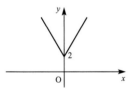

(d) $y = |x^2 - 2|$

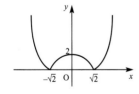

(e) $y = |x^3|$

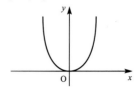

(f) $y = |x^3| - 1$

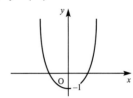

(g) $y = \left|\frac{1}{x}\right|$

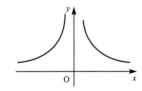

(h) $y = |3 - x|$

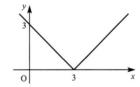

(i) $y = |9 - x^2|$

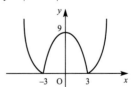

(j) $y = |\sin x|$

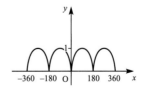

ANSWERS

4 (a)
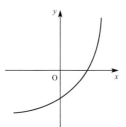
or
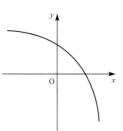

(b)
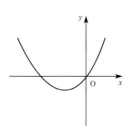
or
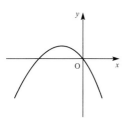

5 (a)
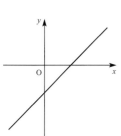

(b)
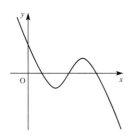

EXERCISE **1H** (page 42)

1 (a) Translation $\begin{pmatrix} -3 \\ 1 \end{pmatrix}$

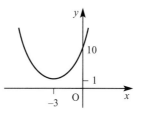

(b) Translation $\begin{pmatrix} 1 \\ 0 \end{pmatrix}$, then reflection in y-axis, then translation $\begin{pmatrix} 0 \\ 2 \end{pmatrix}$

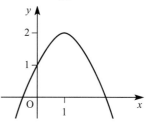

(c) Translation $\begin{pmatrix} 4 \\ 0 \end{pmatrix}$ and stretch parallel to y-axis of sf. 2, in either order

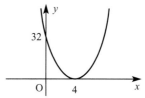

(d) Stretch parallel to x-axis of s.f. $\frac{1}{2}$ and translation $\begin{pmatrix} 0 \\ -4 \end{pmatrix}$, in either order

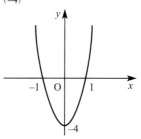

(e) Stretch parallel to y-axis of s.f. 3 and translation $\begin{pmatrix} 0 \\ -5 \end{pmatrix}$, in either order

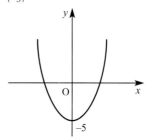

2 (a) Translation $\begin{pmatrix} 90° \\ 0 \end{pmatrix}$
  (b) Stretch parallel to x-axis of s.f. $\frac{1}{3}$
  (c) Stretch parallel to y-axis of s.f. $\frac{1}{2}$
  (d) Stretch parallel to x-axis of s.f. 2
  (e) Stretch parallel to x-axis of s.f. $\frac{1}{3}$ and translation $\begin{pmatrix} 0 \\ 2 \end{pmatrix}$, in either order

3 (a) Translation $\begin{pmatrix} -60° \\ 0 \end{pmatrix}$
  (b) Stretch parallel to y-axis of s.f. $\frac{1}{3}$
  (c) Translation $\begin{pmatrix} 0 \\ 1 \end{pmatrix}$
  (d) Translation $\begin{pmatrix} -90° \\ 0 \end{pmatrix}$, then stretch parallel to x-axis of s.f. $\frac{1}{2}$

4 (i) (a)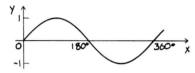
  (b) $y = \sin x$

  (ii) (a)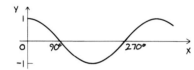
  (b) $y = \cos x$

  (iii) (a)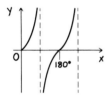
  (b) $y = \tan x$

  (iv) (a)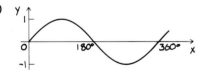
  (b) $y = \sin x$

  (v) (a)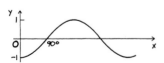
  (b) $y = \cos x$

5 (a) $2(x - 1)^2 + 3$
  (b)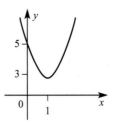

6 (a) $\dfrac{x^2}{9} + \dfrac{y^2}{4} = 1$

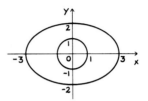

7 (a) $(-1, 5)$; $y = 2$
  (b) $a = -3$, $b = 5$
  (d) $y = 3x^2 + 6x + 2$; $(-1, -1)$

8 (a)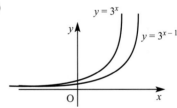

  (b) Translation $\begin{pmatrix} 1 \\ 0 \end{pmatrix}$, stretch parallel to y-axis of s.f. $\frac{1}{3}$

  (c)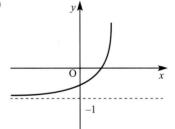

  Translation $\begin{pmatrix} 1 \\ -1 \end{pmatrix}$

9 (a)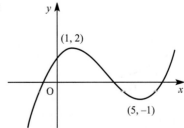

(b)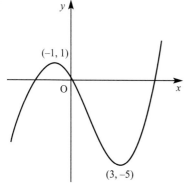

(c)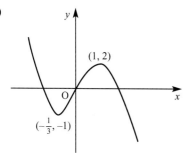

(d)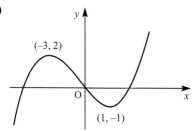

(e)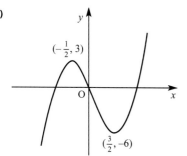

10 (a)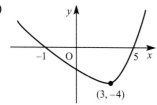

(b)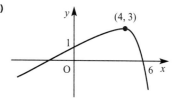

(c)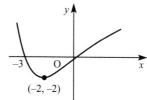

(d)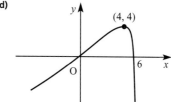

(e)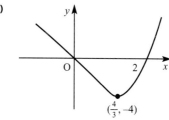

EXERCISE 1I (Page 44)

1 $\dfrac{3x - 7}{(x - 1)(x - 5)}$

2 $-2, \tfrac{1}{2}$

3 $\dfrac{x^2 + x + 1}{x - 1}$

5 $\dfrac{2(x + 1)}{x + 3}$

6 (b) $0 < f(x) < \tfrac{4}{3}$
   (c) $f^{-1}: x \mapsto \tfrac{2}{x} - \tfrac{1}{2}$
   (d) $f^{-1}(x) > 1$

8 (a) $A = -7, B = 1$

9 (a) $3x^2 - 6x + 4 - \dfrac{3}{2x + 3}$
   (b) $(3x^2 - 6x + 4)(2x + 3) - 3$

10 (a) $a = 3, b = 2$
   (b)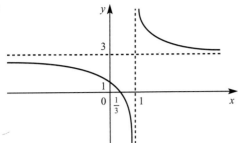

11 (a) $y \geqslant 0$
   (b) $-2, 8$
   (c)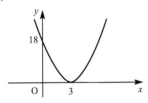
   (d)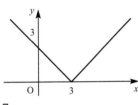
   $-1, 7$

12 (a) $y \geqslant 0$
   (b) $f^{-1}: x \mapsto \dfrac{x-1}{2}$ $g^{-1}: x \mapsto x^2$
   (c)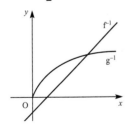
   (d) one solution
   (e) 3.317

13 (a)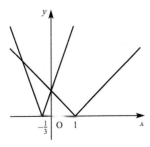
   (b) $-1, 0$

14 (a) $a = 2, b = 1$
   (b)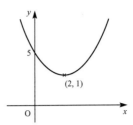
   (c) $x \geqslant 2$
   (d) $f^{-1}: x \mapsto \sqrt{x-1} + 2$
   (e)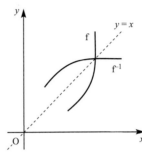
   (f) 3.6

15 (a) $g^{-1}: x \mapsto \dfrac{1}{x} - 1$ $x \neq 0$
   (b) $fg(x) = \dfrac{1-x}{1+x}$
   (c)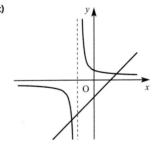
   (d) $-1.3, 0.8$
   (e) $-1.28, 0.781$

16 (a) Y $\left(\dfrac{\pi}{2}, 2\right)$ Z $\left(\dfrac{3\pi}{4}, 0\right)$
   $A = 2$ $B = 2$
   (b) $y = f\left(x + \dfrac{\pi}{2}\right)$
   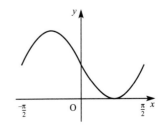

$y = |f(x) - 2|$

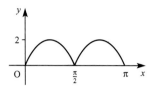

17 (a) $y = f(-x)$

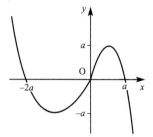

(b) $y = -f(x)$

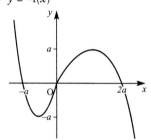

(c) $y = f(x - a)$

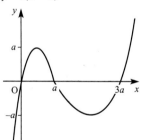

(d) $y = f(x) - a$

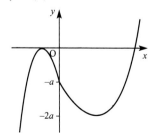

(e) $y = |f(x)|$

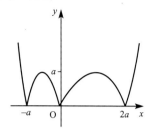

$-a < x < 0 \qquad x > 2a$

18 (a)

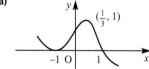

(b)

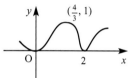

(c)

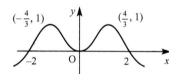

19 (a)

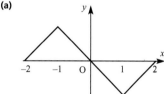

(b) $-\frac{3}{2}, -\frac{1}{2}$

20 (a)

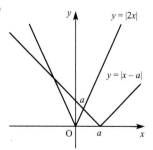

(b) $(a, 0)$ $(0, a)$

(d) $\left(\frac{1}{3}a, \frac{2}{3}a\right)$

21 (a)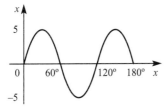

(b) (30, 5) (90, −5) (150, 5)
(c) 10°, 50°, 130°, 170°

22 (a) $a = 2$, $b = \frac{5}{4}$, $c = -\frac{17}{8}$

(b) Translation $\begin{pmatrix} -\frac{5}{4} \\ 0 \end{pmatrix}$, stretch parallel to y-axis of s.f. 2, and translation $\begin{pmatrix} 0 \\ -\frac{17}{8} \end{pmatrix}$, in any order

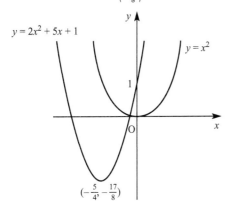

23 (a)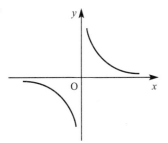

(b) Translation $\begin{pmatrix} -1 \\ 0 \end{pmatrix}$, reflection in x-axis, then translation $\begin{pmatrix} 0 \\ 1 \end{pmatrix}$

(c)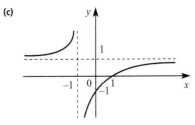

24 (a) $f^{-1}: x \mapsto \sqrt[3]{2 - x}$

(b)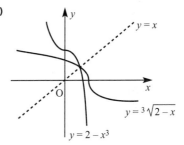

(c) Reflection in $y = x$

25 (a)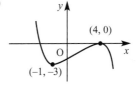

(b)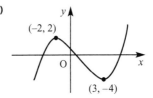

(c)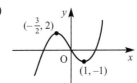

(d)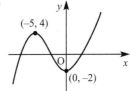

CHAPTER 2

EXERCISE 2A (Page 55)

1 (a) 30°, 150°
(b) 78.5°, 281.5°
(c) $\frac{\pi}{4}, \frac{5\pi}{4}$
(d) $-\frac{\pi}{3}, -\frac{2\pi}{3}, \frac{4\pi}{3}, \frac{5\pi}{3}$
(e) 63.4°, 243.4°

Answers

2
(a) $30°, 150°, 210°, 330°$
(b) $-\frac{3\pi}{2}, -\frac{5\pi}{6}, -\frac{\pi}{6}, \frac{\pi}{2}, \frac{7\pi}{6}, \frac{11\pi}{6}$
(c) $-148.3°, -58.3°, 31.7°, 121.7°$
(d) $-\frac{11\pi}{12}, -\frac{5\pi}{6}, -\frac{5\pi}{12}, -\frac{\pi}{3}, \frac{\pi}{12}, \frac{\pi}{6}, \frac{7\pi}{12}, \frac{2\pi}{3}$
(e) $-\frac{5\pi}{6}, -\frac{\pi}{3}, \frac{\pi}{6}, \frac{2\pi}{3}, \frac{7\pi}{6}, \frac{5\pi}{3}, \frac{13\pi}{6}, \frac{8\pi}{3}$

3
(a) $0°, 60°, 300°, 360°$
(b) $19.5°, 160.5°, 270°$
(c) $63.4°, 45°, 243.4°, 225°$
(d) $\frac{\pi}{2}, \frac{3\pi}{2}$
(e) $6.1°, 23.9°, 66.1°, 83.9°, 126.1°, 143.9°$

4 (a)
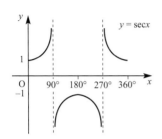

(b)
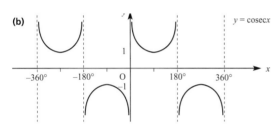

(c)
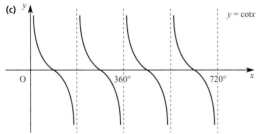

(d)
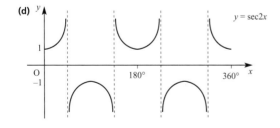

(e)
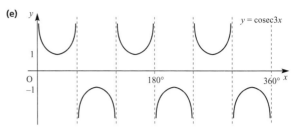

5 (a) -2
(b) -2
(c) $\sqrt{3}$
(d) 2
(e) -2

Exercise 2B (Page 61)

1 (a) $90°, 210°, 330°$
(b) $45°, 63.4°$
(c) $48.2°, 60°, 300°, 311.8°$
(d) $26.6°, 45°, 206.6°, 225°$
(e) $26.6°, 135°, 206.6°, 315°$
(f) $60°, 70.5°$
(g) $33.7°, 45°$
(h) $\frac{\pi}{6}, \frac{5\pi}{6}, \frac{3\pi}{2}$
(i) $0, \frac{2\pi}{3}, \frac{4\pi}{3}, 2\pi$
(j) $63.4°, 123.7°, 243.4°, 303.7°$

3 (a) $B = 60°\ \ C = 30°$
(b) $\sqrt{3}$

4 (a) $L = N = 45°$
(b) $\sqrt{2}, \sqrt{2}, 1$

5 $\frac{1}{2}, -\frac{3}{4}$
$30°, 150°, 228.6°, 311.4°$
$8\sin^2\theta + 2\sin\theta - 3 = 0$
$\theta = 30°$

Exercise 2C (Page 66)

1 (a) $\frac{\sqrt{3}}{2\sqrt{2}} + \frac{1}{2\sqrt{2}}$

(b) $-\frac{1}{\sqrt{2}}$

(c) $\frac{3}{2\sqrt{2}} + \frac{1}{2\sqrt{2}}$

(d) $\frac{\sqrt{3}-1}{\sqrt{3}+1}$

(e) $\frac{\sqrt{3}+1}{\sqrt{3}-1}$

Pure Mathematics: Core 3 — ANSWERS

2 (a) $\frac{1}{\sqrt{2}}(\sin\theta + \cos\theta)$

(b) $\frac{1}{2}(\sqrt{3}\cos\theta + \sin\theta)$

(c) $\frac{1}{2}(\sqrt{3}\cos\theta - \sin\theta)$

(d) $\frac{1}{\sqrt{2}}(\cos 2\theta - \sin 2\theta)$

(e) $\frac{\tan\theta + 1}{1 - \tan\theta}$

(f) $\frac{\tan\theta - 1}{1 + \tan\theta}$

3 (a) $\sin\theta$
(b) $\cos 4\phi$
(c) 0
(d) $\cos 2\theta$

4 (a) $15°$
(b) $157.5°$
(c) $0°$ or $180°$
(d) $111.7°$
(e) $165°$

5 (a) $\frac{\pi}{8}$
(b) 2.79 radians

Exercise 2D (Page 70)

1 (a) $14.5°, 90°, 165.5°, 270°$
(b) $0°, 35.3°, 144.7°, 180°, 215.3°, 324.7°, 360°$
(c) $90°, 210°, 330°$
(d) $30°, 150°, 210°, 330°$
(e) $0°, 138.6°, 221.4°, 360°$

2 (a) $-\pi, 0, \pi$
(b) $-\pi, 0, \pi$
(c) $\frac{-2\pi}{3}, 0, \frac{2\pi}{3}$
(d) $\frac{-3\pi}{4}, \frac{-\pi}{4}, \frac{\pi}{4}, \frac{3\pi}{4}$
(e) $\frac{-11\pi}{12}, \frac{-3\pi}{4}, \frac{-7\pi}{12}, \frac{-\pi}{4}, \frac{\pi}{12}, \frac{\pi}{4}, \frac{5\pi}{12}, \frac{3\pi}{4}$

3 $3\sin\theta - 4\sin^3\theta$,
$\theta = 0, \frac{\pi}{4}, \frac{3\pi}{4}, \pi, \frac{5\pi}{4}, \frac{7\pi}{4}, 2\pi$

4 $51°, 309°$

5 $\cot\theta$

6 $\dfrac{\tan\theta(3 - \tan^2\theta)}{1 - 3\tan^2\theta}$

8 (b) $63.4°$

9 (a) $180°$
(b) $126.8°, 180°$
(c) $0°, 53.1°, 306.9°, 360°$
(d) $97.2°, 262.8°$

Exercise 2E (Page 74)

1 (a) $\sqrt{2}\cos(\theta - 45°)$
(b) $5\cos(\theta - 53.1°)$
(c) $2\cos(\theta - 60°)$
(d) $3\cos(\theta - 41.8°)$

2 (a) $\sqrt{2}\cos\left(\theta + \frac{\pi}{4}\right)$
(b) $2\cos\left(\theta + \frac{\pi}{6}\right)$

3 (a) $\sqrt{5}\sin(\theta + 63.4°)$
(b) $5\sin(\theta + 53.1°)$

4 (a) $\sqrt{2}\sin\left(\theta - \frac{\pi}{4}\right)$
(b) $2\sin\left(\theta - \frac{\pi}{6}\right)$

5 (a) $2\cos(\theta - (-60°))$
(b) $4\cos(\theta - (-45°))$
(c) $2\cos(\theta - 30°)$
(d) $13\cos(\theta - 22.6°)$
(e) $2\cos(\theta - (-150°))$
(f) $2\cos(\theta - (-135°))$

6 (a) $13\cos(\theta + 67.4°)$
(b) Max 13, min -13
(c)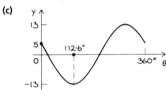

(d) $4.7°, 220.5°$

7 (a) $2\sqrt{3}\sin\left(\theta - \frac{\pi}{6}\right)$
(b) Max $2\sqrt{3}, \theta = \frac{2\pi}{3}$; min $-2\sqrt{3}, \theta = \frac{5\pi}{3}$
(c)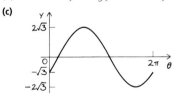

(d) $\frac{\pi}{3}, \pi$

ANSWERS

8 (a) $\sqrt{13}\sin(2\theta + 56.3°)$
   (b) Max $\sqrt{13}$, $\theta = 16.8°$;
       min $-\sqrt{13}$, $\theta = 106.8°$
   (c)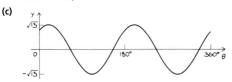
   (d) $53.8°, 159.9°, 233.8°, 339.9°$

9 (a) $\sqrt{3}\cos(\theta - 54.7°)$
   (b) Max $\sqrt{3}$, $\theta = 54.7°$;
       min $-\sqrt{3}$, $\theta = 234.7°$
   (c)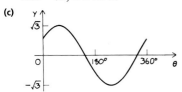
   (d) Max $\dfrac{1}{3 - \sqrt{3}}$, $\theta = 234.7°$;
       min $\dfrac{1}{3 + \sqrt{3}}$, $\theta = 54.7°$

10 (b) $30.6°$ or $82.0°$

EXERCISE 2F (Page 76)

1 (a) $R = 25$, $\alpha = 73.7°$
  (b) $-36.9°, -110.6°$
  (c) 25 when $x = -73.7°$

2 (a) $\cos 2x = 2\cos^2 x - 1$
  (b) $0, \frac{2\pi}{3}, \frac{4\pi}{3}, 2\pi$
  (c) $0, \frac{4\pi}{3}$

3 (a) $\dfrac{1 - \sqrt{3}}{4}$
  (b) $22.5°, 67.5°, 90°, 112.5°, 157.5°, 202.5°, 247.5°, 270°, 292.5°, 337.5°$

4 (a)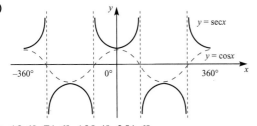
  (b) $18.4°, 71.6°, 198.4°, 251.6°$

5 (a) $R = 10$, $\alpha = 0.93$
  (b) $-10, -0.64$
  (c) 1.34

6 $\dfrac{7\pi}{6}, \dfrac{11\pi}{6}, \dfrac{\pi}{2}$

7 (b) $0.85, 2.29, 4.71$
  (c) $1.70, 4.59$

8 (a) $210°, 330°$
  (b) 2.01

9 (a) $R = 25$, $\alpha = 73.7°$
  (b) $20.6°, 126.8°$
  (c) $\frac{1}{630} \leq f(\theta) \leq \frac{1}{5}$

10 (a) $R = 15$, $\alpha = 53.1°$
   (b) $156.9°, 276.9°$
   (c) $\frac{\pi}{3}, \frac{4\pi}{3}$

11 $0.29, 1.86, 1.08, 2.65$

12 (a) $\frac{\pi}{4}, \frac{\pi}{2}, \frac{3\pi}{4}, \frac{3\pi}{2}$
   (b) $\frac{\pi}{3}, \frac{5\pi}{6}, \frac{4\pi}{3}, \frac{11\pi}{6}$

14 (a) $R = 5.39$, $\alpha = 1.19$
   (b) max 5.39 when $\theta = 1.19$
   (c) max $= 20.39\,°\text{C}$, $t = 4.55$ (or 4 hrs 33 mins)
   (d) $08:30, 24:00$

15 (b) (iii) $26.6°, 206.6°$

CHAPTER 3

EXERCISE 3A (Page 84)

1 (a)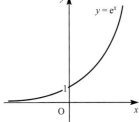

  (b)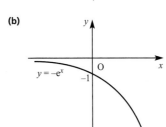

(c)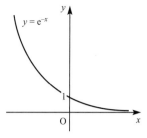

(d)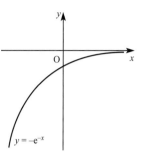

2 (a)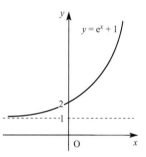

(b)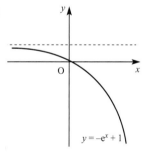

(c)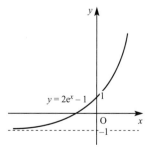

(d)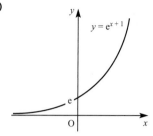

3 (a)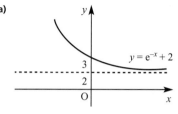

(b)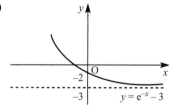

(c)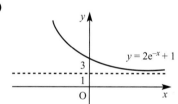

(d)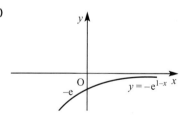

4 (a)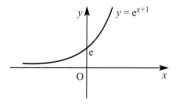

Answers

(b)
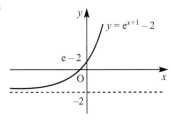

(c)
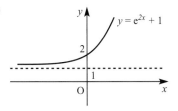

(d)
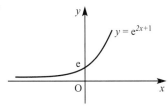

5 (a)
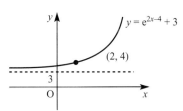

(b)
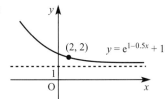

(c)
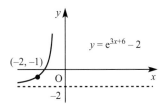

(d)
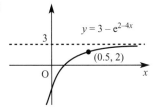

Exercise 3B (page 87)

1 (a)
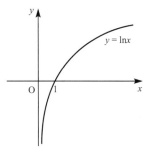

(b)
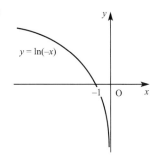

(c)
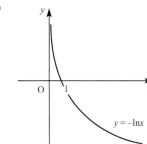

(d)
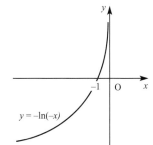

2 (a)
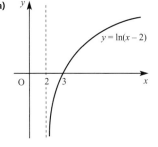

(b)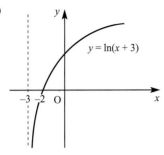

(c)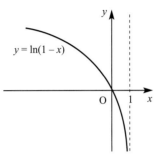

(d)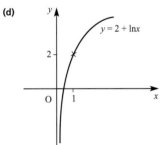

3 (a)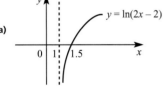

(b)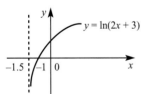

(c)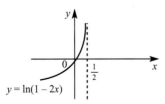

(d)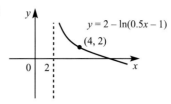

4 (a), (b)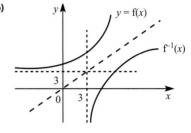

(c) $f^{-1}(x) = 1 + \ln(x - 3)$
(d) $x > 3$

5 (a), (b)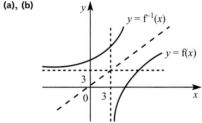

(c) $f^{-1}(x) = \frac{1}{2}(e^x + 6)$
(d) $x \in \mathbb{R}, y > 3$

Exercise 3C (Page 91)

1. (a) 0.693
   (b) 3.08
   (c) 1.61
   (d) −3.91
   (e) 9.90
   (f) −3.00
   (g) 2.69
   (h) 1.26
   (i) 21.2
   (j) 0.231

2. (a) 7.39
   (b) 4.35
   (c) 143
   (d) 2.74
   (e) 1.04
   (f) −0.105

(g) 1.38
(h) 1.45
(i) 52.4
(j) 0.273

3 (a) 100
  (b) 148
  (c) 29.5 hrs

4 (a) $a = 0.50, b = 2.00$
  (b)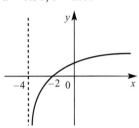

5 (a)
  (b) 100
  (c) 1218
  (d) 185 years

6 (a)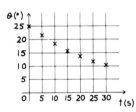
  (b) 25°
  (c) 4.1°
  (d) 22

7 (a)
  (b) 621.5 m
  (c) 8.07 am (to nearest minute)
  (d) Never

8 (a)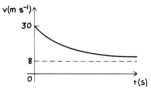
  (b) 30 m s^{-1}, 8 m s^{-1}
  (c) 8.33 m s^{-1}
  (d) 8.7 s

9 (a) 250
  (b) 6.9
  (c)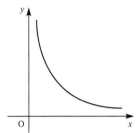

10 (a) Yes
   (b) 1115
   (c) 26 days

Exercise 3D (Page 93)

1 (a) $S = 6000$
  (b) $t = 110$
      $P = P_0 \times 2^{\frac{t}{15}}$
      99.7 hours

2 (a)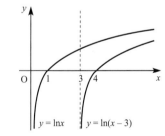
      translation $\binom{3}{0}$
  (b) $y = e^{x+2}$
      stretch scale factor e^2
      parallel to y axis, x axis invariant.

3 (a) Levels off
  (b) 164
  (c) 8.4 weeks
  (d) 260

4 (a) range $\geq k$
  (b) $2k$
  (c) $\ln(x - k)$, domain $x > k$
  (d)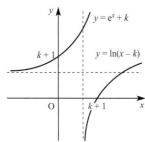

5 (a) $y\ln 2 = x$
  (c) 2.43

6 (a) 330
  (b) 317 to 345

7 (a)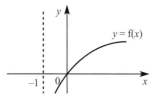
  (b)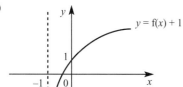
  (c)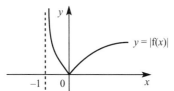
  (d)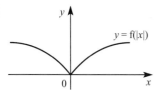

8 (a)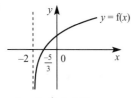
  $x > -2$, $f(x) \in \mathbb{R}$

(b) Yes
(c) $f^{-1}: x \mapsto \frac{1}{3}(e^{4x} - 6)$
(d)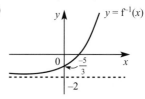
  $x \in \mathbb{R}$, $f^{-1}(x) > -2$

9 (a) (i) 12.4
     (ii) 2.69
  (b)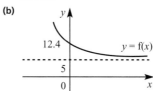
  $x \in \mathbb{R}$, $y > 5$
  (c) $f^{-1}: x \mapsto 2 - \ln(x - 5)$
  (d)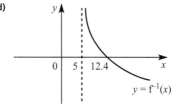
  $x > 5$ $f^{-1}(x) \in \mathbb{R}$

10 (a)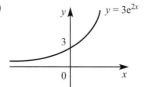
   (b)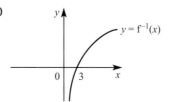
   (c)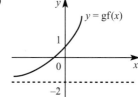

ANSWERS

(d)

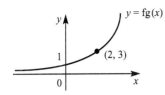

11 (a) 13.2
   (b) 14.7

12 (a) 30°C
   (b) 25°C
   (c) 9 hours 10 minutes
   (d)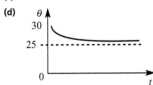

13 (a) $a = 4, b = -2$
   (b) 37.6
   (c), (d)

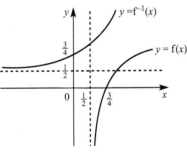

   f(x): domain $x > \frac{1}{2}$
   range f(x) ∈ ℝ

   (e) $f^{-1}: x \mapsto \frac{1}{4}(e^x + 2)$

14 (b) 1650
   (c) 13th day
   (d)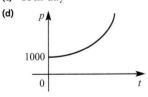

   (e) Not realistic as population continues to grow

15 $x = 1.72, y = 1.77$

CHAPTER 4

Exercise 4A (Page 103)

1 (a) e^x
  (b) $2e^x$
  (c) $-3e^x$
  (d) $2 + 8e^x$
  (e) $0.1e^x - x$
  (f) $\dfrac{1}{2\sqrt{x}} - \dfrac{e^x}{2}$

2 (a) $\frac{1}{x}$
  (b) $\frac{2}{x}$
  (c) $\frac{1}{x}$
  (d) $\frac{4}{x}$
  (e) $\frac{4}{x}$
  (f) $-\frac{1}{x}$
  (g) $-\frac{2}{x}$
  (h) $-\frac{6}{x}$
  (i) $1 - \frac{1}{x}$
  (j) $-\frac{2}{x^2} + \frac{4}{x}$

3 (a) $e + 1$
  (b) $\frac{3}{2} - 4e^2$
  (c) $4 + 2e$
  (d) $2e^4 - \frac{5}{8}$
  (e) $\frac{-5}{4} - 5e^2$
  (f) 3

5 (a) 1030 agents
  (b) 1000 agents/year

6 $y = 4x - 2$

7 (a) $x = \ln 4$
  (b) 4
  (c) $y = 4x - 4\ln 4$

8 (a)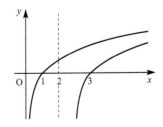

   translation $\binom{2}{0}$

  (b) Using (a) gradient of $\ln x$ is $\frac{1}{x}$
      so gradient of $\ln(x - 2)$ must be $\dfrac{1}{x - 2}$
  (d) $3y = x - 3 - 6\ln 3$

Exercise 4B (Page 111)

1. (a) $3(x + 2)^2$
   (b) $8(2x + 3)^3$
   (c) $6x(x^2 - 5)^2$
   (d) $15x^2(x^3 + 4)^4$
   (e) $-3(3x + 2)^{-2}$
   (f) $\dfrac{-6x}{(x^2 - 3)^4}$
   (g) $3x(x^2 - 1)^{\frac{1}{2}}$
   (h) $3\left(\dfrac{1}{x} + x\right)^2 \left(1 - \dfrac{1}{x^2}\right)$
   (i) $\dfrac{2}{\sqrt{x}}(\sqrt{x} - 1)^3$

2. (a) $9(3x - 5)^2$
   (b) $y = 9x - 17$

3. (a) $8(2x - 1)^3$
   (b) $(\tfrac{1}{2}, 0)$, minimum
   (c)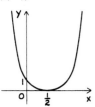

4. (a) $6x(x^2 - 4)^2$
   (b) $(0, -64)$, minimum;
   $(-2, 0)$ point of inflection;
   $(2, 0)$, point of inflection
   (c)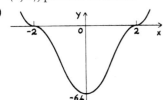

5. (a) $4(2x - 1)(x^2 - x - 2)^3$
   (b) $(-1, 0)$, minimum;
   $(\tfrac{1}{2}, \tfrac{6561}{256})$, maximum;
   $(2, 0)$, minimum
   (c)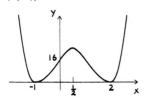

6. 4 cm² s⁻¹

7. -0.015 N s⁻¹

8. 10π cm² day⁻¹

9. (a) $3x(3x - 2)(x^3 - x^2 + 2)^2$
   (b) $\dfrac{dy}{dx} = 0$ when $x = 1$ and when $x = 0$.
   When $x < -1$ (e.g. -1.1) $\dfrac{dy}{dx} > 0$;
   when $-1 < x < 0$ (e.g. -0.5) $\dfrac{dy}{dx} > 0 \Rightarrow$ point of inflection at $x = -1$.
   When x is just greater than 0 (e.g. 0.1) $\dfrac{dy}{dx} < 0 \Rightarrow$ maximum point at $x = 0$.
   (c) $a = \tfrac{2}{3}$
   (d) Gradient at $(1, 8)$ is 12; $y = 12x - 4$

10. $\tfrac{1}{8}$ km

11. (a) $5e^{5x}$
    (b) $2e^{5+x}$
    (c) $-3e^{5-x}$
    (d) $-\tfrac{5}{2}e^{3-5x}$
    (e) $\dfrac{-3}{e^{3x}}$
    (f) $\dfrac{-1}{e^{\frac{1}{2}x}}$
    (g) $2xe^{x^2}$
    (h) $(1 + 2x)e^{1+x+x^2}$
    (i) $2(1 + x)e^{(1+x)^2}$
    (j) $\dfrac{2}{\sqrt{1 + 4x}} e^{\sqrt{1 + 4x}}$

12. (a) $\dfrac{3}{x}$
    (b) $\dfrac{1}{x}$
    (c) $-\dfrac{1}{2 - x}$
    (d) $\dfrac{2}{x}$
    (e) $\dfrac{2}{x} \ln x$
    (f) $\dfrac{4(1 + 2x)}{1 + x + x^2}$
    (g) $-\dfrac{1}{x}$
    (h) $\dfrac{1}{x \ln x}$
    (i) $\dfrac{5x}{x^2 + 1}$
    (j) 1

13. $x = 1$, minimum

14. (a) $\dfrac{dy}{dx} = -\dfrac{3}{e^{3x}} - 2x$

Answers

Exercise 4C (Page 118)

1. (a) $x(5x^3 - 3x + 6)$
   (b) $(x + 1)e^x$
   (c) $2x(6x + 1)(2x + 1)^3$
   (d) $-\dfrac{2}{(3x - 1)^2}$
   (e) $\dfrac{x^2(x^2 + 3)}{(x^2 + 1)^2}$
   (f) $2(2x + 1)(12x^2 + 3x - 8)$
   (g) $\dfrac{2(1 + 6x - 2x^2)}{(2x^2 + 1)^2}$
   (h) $\dfrac{7 - x}{(x + 3)^3}$
   (i) $\dfrac{3x - 1}{2\sqrt{x - 1}}$
   (j) $\dfrac{x^2 + 1}{x} + 2x \ln x$

2. (a) $-\dfrac{1}{(x - 1)^2}$
   (b) $-1;\ y = -x$
   (c) $-1;\ y = -x + 4$
   (d) The two tangents are parallel

3. (a) $3x(x - 2)$
   (b) $(0, 4)$, maximum
   (c) $(2, 0)$, minimum
   (d)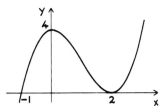

4. (a) $3(4x + 1)(x - 1)^2(2x - 1)^2$
   (b) $x = -1$, point of inflection;
   $x = -\tfrac{1}{4}$, minimum;
   $x = \tfrac{1}{2}$, point of inflection
   (c) $P(-1, 0);\ Q\left(-\tfrac{1}{4}, -\tfrac{729}{512}\right);\ R\left(\tfrac{1}{2}, 0\right)$

5. (a) $\dfrac{\sqrt{x} - 2}{(\sqrt{x} - 1)^2}$
   (b) $\tfrac{1}{4}$
   (c) $(4, 8)$
   (d) Tangent: $y = 8$; normal: $x = 4$
   (e) (i) $Q\left(\tfrac{37}{4}, 8\right)$
       (ii) $R(4, 29)$

6. (a) $-\dfrac{1}{(x - 4)^2}$
   (b) $4y + x = 12$
   (c) $y = x - 3$

   (d) $\dfrac{dy}{dx} \neq 0$ for any value of x
   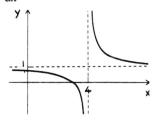

7. (a) $\dfrac{2(x + 1)(x + 2)}{(2x + 3)^2}$
   (b) $(-1, -2);\ (-2, -3)$
   (c) $(-1, -2)$, minimum
   $(-2, -3)$, maximum

8. (a) $\dfrac{2x(x + 1)}{(2x + 1)^2};\ (0, 0)$ and $(-1, -1)$
   (b) $(0, 0)$, minimum
   $(-1, -1)$, maximum

9. (a) $x > 1$
   (b) $\dfrac{dy}{dx} = -\dfrac{5}{2\sqrt{x + 4}\sqrt{(x - 1)^3}}$
   (c) There are no finite values of x for which $\dfrac{dy}{dx} = 0$ so there are no stationary points.

10. (a) $0;\ 2;\ 1.6$
    (b) $(1, 2)$
    (c) $f(x) \to 0$
    (d)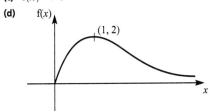
    (e)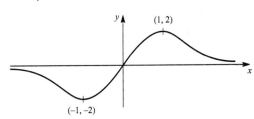

11. (a) $(2x + 1)e^{2x}$
    (b) $\left(-\tfrac{1}{2}, -\dfrac{1}{2e}\right)$
    (c) $4(x + 1)e^{2x}$
    (d) minimum

12 (a) $x(2\ln x + 1)$
   (c) $y = 3ex - 2e^2$
   (d) $\left(\dfrac{2e}{3}, 0\right)$

13 (a) $(2x + 3) x^2 e^x$
   (c) $e^2 x + y + \dfrac{e^4 + 1}{e^2} = 0$

14 (a) $e^x \left[(1 + x) \ln(1 + x) + \dfrac{x}{1 + x}\right]$
   (b) $\ln(1 + x) = -\dfrac{x}{(1 + x)^2}$
   No solutions for $x > 0$

15 (b) $\dfrac{dy}{dx} = \dfrac{e^x(x^2 - 2)}{x^3}$
   (c) both minimum turning points

16 (a) $\dfrac{dy}{dx} = (1 + x)e^x$, $\dfrac{d^2y}{dx^2} = (2 + x)e^x$
   (b) $\left(-1, -\dfrac{1}{e}\right)$

17 (a) odd function since $f(-x) = -f(x)$
   (b) $f'(x) = 2 + \ln(x^2)$, $f''(x) = \dfrac{2}{x}$
   (c) $\left(-\dfrac{1}{e}, \dfrac{2}{e}\right)$ maximum, $\left(\dfrac{1}{e}, -\dfrac{2}{e}\right)$ minimum

18 (a) $\dfrac{e^x(x - 1)}{x^2}$
   (b) $(1, e)$ minimum
   (c)

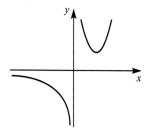

19 (a) $-\dfrac{x^2 + 2x + 5}{(x^2 + 2x - 3)^2}$

20 (b) $\dfrac{3x^2 - 2x - 3}{(x^2 + 1)^2} - \dfrac{3}{(x - 3)^2}$
   (d) $5y + 14x = 8$

EXERCISE 4D (Page 126)

1 0.16 cm min^{-1}

2 (a)

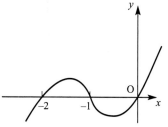

(b)

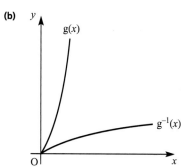

(c) $26; \dfrac{1}{26}$

3 $\dfrac{1}{3\pi}$ cm s^{-1}

4 (a) 3.2π (≈ 10.5) cm^3 s^{-1}
   (b) 0.8 cm s^{-1}

5 (a) $x = 0, 2\dfrac{2}{3}$
   (c)

(d) $\dfrac{4}{13}$

6 (a) (i)

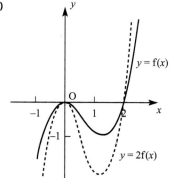

ANSWERS

Pure Mathematics: Core 3

(ii)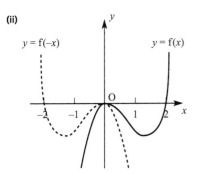

(iii)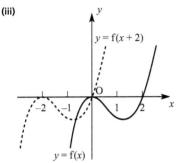

(b) Not one-to-one

(c)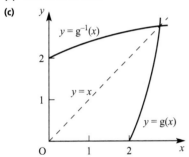

(d) $15; \frac{1}{15}$

8 (a) $x = e^{-y} - 3$

(b) $\frac{dx}{dy} = -e^{-y}$

(c) $\frac{dy}{dx} = -\frac{1}{x+3}$

9 16

10 (a)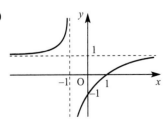

(b) $\frac{2}{(x+1)^2} > 0$

(c) $\frac{9}{2}$

EXERCISE 4E (Page 134)

1 $\frac{\delta y}{\delta x} = \frac{\cos(x + \delta x) - \cos x}{\delta x}$

$= \frac{\cos x \cos \delta x - \sin x \sin \delta x - \cos x}{\delta x}$

As $\delta x \to 0$, $\cos \delta x \to 1 - \frac{1}{2}(\delta x)^2$ and $\sin \delta x \to \delta x$, so

$\lim\limits_{\delta x \to 0} \frac{\delta y}{\delta x} = \lim\limits_{\delta x \to 0} \frac{\cos x(1 - \frac{1}{2}(\delta x)^2) - \sin x \times \delta x - \cos x}{\delta x}$

$= \lim\limits_{\delta x \to 0} -\frac{1}{2}(\cos x)dx - \sin x$

$= -\sin x$

2 (a) $\frac{d}{dx}(\sec x) = \frac{d}{dx}(\cos x)^{-1}$

$= -(\cos x)^{-2} \times -\sin x$

$= \frac{\sin x}{\cos^2 x}$

$= \frac{1}{\cos x} \times \frac{\sin x}{\cos x}$

$= \sec x \tan x$

(b) $\frac{d}{dx}(\csc x) = \frac{d}{dx}(\sin x)^{-1}$

$= -(\sin x)^{-2} \times \cos x$

$= -\frac{1}{\sin x} \times \frac{\cos x}{\sin x}$

$= -\csc x \cot x$

(c) $\frac{d}{dx}(\cot x) = \frac{d}{dx}\left(\frac{\cos x}{\sin x}\right)$

$= \frac{-\sin^2 x - \cos^2 x}{\sin^2 x}$

$= -\frac{1}{\sin^2 x}$

$= -\csc^2 x$

3 (a) $-2\sin x + \cos x$
(b) $\sec^2 x$
(c) $\cos x + \sin x$
(d) $-3\sin 3x$

4 (a) $x\sec^2 x + \tan x$
(b) $\cos^2 x - \sin^2 x = \cos 2x$
(c) $e^x(\sin x + \cos x)$
(d) $2x(x \cos 2x + \sin 2x)$

5 (a) $\frac{x \cos x - \sin x}{x^2}$

(b) $\frac{e^x(\cos x + \sin x)}{\cos^2 x}$

(c) $\frac{\sin x(1 - \sin x) - \cos x(x + \cos x)}{\sin^2 x}$

(d) $-\csc^2 x$

6 (a) $2x\sec^2(x^2+1)$
 (b) $-\sin 2x$
 (c) $\cot x$
 (d) $2x\cos(x^2+1)$

7 (a) $-\dfrac{\sin x}{2\sqrt{\cos x}}$
 (b) $e^x(\tan x + \sec^2 x)$
 (c) $8x\cos 4x^2$
 (d) $-2\sin 2x\, e^{\cos 2x}$
 (e) $\dfrac{1}{(1+\cos x)}$
 (f) $\dfrac{1}{(\sin x \cos x)} = 2\csc 2x$

8 (a) $3\sec 3x \tan x$
 (b) $-20\csc 4x \cot 4x$
 (c) $-3\cot^2 x \csc^2 x$
 (d) $\sec x\,(2\sec^2 x - 1)$

9 (a) $\cos x - x\sin x$
 (b) -1
 (c) $y = -x$
 (d) $y = x - 2\pi$

10 (a) $3\cos x \sin^2 x$
 (b) $(-\pi, 0)$ point of inflection, $(-\tfrac{1}{2}\pi, -1)$ min,
 $(0, 0)$ point of inflection, $(\tfrac{1}{2}\pi, 1)$ max,
 $(\pi, 0)$ point of inflection

11 (a) $1 + 2\cos 2x$
 (b) $\left(\dfrac{\pi}{3}, \dfrac{\pi}{3} + \dfrac{\sqrt{3}}{2}\right)$ max, $\left(\dfrac{2\pi}{3}, \dfrac{2\pi}{3} - \dfrac{\sqrt{3}}{2}\right)$ min,
 $\left(\dfrac{4\pi}{3}, \dfrac{4\pi}{3} + \dfrac{\sqrt{3}}{2}\right)$ max, $\left(\dfrac{5\pi}{3}, \dfrac{5\pi}{3} - \dfrac{\sqrt{3}}{2}\right)$ min,

13 (a) $e^{-x}(\cos x - \sin x)$
 (b) $(0.79, 0.32), (-2.4, -7.5)$

14 (a) (ii) $2\cos 2x$
 (b) (i) $1 - \tfrac{1}{2}\sin^2 2x$

15 (a) $2e^{-x}\cos 2x - e^{-x}\sin 2x$
 (c) $x = 0.55$ or 2.12
 (d) $r = \sqrt{5}, \alpha = 0.46$

16 $e^{-2x}\sec^2 x - 2e^{-2x}\tan x$

Exercise 4F (page 136)

1 (a) $h = \ln 4$
 (b) $y = 4x + 3 - 4\ln 4$

2 (b) $15y = 4x - 121$

3 (b) $y = 2x + 3$

4 Tangent: $y = (e+2)x + 1$
 Normal: $x + (e+2)y = e^2 + 5e + 7$

5 (b) $\dfrac{1}{x} - \tfrac{1}{2}e^x$
 (c) (i) $2y = (2-e)x + 18 + 2\ln 3$
 (ii) $9 + \ln 3$

6 (b) $\tfrac{5}{4} + \tfrac{1}{2}\ln 2$

7 (a) (i) $2x(x\cos 2x + \sin 2x)$
 (ii) $-\dfrac{1}{(x+1)^{\frac{1}{2}}(x-1)^{\frac{3}{2}}}$
 (iii) $(1 + \ln x)e^{x\ln x}$
 (b) $25\ln 5$

8 (a) $\dfrac{x\cos x - \sin x}{x^2}$
 (b) $-\dfrac{2x}{x^2+9}$

9 (a) $\tfrac{1}{15}$

10 (a) $xe^{3x}(2+3x)$
 (c) $x = 0$, minimum;
 $x = -\tfrac{2}{3}$, maximum

11 (a) $\dfrac{dy}{dx} = 2\cos 2x - 2\sin x$
 $\dfrac{d^2y}{dx^2} = -4\sin 2x - 2\cos x$
 (b) $\tfrac{\pi}{6}, \tfrac{5\pi}{6}, \tfrac{3\pi}{2}$
 (c) maximum, minimum, point of inflection
 (d)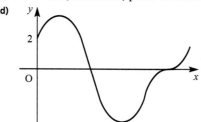

12 (a) $0, \tfrac{\pi}{2}, \pi$
 (b) $\left(\dfrac{3\pi}{8}, \dfrac{1}{\sqrt{2}}e^{\frac{3\pi}{4}}\right), \left(\dfrac{7\pi}{8}, -\dfrac{1}{\sqrt{2}}e^{\frac{7\pi}{4}}\right)$
 (d) $x = \dfrac{3\pi}{8}$, max; $x = \dfrac{7\pi}{8}$, min

13 $|x| > \sqrt{2}$

14 (a) $\dfrac{1}{y(1+2\ln y)}$
 (b) $\dfrac{1}{3e}$

CHAPTER 5

EXERCISE 5A (Page 146)

1 −1.62, 1.28

2 (a) [−2, −1]; [1, 2]; [4, 5]
   (b)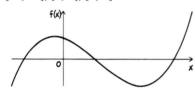
   (c) −1.51, 1.24, 4.26
   (d) $a = -1.51171875$, $n = -8$
       $a = 1.244386172$, $n = -12$
       $a = 4.262695313$, $n = -10$

3 (a) [1, 2]; [4, 5]
   (b) 1.857, 4.536

4 (b)
   (c) 1.154

5 (a) 2
   (b) [0, 1]; [1, 2]
   (c) 0.62, 1.51

6 (a)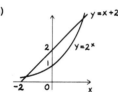
   (b) 2 roots
   (c) 2, −1.690

7 −1.88, 0.35, 1.53

8 (a) (ii) No root
       (iii) Convergence to a non-existent root
   (b) (ii) $x = 0$
       (iii) Success
   (c) (ii) $x = 0$
       (iii) Failure to find root

9 $a = 4, 0.8$

10 −3, 0.5, 2; 0.53

EXERCISE 5B (Page 154)

1 (c) 1.521

2 (c) 2.120

3 (c) $x = \sqrt[3]{3 - x}$
   (d) 1.2134

4 (b) 1.503

5 (a)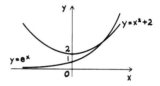
   (b) Only one point of intersection
   (d) 1.319 (to 3 d.p.)

6 (a)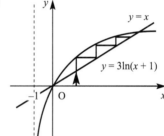
   (b) 5.711

7 (a)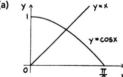
   (b) 0.73909

8 (a) $x^2 - 3x + 1 = 0$
   (b) 0.382
   (c)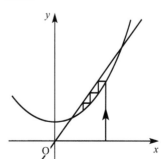
   (d) 2.62

9 (a)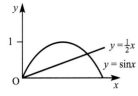
(c) 1.895

10 (a)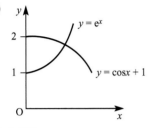
(c) 0.601

EXERCISE 5C (Page 155)

1 (b) $N = 23$
(c) $[-3, -2]$

2 (b) $N = -3$
(c) 1

3 (b)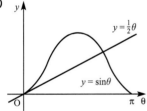
(c) $[1.89, 1.90]$

4 (a) $a = 2, b = 5, 3.449$
(b) -1.449

5 $p = 5, q = 6, r = 5; 1.708$

6 (a)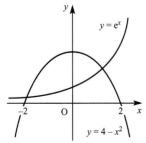
(b) -1.9646
(c) No real solution

7 (a) $a = 6, b = 2$
(b) 2.60

8 (a) $x_5 \approx 1.5607$
(b) $x_1 \approx 1.6330$ (4 d.p.)
$x_2 \approx 1.4663$
$x_3 \approx 1.6731$
$x_4 \approx 1.4144$
oscillating between two solutions.

9 (a) $q = -\frac{1}{4}$
(b) $x_4 \approx 0.1888$

10 (b) $x_3 \approx 1.327$
(c) $p = -1.25$
(d) $x_3 \approx -2.642$

11 (a) $f'(x) = 1 + x + (1 + 2x)\ln x$
(c) $x_3 \approx 0.4685$
(d) $(0.47, -0.52)$

12 (a) 2.422
(b) $f(2.4220) \approx -2.6 \times 10^{-4} < 0$
$f(2.4225) \approx 4.5 \times 10^{-4} > 0$
∴ 2.422 is accurate to 3 d.p.

13 (a) $f'(x) = (-x^3 + 3x^2 + 2x - 2)e^{-x}$
(c) $x \approx -1.455$

14 (a) $2y = x + 1$
(b) from $\frac{1}{2}e^x - 2x = 0$
(c) $x_3 \approx 2.1530$

15 (a)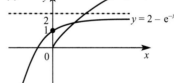

(b) One point of intersection
(c) $f(3) < 0$
$f(4) > 0$
(d) $x \approx 3.921$

Answers (Core 4)

Chapter 6

Exercise 6A (Page 163)

1. $\dfrac{x-3}{x-2}$

2. $\dfrac{1}{(x+2)(x+3)}$

4. $\dfrac{2x}{(x-5)}$

5. $\dfrac{x+1}{(x-1)(x-2)^2}$

Exercise 6B (Page 166)

1. $\dfrac{1}{(x-2)} - \dfrac{1}{(x+3)}$

2. $\dfrac{1}{x} - \dfrac{1}{(x+1)}$

3. $\dfrac{2}{(x-4)} - \dfrac{2}{(x-1)}$

4. $\dfrac{2}{(x-1)} - \dfrac{1}{(x+2)}$

5. $\dfrac{1}{(x+1)} + \dfrac{1}{(2x-1)}$

6. $\dfrac{2}{(x-2)} - \dfrac{2}{x}$

7. $\dfrac{1}{(x-1)} - \dfrac{3}{(3x-1)}$

8. $\dfrac{3}{5(x-4)} + \dfrac{2}{5(x+1)}$

9. $\dfrac{5}{(2x-1)} - \dfrac{2}{x}$

10. $\dfrac{2}{(2x-3)} - \dfrac{1}{(x+2)}$

11. $\dfrac{8}{13(2x-5)} + \dfrac{9}{13(x+4)}$

12. $\dfrac{19}{24(3x-2)} - \dfrac{11}{24(3x+2)}$

13. $\dfrac{3}{5(x-3)} - \dfrac{1}{(x-1)} + \dfrac{2}{5(x+2)}$

14. $\dfrac{1}{(x-1)} - \dfrac{5}{2(x-2)} + \dfrac{3}{2(x-4)}$

15. $\dfrac{1}{5(2x+1)} + \dfrac{1}{(2x-1)} - \dfrac{9}{5(3x-1)}$

Exercise 6C (Page 168)

1. $\dfrac{9}{(1-3x)} - \dfrac{3}{(1-x)} - \dfrac{2}{(1-x)^2}$

2. $\dfrac{8}{3(2x-1)} - \dfrac{4}{3(x+1)} + \dfrac{1}{(x+1)^2}$

3. $\dfrac{1}{(x-1)^2} - \dfrac{1}{(x-1)} + \dfrac{1}{(x+2)}$

4. $\dfrac{1}{4(x-2)} - \dfrac{1}{4(x+4)} + \dfrac{1}{2(x+4)^2}$

5. $\dfrac{1}{(2x-3)^2} - \dfrac{2}{2x-3} + \dfrac{1}{x+2}$

6. $\dfrac{2}{x} - \dfrac{1}{x^2} - \dfrac{3}{(2x+1)}$

7. $\dfrac{1}{x} - \dfrac{3}{3x-1} + \dfrac{4}{(3x-1)^2}$

8. $\dfrac{4}{(2x+1)^2} - \dfrac{5}{(2x+1)} + \dfrac{3}{x+1}$

9. $\dfrac{8}{(2x-1)} - \dfrac{4}{(2x-1)^2} - \dfrac{3}{x}$

10. $\dfrac{1}{(1-x)} + \dfrac{1}{(1+x)} + \dfrac{2}{(1+x)^2}$

11 $\dfrac{16}{9(2x-1)} - \dfrac{8}{9(1+x)} - \dfrac{1}{3(1+x)^2}$

12 $\dfrac{7}{36(4-x)} + \dfrac{7}{36(2+x)} + \dfrac{1}{6(2+x)^2}$

13 $\dfrac{7}{48(2-x)} - \dfrac{5}{3(1+x)} + \dfrac{87}{48(2+x)} + \dfrac{9}{4(2+x)^2}$

14 $\dfrac{4}{x-2} - \dfrac{4}{x-3} + \dfrac{5}{(x-3)^2}$

15 $\dfrac{1}{9(x+1)} - \dfrac{1}{9(x-2)} + \dfrac{1}{3(x-2)^2}$

EXERCISE 6D (Page 170)

1 $1 + \dfrac{1}{2(x-1)} + \dfrac{13}{2(x-3)}$

2 $x + \dfrac{12}{5(x-2)} + \dfrac{18}{5(x+3)}$

3 $1 + \dfrac{3}{4(x-2)} - \dfrac{3}{4(x+2)}$

4 $1 + \dfrac{3}{x+2} - \dfrac{5}{x+3}$

5 $x + 7 + \dfrac{130}{3(x-5)} - \dfrac{10}{3(x-2)}$

6 $1 - \dfrac{3}{2(x+1)} + \dfrac{x-3}{2(x^2+1)}$

7 $x + \dfrac{5}{4(x-2)} - \dfrac{1}{4(x+2)} + \dfrac{3}{(x+2)^2}$

8 $1 + \dfrac{1}{12(x+1)} + \dfrac{3}{16(x+3)} + \dfrac{35}{48(x-5)}$

9 $2 - \dfrac{2}{7(2x+1)} + \dfrac{36}{7(x-3)}$

10 $1 + \dfrac{x}{x^2+1} + \dfrac{1}{x-1} + \dfrac{1}{(x-1)^2}$

EXERCISE 6E (Page 171)

1 (a) $\dfrac{20}{x+4} - \dfrac{15}{x-6}$

  (b) $f'(x) = -\dfrac{20}{(x+4)^2} + \dfrac{15}{(x-6)^2}$

2 (a) $A = 4, B = -1$

3 (a) $\dfrac{2}{x+2} - \dfrac{3}{2x-1}$

  (b) $5\tfrac{1}{2}$

4 (a) $\dfrac{-7x^2 - 26x - 47}{(x^2 + 2x - 3)^2}$

  (b) $f(x) = \dfrac{2}{x+3} + \dfrac{5}{x-1}$

  $f'(x) = -\dfrac{2}{(x+3)^2} - \dfrac{5}{(x-1)^2}$

5 (a) $\dfrac{3}{2-x} + \dfrac{4}{4-x} - \dfrac{2}{1-x}$

6 (a) $A = -1, B = 3, C = 2$

  (b) $f'(x) = \dfrac{1}{(x+2)^2} - \dfrac{3}{(x+1)^2} - \dfrac{2}{(x-1)^2}$

7 (a) $\dfrac{5}{4-x} - \dfrac{3}{x+1} + \dfrac{2}{(x+1)^2}$

  (c) $\tfrac{41}{8}$

8 (a) $A = -4, B = -9, C = 2$

9 (a) $A = 1, B = 2, C = 1, D = -1$

  (c) 1

10 (a) $Q = 2x - 3, R = -2x - 16$

  (b) $2x - 3 + \dfrac{2}{x+3} - \dfrac{4}{x-2}$

  (c) $f'(x) = 2 - \dfrac{2}{(x+3)^2} + \dfrac{4}{(x-2)^2}$

EXERCISE 6F (Page 173)

1 $\dfrac{1}{x+1} + \dfrac{2}{x+2} - \dfrac{2}{x+3}$

2 $\dfrac{3}{2(x-1)} - \dfrac{3}{2(x+1)} - \dfrac{2}{(x+1)^2}$

3 (b) $2 + \dfrac{12}{5(x-2)} - \dfrac{9}{5(2x+1)}$

  (c) As $x \to \pm\infty$, $f(x) \to 2$

4 (a) $1 + \dfrac{1}{2(x-1)} - \dfrac{1}{2(x+1)}$

5 $A = 2, B = 4, C = -1$

6 $\dfrac{2}{x-1} + \dfrac{1}{2x-1} - \dfrac{3}{3x+1}$

7 (a) $\dfrac{4}{x-2} + \dfrac{3}{(x+2)^2}$

  (c) $-\tfrac{11}{3}$

8 $A = 1, B = 3, C = -1$

9 $\dfrac{1}{x-1} - \dfrac{4}{x+1} + \dfrac{3}{x+2} + \dfrac{1}{(x+2)^2}$

10 $A = 2, B = \tfrac{11}{5}, C = \tfrac{2}{5}, D = 10$

Answers

Exercise 7A (Page 183)

1 (a) (i) $y = \dfrac{x^2}{4}$

(ii)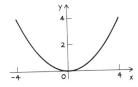

(b) (i) A segment of $y = \dfrac{1-x}{2}$, where $-1 \leq x \leq 1$ and $0 \leq y \leq 1$

(ii)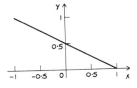

(c) (i) $y^2 = x^3$

(ii)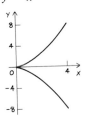

(d) (i) Part of $(y-1)^2 = 4x$, where $0 \leq x \leq 1$ and $-1 \leq y \leq 3$

(ii)

(e) (i) $x^2 - y^2 = 4$

(ii)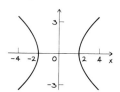

(f) (i) Part of $y^2 = \tfrac{9}{2}(2-x)$ where $0 \leq x \leq 2$ and $-3 \leq y \leq 3$

(ii)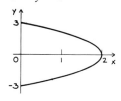

(g) (i) $y = 1 - x^2$

(ii)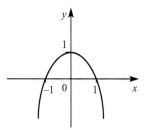

(h) (i) Part of $y = 1 - 2x^2$ where $-1 \leq x \leq 1$ and $-1 \leq y \leq 1$

(ii)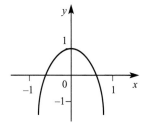

(i) (i) $y = \dfrac{x}{1-2x}$

(ii)

2 (a) $x^2 + y^2 = 25$

(b) $\dfrac{x^2}{9} + \dfrac{y^2}{4} = 1$

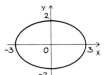

(c) Segment of $y = x - 3$, where $1 \leq x \leq 7$ and $-2 \leq y \leq 4$

(d) $(x + 1)^2 + (y - 3)^2 = 4$

3 (a) (i) $xy = 1$, $xy = 16$
  (b)

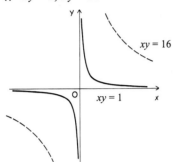

  (c) $xy = 16$ is a stretch of $xy = 1$, parallel to y-axis

4 (a) $\dfrac{x^2}{25} + \dfrac{y^2}{9} = 1$

  (b) Inscribed circle $x = 3\cos\theta$, $y = 3\sin\theta$; circumscribing circle $x = 5\cos\theta$, $y = 5\sin\theta$

5 (a) $\dfrac{x^2}{9} + \dfrac{y^2}{16} = 1$
  (b) $x = 3\cos\theta$, $y = 4\sin\theta$

6 $x = 2t$, $y = \dfrac{2}{t}$

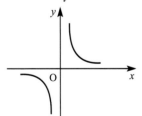

7 (a) $y = 8x^2(x - 1)$
  (b)

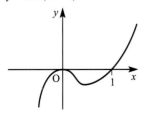

8 (a)

t	−4	−3	−2	−1	0	1	2	3	4
x	9	4	1	0	1	4	9	16	25
y	−5	−4	−3	−2	−1	0	1	2	3

  (b)

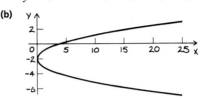

  (c) $y = -2$
  (d) $x = (y + 2)^2$

9 $x^{\frac{2}{3}} + y^{\frac{2}{3}} = 1$

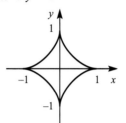

10 (b)

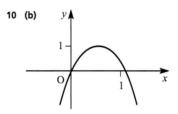

EXERCISE 7B (Page 187)

1 (a) $\dfrac{448}{3}$
  (b) $\dfrac{4}{3}$
  (c) $\dfrac{15}{8} - 2\ln 2 \approx 0.489$
  (d) $74\dfrac{2}{3}$

2 (a) $\dfrac{3\pi}{2}$
  (b) $\dfrac{1}{2}\pi r^2$
  (c) $\dfrac{1}{8}(3\sqrt{3} - 1) \approx 0.525$
  (d) $\dfrac{1}{2}(\pi^2 + 3\pi + 4) \approx 11.647$

3 (b) 583.2

ANSWERS

4 (a)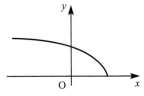
 (b) $t = 1$
 (c) $\frac{4}{3}$

5 (a)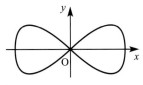
 (b) $\frac{8}{3}$

EXERCISE 7C (Page 188)

1 (a) $y = \frac{2}{9}x^2 - 1$
 (b)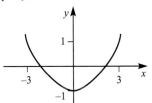

2 (a) $y = x^2 + 3x + 2$
 (b)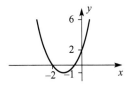

3 (a) $t = -\frac{\pi}{4} \Rightarrow x = -\sqrt{2}$
   $t = \frac{\pi}{4} \Rightarrow x = \sqrt{2}$
 (b) $y = 1 - \frac{1}{2}x^2$
 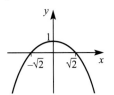

4 (a) (i) $\dfrac{x^2}{9} + \dfrac{y^2}{16} = 1$
 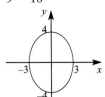

 (ii) $x^2 + y^2 = 16$
 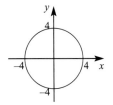

 One-way stretch, scale factor $\frac{4}{3}$, parallel to x-axis, y-axis invariant

 (b) $x = 5\cos t$, $y = 5\sin t$

5 (a) $y = \dfrac{x + 3}{x - 1}$
 (b) $5y = 2x + 9$
 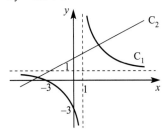
 (c) $(-4, 0.2)$, $(3, 3)$

6 46.9

7 (a) $y^2 = 4x^2(9 - x^2)$
 (b) $A = -27$
 (c) -17
 (d) 40

8 (a) 20π
 (c) 0.345

9 (a) $-\frac{1}{8}$
 (b) $y = 2 - x$
 (c)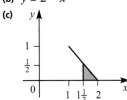

10 (a) $7\frac{1}{5}$

Chapter 8

Exercise 8A (Page 195)

1. $243x^5 - 810x^4 + 1080x^3 - 720x^2 + 240x - 32$

2. $x^8 + 8x^6 + 28x^4 + 56x^2 + 70 + \dfrac{56}{x^2} + \dfrac{28}{x^4} + \dfrac{8}{x^6} + \dfrac{1}{x^8}$

3. $1024x^{10} + 2560x^9 + 2880x^8 + \ldots$

4. $3\,294\,720$

5. $1 + nx + \dfrac{1}{2!}n(n-1)x^2 + \dfrac{1}{3!}n(n-1)(n-2)x^3 + \ldots + nx^{n-1} + x^n$

Exercise 8B (Page 202)

1. (a) $1 - 2x + 3x^2$
   (b) $|x| < 1$

2. (a) $1 - 2x + 4x^2$
   (b) $|x| < \tfrac{1}{2}$

3. (a) $1 - \dfrac{x^2}{2} - \dfrac{x^4}{8}$
   (b) $|x| < 1$

4. (a) $1 + 4x + 8x^2$
   (b) $|x| < \tfrac{1}{2}$

5. (a) $\dfrac{1}{3} - \dfrac{x}{9} + \dfrac{x^2}{27}$
   (b) $|x| < 3$

6. (a) $2 - \dfrac{7x}{4} - \dfrac{17x^2}{64}$
   (b) $|x| < 4$

7. (a) $-\dfrac{2}{3} - \dfrac{5x}{9} - \dfrac{5x^2}{27}$
   (b) $|x| < 3$

8. (a) $\dfrac{1}{2} - \dfrac{3x}{16} + \dfrac{27x^2}{256}$
   (b) $|x| < \tfrac{4}{3}$

9. (a) $1 + 6x + 20x$
   (b) $|x| < \tfrac{1}{2}$

10. (a) $1 + 2x^2 + 2x^4$
    (b) $|x| < 1$

11. (a) $1 + \dfrac{2x^2}{3} - \dfrac{4x^4}{9}$
    (b) $|x| < \dfrac{1}{\sqrt{2}}$

12. (a) $1 - 3x + 7x^2$
    (b) $|x| < \tfrac{1}{2}$

13. (a) $1 + 3x + 3x^2 + x^3$
    (b) $1 + 4x + 10x^2 + 20x^3$ for $|x| < 1$
    (c) $a = 25, b = 63$

14. (a) $16 - 32x + 24x^2 - 8x^3 + x^4$
    (b) $1 - 6x + 24x^2 - 80x^3$ for $|x| < \tfrac{1}{2}$
    (c) $a = -128, b = 600$

15. (a) $1 + x + x^2 + x^3$ for $|x| < 1$
    (b) $1 - 4x + 12x^2 - 32x^3$ for $|x| < \tfrac{1}{2}$
    (c) $1 - 3x + 9x^2 - 23x^3$ for $|x| < \tfrac{1}{2}$

16. (b) $1 + \dfrac{x}{8} + \dfrac{3x^2}{128}$ for $|x| < 4$
    (c) $1 + \dfrac{9x}{8} + \dfrac{19x^2}{128}$

17. (a) $1 - y + y^2 - y^3 \ldots$
    (b) $1 - \dfrac{2}{x} + \dfrac{4}{x^2} + \dfrac{8}{x^3}$
    (c) $\dfrac{x}{2}\left(1 + \dfrac{x}{2}\right)^{-1}$
    (d) $\dfrac{x}{2} - \dfrac{x^2}{4} + \dfrac{x^3}{8} - \dfrac{x^4}{16}$
    (e) $x < -2$ or $x > 2$; $-2 < x < 2$; no overlap in range of validity

18. (a) $1 - x - x^2 - \tfrac{5}{3}x^3$
    (b) $9.989\,989\,98$

Exercise 8C (Page 204)

1. (a) $\dfrac{2}{(2x-1)} - \dfrac{3}{(x+2)}$
   (b) $1 + 2x + 4x^2 \ldots a = 1, b = 2, c = 4$, for $|x| < \tfrac{1}{2}$
   (c) $\dfrac{1}{2} - \dfrac{x}{4} + \dfrac{x^2}{8}$ for $|x| < 2$
   (d) $-\dfrac{7}{2} - \dfrac{13x}{4} - \dfrac{67x^2}{8}$

2. $1 + x^2 + x^4, |x| < 1$

3. $\dfrac{1}{3} + \dfrac{2x}{9} + \dfrac{5x^2}{24}, |x| < 2$

4. $2 - 5x + \dfrac{23x^2}{2}, |x| < \tfrac{1}{2}$

5. $\dfrac{5}{3} + \dfrac{7x}{9} + \dfrac{29x^2}{27}, |x| < 1$

6. $4 + 20x + 72x^2, |x| < \tfrac{1}{3}$

Answers

Pure Mathematics: Core 4

7 $-5 - 2x - 15x^2$, $|x| < \frac{1}{2}$

8 $\frac{5}{2} + \frac{11x}{4} + \frac{33x^2}{8}$, $|x| < 1$

EXERCISE 8D (Page 205)

1 (a) $1 + x - \frac{3}{2}x^2 + \frac{7}{2}x^3$
  (b) $1.009\,854$

2 (a) $1 - 6x + 24x^2 - 80x^3$
  (b) $a = -7$, $b = -104$

3 (a) $10 + \frac{x}{300} - \frac{x^2}{900\,000}$
  (b) $10.009\,99$

4 (a) $1 - \frac{1}{2}x - \frac{1}{8}x^2 - \frac{1}{16}x^3$
  (c) $1.414\,21$

5 (a) $\frac{1}{2} + \frac{1}{4}x + \frac{1}{8}x^2 + \frac{1}{16}x^3$
  (b) $|x| < 2$

6 (a) $1 + \frac{1}{3}x - \frac{1}{9}x^2$
  (b) $a = 3$, $b = -\frac{2}{3}$

7 (a) $1 + 2x + 2x^2 + 2x^3$
  (b) $|x| < 1$

8 (a) $1 - 4x + 12x^2$, $|x| < \frac{1}{2}$

9 (a) $-\frac{1}{2}, \frac{3}{2}$
  (b) $-\frac{3}{2}, \frac{9}{2}$

10 (a) $1 - 3x + 9x^2 - 27x^3$
   (b) $x = 0.01$, $\frac{101}{103} \approx 0.980\,582$

11 (a) $\frac{3}{1+x} + \frac{2}{2-x}$

12 (a) $A = 2$, $B = 16$
   (b) $10 + 10x^2 + 15x^3$

13 (a) $A = 2$, $B = 1$, $C = 3$, $D = 1$
   (b) $\frac{1}{2} + \frac{17}{12}x + \frac{1}{72}x^2$
   (c) $|x| < 2$

14 (a) $\frac{1}{2} - \frac{1}{4}x + \frac{1}{8}x^2 - \frac{1}{16}x^3 + \frac{1}{32}x^4 - \frac{1}{64}x^5$
   (b) $\frac{1}{2} + \frac{1}{4}x + \frac{1}{8}x^2 + \frac{1}{16}x^3 + \frac{1}{32}x^4 + \frac{1}{64}x^5$
   (c) $\frac{1}{4}\left(\frac{1}{2+x} + \frac{1}{2-x}\right)$
   (d) $\frac{1}{4} + \frac{1}{16}x^2 + \frac{1}{64}x^4$

15 (a) $A = 3$, $B = 1$, $C = 2$
   (b) $a = -\frac{14}{3}$, $b = -\frac{242}{27}$
   (c) $|x| < 1$

CHAPTER 9

EXERCISE 9A (Page 214)

1 (a) $4y^3 \dfrac{dy}{dx}$
  (b) $2x + 3y^2 \dfrac{dy}{dx}$
  (c) $x\dfrac{dy}{dx} + y + 1 + \dfrac{dy}{dx}$
  (d) $-\sin y \dfrac{dy}{dx}$
  (e) $e^{(y+2)} \dfrac{dy}{dx}$
  (f) $y^3 + 3xy^2 \dfrac{dy}{dx}$
  (g) $4xy^5 + 10x^2y^4 \dfrac{dy}{dx}$
  (h) $1 + \dfrac{1}{y}\dfrac{dy}{dx}$
  (i) $xe^y \dfrac{dy}{dx} + e^y + \sin y \dfrac{dy}{dx}$
  (j) $\dfrac{x^2}{y}\dfrac{dy}{dx} + 2x\ln y$
  (k) $e^{\sin y} + x\cos y\, e^{\sin y}\dfrac{dy}{dx}$
  (l) $\tan y + (x\sec^2 y)\dfrac{dy}{dx} - (\tan x)\dfrac{dy}{dx} - y\sec^2 x$

2 $\frac{1}{5}$

3 0

4 (a) -1
  (b) $x + y - 2 = 0$

5 $(1, -2)$ and $(-1, 2)$

6 (b) Max $(4, 8)$, min $(-4, -8)$

7 (a) $\dfrac{y+4}{6-x}$
  (b) $x - 2y - 11 = 0$
  (c) $(2, -4\frac{1}{2})$
  (d)
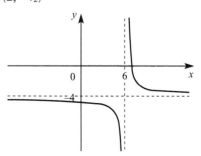

Asymptotes $x = 6$, $y = -4$

Pure Mathematics: Core 4 — Answers

8 (b) $8x + 7y = 22$

9 (a) $-\dfrac{x+y}{x+3y}$

(b) $-\tfrac{1}{9}, -\tfrac{5}{9}$

10 (a) $\dfrac{3y - 2x}{2y - 3x}$

(b) $x = 2$

(c) $(3, 2), (3, 7)$

(d) $x + 3y = 24$

EXERCISE 9B (Page 218)

1 (a) t

(b) $\dfrac{1 + \cos\theta}{1 + \sin\theta}$

(c) $\dfrac{t^2 + 1}{t^2 - 1}$

(d) $-\tfrac{2}{3}\cot\theta$

(e) $\dfrac{t - 1}{t + 1}$

(f) $-\tan\theta$

(g) $\dfrac{1}{2e^t}$

(h) $\dfrac{(1 + t)^2}{(1 - t)^2}$

2 (a) 6

(b) $y = 6x - \sqrt{3}$

(c) $3x + 18y - 19\sqrt{3} = 0$

3 (a) $(\tfrac{1}{4}, 0)$

(b) 2

(c) $y = 2x - \tfrac{1}{2}$

(d) $(0, -\tfrac{1}{2})$

4 (a) $x - ty + at^2 = 0$

(b) $tx + y = at^3 + 2at$

(c) Normal cuts x-axis $(at^2 + 2a, 0)$, y-axis $(0, at^3 + 2at)$

6 (a) $-\dfrac{b}{at^2}$

(b) $at^2y + bx = 2abt$

(c) X$(2at, 0)$, Y$\left(0, \dfrac{2b}{t}\right)$

(d) area = $2ab$

7 (a)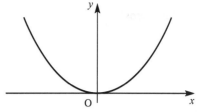

(c) $y = tx - 2t^2$

(d) $[2(t_1 + t_2), 2t_1t_2]$

(e) $x = 4$

8 (a) $t = 1$

(c) $x + y = 3$

(e) $(-8, -5)$

9 (a) $t = -2$

(c) $y = 2x - 6$

(d) $(-5, 9)$

10 (a) $-\dfrac{3\cos t}{4\sin t}$

(b) $3x\cos t + 4y\sin t = 12$

(c) $t = 0.6435 + n\pi$

11 (a) $x\cos\theta + y\sin\theta = 3\sin\theta + 3\cos\theta + 2$

(c) 2.85, 5.01 radians

(d)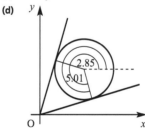

12 (a) $-\dfrac{\cos\theta}{\sin\theta}$

(b) $y\cos\theta - x\sin\theta = 5\cos\theta - 2\sin\theta$

13 (a) $\dfrac{x^2}{9} + \dfrac{y^2}{4} = 1$

(b)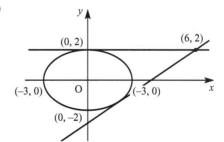

(c) $-\dfrac{2\cos\theta}{3\sin\theta}$

(e) $\theta = 1.57$ or 5.64 (2 d.p.)

ANSWERS

14 (a) $\frac{dx}{dt} = -2\sin t + 2\cos 2t, \frac{dy}{dt} = -\sin t - 4\cos 2t$
 (b) $\frac{1}{2}$
 (c) $y + 2x = \frac{5\sqrt{2}}{2}$

15 (a) $3y = 2x + 1$
 (b) $(8, 4)$
 (c) $2t^2 + t + 2 = 0$ has no real roots.

EXERCISE 9C (Page 224)

1 (a) $3^x \ln 3$
 (b) $b^x \ln b$
 (c) $2^{\sin x} \times \ln 2 \times \cos x$
 (d) $12^x \ln 12$
 (e) $10^{x^2} \times \ln 10 \times 2x$
 (f) $x^x(1 + \ln x)$
 (g) $(\tan x)^x \left(\frac{x}{\sin x \cos x} + \ln(\tan x) \right)$
 (h) $\sec x \tan x e^{\sec x}$
 (i) 3×8^x

2 (a) $71\,000$ mm^2/day
 (b) 7.2 days

3 (a) 750
 (b) 830 people/year

4 (a) 6400
 (b) -4600 cells/day
 (c) decreases

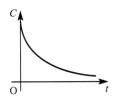

5 (a)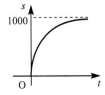
 (b) no
 (c) 355 m/hour
 (d) 78 m/hour, 898 m

EXERCISE 9D (Page 228)

2 $\frac{ds}{dt} = \frac{k}{s^2}$

3 $\frac{dh}{dt} = k\ln(H - h)$

4 $\frac{dm}{dt} = \frac{k}{m}$

5 $\frac{dP}{dt} = k\sqrt{P}$

6 $\frac{de}{d\theta} = k\theta$

7 $\frac{d\theta}{dt} = -\frac{(\theta - 15)}{160}$

8 $\frac{dN}{dt} = \frac{N}{20}$

9 $\frac{dv}{dt} = \frac{4}{\sqrt{v}}$

10 $\frac{dA}{dt} = \frac{2k\sqrt{\pi}}{\sqrt{A}} = \frac{k'}{\sqrt{A}}$

11 $\frac{d\theta}{ds} = -\frac{s}{4}$

12 $\frac{dh}{dt} = -\frac{1}{8\pi h^2}$

13 $\frac{dV}{dt} = -\frac{2V}{1125\pi}$

14 $\frac{dh}{dt} = -\frac{(2 - k\sqrt{h})}{100}$

EXERCISE 9E (Page 230)

1 (a) $(1, -7)$
 (b) $-\frac{4x + 5y}{5x + 2y}$
 (c) $14y = 9x + 19$

2 $8y = 19x - 46$

3 (a) $(3, -8)$
 (b) $7x + 10y = 41$

4 $x + 4y + 4 = 0$

5 (b) $(-2, 1), (2, -1)$

6 (b) 2

7 (b) $4y = 6x - 1$

Pure Mathematics: Core 4 — ANSWERS

8 (a) $\frac{\pi}{6}$

(b) $-\frac{3}{16\sin t}$

(d) $\frac{-123}{64}$ (−1.92)

9 (b) $5y = 3x - 15$

10 (a) $\left(\pm\frac{5\sqrt{3}}{2}, 0\right)$, (0, 2), (0, −6)

(b)
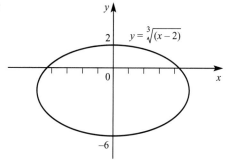

12 (a) $A = 5 \times 2^t$

14 $V = \frac{\pi}{27}h^3$, $\frac{dh}{dt} = \frac{16}{h^2}$

15 (a) $-4\pi\ln 2$

(b) $-\frac{\pi}{128}\ln 2$

Chapter 10

Exercise 10A (Page 237)

1 (a) $-\frac{10}{3}x^{-3} + c$
(b) $x^2 + x^{-3} + c$
(c) $2x + \frac{1}{4}x^4 - \frac{5}{2}x^{-2} + c$
(d) $2x^3 + 7x^{-1} + c$
(e) $4x^{\frac{3}{4}} + c$
(f) $-\frac{1}{3x^3} + c$
(g) $\frac{2x^{\frac{3}{2}}}{3} + c$
(h) $\frac{2x^5}{5} + \frac{4}{x} + c$

2 (a) $2\frac{1}{4}$
(b) $\frac{3}{4}$
(c) 41.4
(d) $-2\frac{2}{3}$
(e) 3.68
(f) $10\frac{2}{3}$

3 $21\frac{1}{3}$

4 (a) P(−4, 3); Q(−2, 0); R(2, 0); S(4, 3)
(b) 8

5 (a) (4, 2)
(b) and (c)
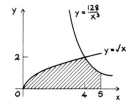
(d) 6.77 (to 3 s.f.)

6 (b) B(1, −2)
(c) $1\frac{1}{4}$

7 (a) $y = -\frac{2}{x} - 3x + c$
(b) $y = -\frac{2}{x} - 3x + 17$

8 $y = \frac{2}{3}x^{\frac{3}{2}} + c$; $y = \frac{2}{3}x^{\frac{3}{2}} + 2$

9 (a) $y = 2x^{\frac{5}{2}} - 2x^{\frac{1}{2}} + c$
(b) $y = 2x^{\frac{5}{2}} - 2x^{\frac{1}{2}} + 8$

10 (a)
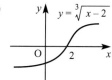

(b) $\frac{23}{4}$

Exercise 10B (Page 241)

1 (a) $e^x + c$
(b) $2e^x + c$
(c) $3e^x + c$
(d) $e^x + 2x + c$
(e) $20x + 15e^x + c$
(f) $4x - e^x + c$
(g) $x^2 + 3e^x + c$
(h) $\frac{1}{2}e^x + \frac{2}{3}x^{\frac{3}{2}} + c$
(i) $4e^x - \frac{1}{x^2} + c$
(j) $e^{x+2} + c$

2 (a) 1.72
(b) 12.8
(c) 1.63
(d) −47.2
(e) −0.198
(f) −255
(g) 32.8

Answers

Pure Mathematics: Core 4

(h) 33.1
(i) 6.39

4 (a)

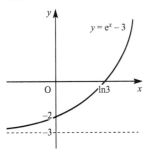

Exercise 10C (Page 244)

1 (a) $2\ln x + c$
  (b) $\frac{1}{2}\ln x + c$
  (c) $\frac{3}{4}\ln x + 5x + c$
  (d) $\ln x + e^x + c$
  (e) $\frac{1}{2}\ln x + \sqrt{x} + c$
  (f) $-\frac{1}{x} - \ln x + c$
  (g) $4\ln x + x + c$
  (h) $5\ln x + \frac{1}{10}x^2 + c$
  (i) $\frac{1}{2}x^2 + 2x + \ln x + c$
  (j) $\frac{1}{2}x^2 - \ln x + c$

2 (a) 0.693
  (b) 4.16
  (c) −3.97
  (d) −1.10
  (e) 2.81
  (f) 0.189
  (g) 10.7
  (h) 10.5
  (i) 3.64
  (j) −3.36

3 (a)

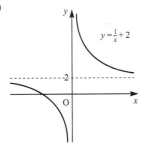

  (b) $2 + \ln\left(\frac{3}{2}\right)$

5 −1.616

Exercise 10D (Page 248)

1 (a) $\frac{1}{5}e^{5x} + c$
  (b) $-\frac{1}{4}e^{3-4x} + c$
  (c) $\frac{2}{3}e^{3x} + c$
  (d) $16e^{\frac{x}{4}} + c$
  (e) $-\dfrac{3}{e^{x-2}} + c$
  (f) $\frac{5}{3}e^{3x-1} + c$

2 (a) $e - 1$
  (b) $e^2(e^2 - 1)$
  (c) $3e(e - 1)$
  (d) $\frac{1}{2}\left(1 - \frac{1}{e^4}\right)$
  (e) $2e(e^4 - 1)$
  (f) $\dfrac{3(e^4 - 1)}{4e^9}$

3 (a) $\ln|x + 3| + c$ or $\ln k|x + 3|$
  (b) $\frac{1}{3}\ln|3x + 2| + c$
  (c) $-\frac{4}{5}\ln|3 - 5x| + c$
  (d) $x - \ln|1 + x| + c$
  (e) $\frac{1}{15}\ln|2 + 3x| + c$
  (f) $\frac{1}{6}\ln|5 + 6x| + c$

4 (a) $\ln 2$
  (b) $\ln 5$
  (c) $2\ln 2$
  (d) $\frac{1}{6}\ln\frac{5}{2}$
  (e) $3\ln 9$
  (f) $\frac{1}{3}\ln\frac{11}{8}$

5 (a)

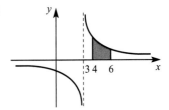

  (b) $2\ln 3$
  (c) discontinuity at $x = 3$

7 (a) $\dfrac{1}{x - 1} + \dfrac{1}{2x + 1}$

10 (a, b)

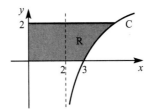

Exercise 10E (Page 250)

1. (a) $\frac{1}{2}\ln|x^2 - 1| + c$
   (b) $-2\ln|3 - x^2| + c$
   (c) $\frac{1}{2}\ln|x^2 + 4x + 1| + c$
   (d) $\ln|1 + \sin x| + c$
   (e) $\frac{1}{3}\ln|1 + e^{3x}| + c$
   (f) $\ln|\ln x| + c$

2. (a) $\frac{1}{3}\ln 2$
   (b) $2\ln 2$
   (c) $2\ln 5$
   (d) $\frac{1}{3}\ln 2$
   (e) $\ln(e + 1)$
   (f) $\frac{3}{2}\ln 3$

3. $\frac{1}{2}\ln 10$

4. $\frac{1}{2}\ln 10$

5. (a) $y = \frac{1}{4}\ln|x^4 + 1| + c$ or $y = \frac{1}{4}\ln k|x^4 + 1|$
   (b) $y = \frac{1}{4}\ln|8(x^4 + 1)|$

Exercise 10F (Page 252)

1. (a) $-\frac{1}{2}\cos 2x + c$
   (b) $\frac{2}{3}\sin 3x + c$
   (c) $5\sec x + c$
   (d) $-4\cot x + c$
   (e) $2\ln|\sec\frac{1}{2}x| + c$
   (f) $2\tan 3x + c$

2. (a) $\dfrac{\sqrt{3} - 1}{2}$
   (b) $2 - \sqrt{2}$
   (c) 1
   (d) $\frac{1}{2}$
   (e) $\frac{1}{2}\ln 2$
   (f) $\dfrac{\sqrt{3} - 1}{\sqrt{3}}$

3. 2

4. (a) $-3\cos x + 4\sin x + c$
   (b) $\dfrac{4\sqrt{2} - 1}{\sqrt{2}} \approx 3.293$
   (c) $5\sin(x + 0.927)$

Exercise 10G (Page 259)

1. (a) $\frac{1}{4}(x + 1)^4 + c$
   (b) $\frac{2}{3}(2x - 1)^{\frac{3}{2}} + c$
   (c) $\frac{1}{8}(x^3 + 1)^8 + c$
   (d) $\frac{1}{6}(x^2 + 1)^6 + c$
   (e) $\frac{1}{5}(x^3 - 2)^5 + c$
   (f) $\frac{1}{6}(2x^2 - 5)^{\frac{3}{2}} + c$
   (g) $\frac{1}{15}(2x + 1)^{\frac{3}{2}}(3x - 1) + c$
   (h) $\frac{2}{3}(x + 9)^{\frac{1}{2}}(x - 18) + c$
   (i) $2e^{3x^2} + c$
   (j) $-\frac{1}{3}\cos(x^3) + c$

2. (a) 205
   (b) $928\,000$
   (c) $5\frac{1}{3}$
   (d) 30
   (e) $222\,000$
   (f) 586
   (g) 18.1
   (h) 0.0183
   (i) 0.585
   (j) 0.653

3. (a) $\frac{1}{2}(e - 1)$
   (b) $\frac{1}{2}(e^4 - 1)$
   (c) $\frac{1}{2}(e + e^4) - 1 \approx 27.7$ (to 3 s.f.)

4. (a) $(1 - 2x^2)e^{-x^2}$
   (b) $\dfrac{1}{\sqrt{2}}$
   (c) 0.294

5. $0.490, 0.314$

6. (a) $A(-1, 0), x \geq -1$

7. (a) (i) $\dfrac{(1 + x)^4}{4} + c$
   (ii) $2\frac{2}{5}$
   (b) $\frac{1}{3}(2\sqrt{2} - 1) \approx 0.609$

8. (a) $\ln\left(\dfrac{e^2 + 1}{2}\right) \approx 1.434$
   (b) $\ln\left(\dfrac{e^2 + 1}{2}\right) \approx 1.434$
   (c) The same. The substitution $e^x = t^2$ transforms the integral in (a) into that in (b).

9. (a) $\dfrac{1 - x^2}{(x^2 + 1)^2}$
   (b) $(-1, -\frac{1}{2})$ and $(1, \frac{1}{2})$
   (c) $\frac{1}{2}\ln|x^2 + 1| + c$
   (d) 20

10. (b) $\frac{1}{2}\ln 2$
    (c) $-\frac{1}{2}\ln 2$
    (d) translation $\begin{pmatrix} -1 \\ 0 \end{pmatrix}$
    (e) $-\frac{1}{2}\ln 2$

11. $\frac{1}{14}$

ANSWERS

12 $\frac{1}{4}\int u^5 - u^4 \, du$, $\frac{1}{120}(2x+1)^5(10x-1) + c$

13 1

14 (a) 0.02
    (b) 0.2
    (c) 0.02

EXERCISE 10H (Page 267)

1 (a) $u = x$, $\frac{dv}{dx} = e^x$
    (b) $xe^x - e^x + c$

2 (a) $u = x$, $\frac{dv}{dx} = \cos 3x$
    (b) $\frac{1}{3}x\sin 3x + \frac{1}{9}\cos 3x + c$

3 (a) $u = 2x + 1$, $\frac{dv}{dx} = \cos x$
    (b) $(2x + 1)\sin x + 2\cos x + c$

4 (a) $u = x$, $\frac{dv}{dx} = e^{-2x}$
    (b) $-\frac{1}{2}xe^{-2x} - \frac{1}{4}e^{-2x} + c$

5 (a) $u = x$, $\frac{dv}{dx} = e^{-x}$
    (b) $-xe^{-x} - e^{-x} + c$

6 (a) $u = x$, $\frac{dv}{dx} = \sin 2x$
    (b) $-\frac{1}{2}x\cos 2x + \frac{1}{4}\sin 2x + c$

7 $\frac{1}{4}x^4 \ln x - \frac{1}{16}x^4 + c$

8 $x^2 e^x - 2xe^x + 2e^x + c$

9 $(2 - x)^2 \sin x - 2(2 - x)\cos x - 2\sin x + c$

10 $\frac{1}{3}x^3 \ln 2x - \frac{1}{9}x^3 + c$

11 $\frac{2}{15}(1 + x)^{\frac{3}{2}}(3x - 2) + c$

12 $\frac{1}{15}(x - 2)^5(5x + 2) + c$

13 (a) $x\ln 3x - x + c$

EXERCISE 10I (Page 269)

1 $\frac{2}{9}e^3 + \frac{1}{9}$

2 -2

3 $2e^2$

4 $3\ln 2 - 1$

5 $\frac{1}{8}\pi^2 - \frac{1}{2}$

6 $\frac{64}{3}\ln 4 - 7$

7 (a) $(2, 0)$, $(0, 2)$
    (c) $e^{-2} + 1$

8 (b) π

9 $5\ln 5 - 5$

10 $\frac{1}{2}\pi^2 - 4$

11 $\frac{-4}{15}$

12 $\frac{1}{2}$, 0.134

14 π

15 (b) $-x^2 \cos x + 2x\sin x + 2\cos x + c$
    (c) $y = -x^2 \cos x + 2x\sin x + 2\cos x + 2 - \pi^2$

16 (a) $45\pi^2$
    (b) $190(\pi - 1)$
    (c) 13.6%

EXERCISE 10J (Page 274)

1 (a) $\ln \left| \dfrac{3x - 2}{1 - x} \right| + c$

    (b) $\dfrac{1}{1 - x} + \ln \left| \dfrac{x - 1}{2x + 3} \right| + c$

    (c) $\frac{1}{6}\ln \left| \dfrac{(x - 1)^4}{2x + 1} \right| + c$

    (d) $\ln \left| \dfrac{(x - 1)^2}{\sqrt{2x + 1}} \right| + c$

    (e) $\ln \left| \dfrac{x(2 - x)}{(1 - x)^2} \right| + c$

    (f) $\frac{1}{2}\ln \left| \dfrac{x + 1}{x + 3} \right| + c$

    (g) $x + 3\ln|x - 1| + \frac{5}{2}\ln|2x + 1| + c$

    (h) $\ln \left| \dfrac{2x + 1}{x + 2} \right| + \dfrac{1}{2(2x + 1)} + c$

2 (a) (i) $\dfrac{2}{1 - 2x} + \dfrac{1}{1 + x}$
      (ii) $\ln\left(\frac{11}{8}\right) = 0.31845$
    (b) (i) $3 + 3x + 9x^2 + \ldots$
      (ii) 0.31800, 0.14%

3 (a) $A = 1$, $B = 3$, $C = -2$
    (b) $2 + \ln\frac{125}{3} \approx 5.730$

4 $A = 8$, $C = 1$

5 (a) $\dfrac{2}{1 + 2x} - \dfrac{1}{2 + x}$

6 (b) 2.80

Pure Mathematics: Core 4

ANSWERS

7 (a) $A = \frac{1}{2}, B = 2, C = -1$
  (b) $\frac{1}{2}\ln|x| + 2\ln|x-1| + \frac{1}{x-1} + c$

8 (a) $\frac{2}{3+2x} + \frac{10}{(3+2x)^2} + \frac{1}{1-x}$
  (b) $\ln\left|\frac{3+2x}{1-x}\right| - \frac{5}{3+2x} + c$

EXERCISE 10K (Page 278)

1 (a) $-\cos x - 2\sin x + c$
  (b) $3\sin x - 2\cos x + c$
  (c) $-5\cos x + 4\sin x + c$

2 (a) $\frac{1}{3}\sin 3x + c$
  (b) $\cos(1-x) + c$
  (c) $-\frac{1}{4}\cos^4 x + c$
  (d) $\ln|2 - \cos x| + c$
  (e) $-\ln|\cos x| + c$
  (f) $-\frac{1}{6}(\cos 2x + 1)^3 + c$

3 (a) $-\cos(x^2) + c$
  (b) $e^{\sin x} + c$
  (c) $\frac{1}{2}\tan^2 x + c$ or $\frac{1}{2}\sec^2 x + c'$
  (d) $\frac{-1}{\sin x} + c$

4 (a) 1
  (b) $\frac{1}{16}$
  (c) 1
  (d) $e - 1$
  (e) $\ln 2$

5 (b) $\frac{8}{3}$

6 (a) $\frac{1}{4}\sin 2x + \frac{1}{2}x + c$
  (b) $-\cos x + \frac{1}{3}\cos^3 x + c$
  (c) $\frac{3}{8}x - \frac{1}{4}\sin 2x + \frac{1}{32}\sin 4x + c$
  (d) $\sin x - \frac{2}{3}\sin^3 x + \frac{1}{5}\sin^5 x + c$

7 (a) $\sec^2 x - 1$
  (b) $\tan x$
  (c) $\frac{3}{16}(4 - \pi)$

8 (b) $\frac{1}{64}(\pi^2 - 4\pi + 8)$

9 (a) $\frac{1}{4}\ln|\sec 4x| + c$
  (b) $\frac{1}{4}\tan 4x - x + c$
  (d) $\frac{1}{12}\tan^3 4x - \frac{1}{4}\tan 4x + x + c$

10 (a) $\frac{1}{2}$
  (b) $\frac{3\pi}{16}$

EXERCISE 10L (Page 282)

1 (a)
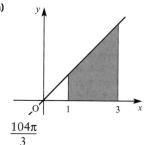
$\frac{104\pi}{3}$

(b)
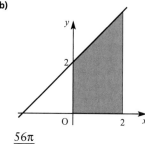
$\frac{56\pi}{3}$

(c)
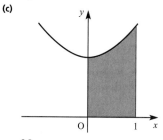
$\frac{28\pi}{15}$

(d)
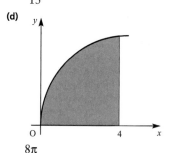
8π

(e)
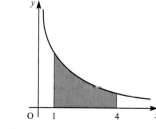
$\pi\ln 4$

Answers

(f)
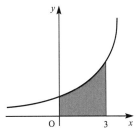
$(e^3 - 1)\pi$

(g)
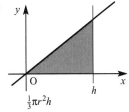
$\frac{1}{3}\pi r^2 h$
Volume of a cone

(h)
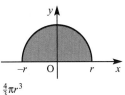
$\frac{4}{3}\pi r^3$
Volume of a sphere

2 (a) $A(-3, 4)$ $B(3, 4)$
  (b) 36π

3 (a)
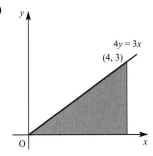
  (b) 12π

4 (a)
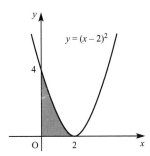
  (b) $2\frac{2}{3}$

  (d) $\frac{32\pi}{5}$

5 (a) $\frac{726\pi}{5}$
  (b) $x = 1, x = 3$

7 (b) $\frac{\pi}{12}(2\pi + 3)$

8 $\frac{1}{2}\pi^2$

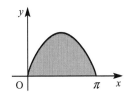

9 (c) 0.796

10 $\frac{592}{5}\pi$

Exercise 10M (page 286)

1 120π

2 $\frac{\pi}{4}$

3 24π

4 3.57π

5 $\frac{352}{81}\pi$

6 $\frac{\pi}{6}$

7 $(2\pi - \frac{35}{3})\pi$

8 $\frac{16\pi}{3}$

9 $\frac{\pi}{5}$

Exercise 10N (Page 290)

1 (a) $y = \frac{1}{3}x^3 + c$
  (b) $y = \sin x + c$
  (c) $y = e^x + c$
  (d) $y = \frac{2}{3}x^{\frac{3}{2}} + c$

2 (a) $y = -\dfrac{2}{(x^2 + c)}$
  (b) $y^2 = \frac{2}{3}x^3 + c$
  (c) $y = Ae^x$
  (d) $y = \ln|e^x + c|$
  (e) $y = Ax$
  (f) $y = (\frac{1}{4}x^2 + c)^2$
  (g) $y = -\dfrac{1}{(\sin x + c)}$
  (h) $y^2 = A(x^2 + 1) - 1$
  (i) $y = -\ln(c - \frac{1}{2}x^2)$
  (j) $y^3 = \frac{3}{2}x^2\ln x - \frac{3}{4}x^2 + c$

Pure Mathematics: Core 4 — Answers

EXERCISE 10O (Page 292)

1. (a) $y = \frac{1}{3}x^3 - x - 4$
   (b) $y = e^{\frac{x^3}{3}}$
   (c) $y = \ln(\frac{1}{2}x^2 + 1)$
   (d) $y = \frac{1}{(2-x)}$
   (e) $y = e^{(x^2-1)/2} - 1$
   (f) $y = \sec x$

2. (a) $\theta = 20 - Ae^{-2t}$
   (b) $\theta = 20 - 15e^{-2t}$
   (c) $t = 1.01$ seconds

3. (a) $N = Ae^t$
   (b) $N = 10e^t$
   (c) N tends to ∞, which would never be realised because of the combined effects of food shortage, predators and human controls

4. $s = \sqrt{4t + k}$

5. (a) $\frac{1}{3y} + \frac{1}{3(3-y)}$
   (b) $\frac{1}{3}\ln\left|\frac{y}{3-y}\right| + c$ or $\frac{1}{3}\ln\left|\frac{Ay}{3-y}\right|$
   (c) $y = \frac{3x^3}{(4+x^3)}$

6. $T = T_0 e^{(1-\cos kt)}$

7. (a) $N = 1500 e^{0.0347t} = 1500 \times 2^{\frac{t}{20}}$
   (b) $N = 24\,000$
   (c) time taken = 11 hours 42 minutes

8. (b) $\frac{1}{x-1} - \frac{1}{x+1}$
   (c) $y = \frac{x+1}{2(x-1)} e^{3-x}$

10. $\frac{dx}{dt} = \frac{(2-3x)}{100}$, $x = \frac{1}{3}(2 - 2e^{\frac{-3t}{100}})$,
    time taken = 40.8 seconds

11. (d) $N_0 e^{\frac{k}{a}}$

12. (a) $\frac{3}{(3x-1)} - \frac{1}{x}$
    (c) $t = 1.967$ (3 d.p.)
    (d) 500 and 3550

EXERCISE 10P (Page 297)

1. (a) $\frac{1}{3}\sin(3x-1) + c$
   (b) $-\frac{1}{x^2+x-1} + c$
   (c) $\frac{1}{3}\tan^3 x + c$
   (d) $-e^{1-x} + c$
   (e) $-\frac{1}{2}x^2\cos 2x + \frac{1}{2}x\sin 2x + \frac{1}{4}\cos 2x + c$
   (f) $\frac{1}{4}\sin 2x + \frac{1}{2}x + c$
   (g) $x\ln 2x - x + c$
   (h) $\frac{-1}{4(x^2-1)^2} + c$
   (i) $\frac{1}{3}(2x-3)^{\frac{3}{2}} + c$
   (j) $\ln\left|\frac{x-1}{x+2}\right| - \frac{1}{x-1} + c$
   (k) $-\frac{1}{2}\cos 2x + \frac{1}{6}\cos^3 2x + c$
   (l) $\frac{1}{4}x^4\ln x - \frac{1}{16}x^4 + c$
   (m) $\ln\left|\frac{x-3}{2x-1}\right| + c$
   (n) $\frac{1}{2}e^{x^2+2x} + c$

2. (a) $\frac{8}{3}$
   (b) $\frac{1}{3}\ln 4$
   (c) $48 + 8\ln 4$
   (d) $\frac{5}{24}$
   (e) $\frac{8}{3}\ln 2 - \frac{7}{9}$

3. $\frac{4}{3}$

4. $\frac{1}{3}(2\sqrt{2} - 1)$

5. (a) $\frac{\pi}{4}$
   (b) 0.24

6. $\frac{1}{8}\pi - \frac{1}{4}$

7. (a) $-\frac{1}{2}xe^{-2x} - \frac{1}{4}e^{-2x} + c$
   (b) 0.112

8. (a) $-\frac{1}{2}\cos(2x-3) + c$
   (b) $\frac{3}{4}e^4 + \frac{1}{4}$
   (c) $\frac{1}{2}\ln|x^2 - 9| + c$

9. (a) $\frac{38}{9}$
   (b) $\frac{1}{4} - \frac{3}{4e^2}$

10. $\frac{1}{4}$, $\frac{1}{4} - \frac{3}{4e^2}$

Answers

Pure Mathematics: Core 4

Exercise 10Q (Page 300)

1 (b)

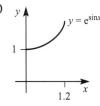

(c) 2.118, 2.119
(d) overestimates
(e) 3.3×10^{-4}

2 (a) (i) 3.2425
(ii) 3.2821
(iii) 3.2924
(b) 3.2924
(c) 3.43×10^{-3}

3 9.64×10^{-4}

4 (a) $\frac{\pi}{4}$
(b) 0.7828
(c) 0.00260
(d) 3.1312, 0.0104

5 3.1454, 0.02524

6 (a) 0.916, 0.9204, 0.9213
(b) 0.9213
(c) 3.114×10^{-4}

7 (b)

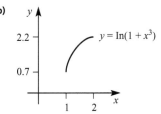

(c) 1.461, 1.464
(d) 6.56×10^{-4}

8 0.7854, 0.9720, 1.0969
1.0969
4.2×10^{-2}

9 0.3519, 0.3896, 0.3988; more strips

10 (b) 0.35901, 0.34976
(c) 0.69952, 0.006

Exercise 10R (Page 302)

1 $e^x(x^2 - 2x + 2) + c$
$e^{-2y} = -2e^x(x^2 - 2x + 2) + k$

2 (a) $\frac{4}{2x-3} - \frac{3}{x+1}$
(b) $y = \frac{108(2x-3)^2}{(x+1)^3}$

3 (a) $\frac{1}{3}(x+2)\sqrt{2x+1}$
(b) $\frac{16}{3}$

4 (b) $\frac{1}{2}(x - \sin x \cos x) + c$
(c) $\frac{1}{4}(x^2 - 2x\sin x \cos x + \sin^2 x) + c$

5 (a) $x = \frac{1}{4}$
(b) $V = \frac{\pi}{6}(1 - 5e^{-4})$

6 (a) $\sin y \cos y + c$
(b) $x \sin x + \cos x + c$
(c) $\sin y \cos y = x \sin x + \cos x + c$

7 (a) $\frac{1}{12}(x^2 + 3)^6 + c$

8 (a) $\theta = 20 - 15e^{-0.2t}$
(b) 10 minutes

9 $16\pi(\frac{1}{4}\ln 3 - \frac{1}{6})$

10 (a) $2\ln 2 - 1$
(b) $2\ln 2 - \frac{3}{4}$
(c) $\frac{1}{9}(24\ln 2 - 7)$

11 (a) $A = \frac{1}{2}, B = 2, C = -1$
(b) $\frac{1}{2}\ln x + 2\ln(x-1) + \frac{1}{x-1} + c$

12 (b) $3\ln(x-4) - \frac{1}{2}\ln(x^2+1) + c$
(c) $y = -\frac{2(x-4)^3}{\sqrt{x^2+1}}$

13 (a) $2x - \frac{2}{3}\ln(1+3x) + c$
(b) $\frac{2}{3}\left(\ln(1+3x) + \frac{1}{1+3x}\right) + c$

14 (b) $\frac{1}{8}\sin 4x + \frac{1}{4}\sin 2x + c$
(c) $\frac{\pi}{12} - \frac{\sqrt{3}}{8}$

15 (c) $k = 9, q = 8$

16 (a) $\frac{80\pi}{3}$
(b) $a = 5, b = -5$

17 (a) $\frac{2}{5}\pi a^3$
(b) $y = \frac{x^2}{4a}$

18 (a) 0, 29.05, 33.46
(b) 33.46, 33.76
(c) 359.65

19 (a) 7.80, 7.80, 6.13
(b) 55.72 m²
(c) 84.28 m²
(d) Overestimate
(e) Use more strips

20 (a) 4, 4.84, 7.06
(b) 4.93

Chapter 11

Exercise 11A (Page 315)

1 (a) $3i + 2j$
(b) $5i - 4j$
(c) $3i$
(d) $-3i - j$
(e) $2j$

2 (a)
($\sqrt{13}$, 56.3°)

(b)
($\sqrt{13}$, −33.7°)

(c)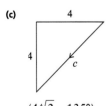
($4\sqrt{2}$, −135°)

(d)
2 $\diagdown$ d
    1
($\sqrt{5}$, 116.6°)

(e)
(5, −53.1°)

3 (a)
3.54i + 3.54j

(b)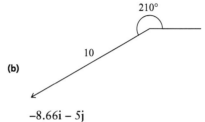
−8.66i − 5j

(c)
↑
4

4j

(d) ────── 8 ─→
8i

(e)
−2.83i − 2.83j

4 (a) $2i - 2j$
(b) $2i$
(c) $-4j$
(d) $4j$
(e) $-i + j$
(f) $i - j$
(g) $8i$
(h) $-8i$
(i) $2i - 4j$
(j) $4i + 4j$

Answers

Pure Mathematics: Core 4

5 (a) A: $2i + 3j$, C: $-2i + j$
   (b) $\overrightarrow{AB} = -2i + j$, $\overrightarrow{CB} = 2i + 3j$
   (c) $\overrightarrow{AB} = \overrightarrow{OC}$, $\overrightarrow{CB} = \overrightarrow{OA}$
   (d) a parallelogram

Exercise 11B (Page 321)

1 (a) $\begin{pmatrix} 6 \\ 8 \end{pmatrix}$
  (b) $\begin{pmatrix} 1 \\ 1 \end{pmatrix}$
  (c) $\begin{pmatrix} 0 \\ 0 \end{pmatrix}$
  (d) $\begin{pmatrix} 8 \\ -1 \end{pmatrix}$
  (e) $-3j$

2 (a) $2i + 3j$
  (b) i
  (c) j
  (d) $3i + 2j$
  (e) 0

3 (a) (i) b
      (ii) $a + b$
      (iii) $-a + b$
  (b) (i) $\tfrac{1}{2}(a + b)$ (ii) $\tfrac{1}{2}(-a + b)$
  (c) PQRS is any parallelogram and $\overrightarrow{PM} = \tfrac{1}{2}\overrightarrow{PR}$, $\overrightarrow{QM} = \tfrac{1}{2}\overrightarrow{QS}$

4 (a) (i) i
      (ii) $2i$
      (iii) $i - j$
      (iv) $-i - 2j$
  (b) $|\overrightarrow{AB}| = |\overrightarrow{BC}| = \sqrt{2}$, $|\overrightarrow{AD}| = |\overrightarrow{CD}| = \sqrt{5}$

5 (a) $-p + q$, $\tfrac{1}{2}p - \tfrac{1}{2}q$, $-\tfrac{1}{2}p$, $-\tfrac{1}{2}q$
  (b) $\overrightarrow{NM} = \tfrac{1}{2}\overrightarrow{BC}$, $\overrightarrow{NL} = \tfrac{1}{2}\overrightarrow{AC}$, $\overrightarrow{ML} = \tfrac{1}{2}\overrightarrow{AB}$

6 (a) $\begin{pmatrix} \frac{2}{\sqrt{13}} \\ \frac{3}{\sqrt{13}} \end{pmatrix}$
  (b) $\tfrac{3}{5}i + \tfrac{4}{5}j$
  (c) $\begin{pmatrix} \frac{-1}{\sqrt{2}} \\ \frac{-1}{\sqrt{2}} \end{pmatrix}$
  (d) $\tfrac{5}{13}i + \tfrac{12}{13}j$
  (e) i
  (f) $\begin{pmatrix} \frac{-1}{\sqrt{5}} \\ \frac{2}{\sqrt{5}} \end{pmatrix}$
  (g) $\begin{pmatrix} \frac{-1}{\sqrt{5}} \\ \frac{2}{\sqrt{5}} \end{pmatrix}$
  (h) $\begin{pmatrix} \frac{1}{\sqrt{5}} \\ \frac{2}{\sqrt{5}} \end{pmatrix}$
  (i) $\begin{pmatrix} \cos\alpha \\ \sin\alpha \end{pmatrix}$
  (j) $\begin{pmatrix} \cos\beta \\ \sin\beta \end{pmatrix}$

Exercise 11C (Page 325)

1 (a) $\begin{pmatrix} 4 \\ 1 \\ 9 \end{pmatrix}$
  (b) $\begin{pmatrix} -2 \\ 5 \\ 1 \end{pmatrix}$
  (c) $\begin{pmatrix} 1 \\ 10 \\ 8 \end{pmatrix}$
  (d) $\begin{pmatrix} 1 \\ -5 \\ -6 \end{pmatrix}$
  (e) $\begin{pmatrix} 1 \\ 2 \\ 1 \end{pmatrix}$
  (f) $\begin{pmatrix} x \\ y \\ z \end{pmatrix}$

2 (a) $6i + 4j + 12k$
  (b) $8i + 4k$
  (c) $17i + 18j + 38k$
  (d) $2i + 20j + 24k$
  (e) $17i + 2j + 2k$
  (f) $(4a - b + 3c)i + (-2a + 2b + 4c)j + (3a + 4b + 5c)k$

3 (a) (i) 3 (ii) $\tfrac{1}{3}\begin{pmatrix} 2 \\ -1 \\ 2 \end{pmatrix}$
  (b) (i) 13 (ii) $\tfrac{1}{13}\begin{pmatrix} 4 \\ 12 \\ -3 \end{pmatrix}$
  (c) (i) 6 (ii) $\tfrac{1}{6}\begin{pmatrix} -4 \\ 4 \\ -2 \end{pmatrix}$
  (d) (i) 7 (ii) $\tfrac{1}{7}(6i + 3j + 2k)$
  (e) (i) 11 (ii) $\tfrac{1}{11}(6i - 7j + 6k)$
  (f) (i) $\sqrt{3}$ (ii) $\tfrac{1}{\sqrt{3}}(i + j + k)$

Pure Mathematics: Core 4

ANSWERS

EXERCISE 11D (Page 328)

1. (a) (i) $2i + 8j$ (ii) $\sqrt{68}$ (iii) $3i + 7j$
   (b) (i) $-4i - 3j$ (ii) 5 (iii) $2i + 1.5j$
   (c) (i) $6i + 8j$ (ii) 10 (iii) $i + 3j$
   (d) (i) $6i - 8j$ (ii) 10 (iii) 0
   (e) (i) $5i + 12j$ (ii) 13 (iii) $-7.5i - 2j$
   (f) (i) $2i - 6j - 2k$ (ii) $\sqrt{44}$ (iii) $2i + j + k$
   (g) (i) $4i + 4j + 2k$ (ii) 6 (iii) $4i - j + 2k$
   (h) (i) $4i - 2j - 2k$ (ii) $\sqrt{24}$ (iii) $2i + 4j - 4k$
   (i) (i) $-4i - 4j + 8k$ (ii) $\sqrt{96}$ (iii) $4i + k$
   (j) (i) $2i - j - k$ (ii) $\sqrt{6}$ (iii) $2i + 1.5j + 2.5k$

2. (a) 5
   (b) 13
   (c) 25
   (d) $4\sqrt{2}$
   (e) 4
   (f) 3
   (g) 7
   (h) 11
   (i) $2\sqrt{3}$
   (j) $\sqrt{3}$

3. $AB = \sqrt{65}$, $BC = \sqrt{65}$, $CD = \sqrt{21}$, $AD = \sqrt{59}$

4. $a = 1$ or 9

5. $p = 1$ or 2

EXERCISE 11E (Page 335)

Note: Many of these answers are not unique.

1. (a) $\mathbf{r} = \begin{pmatrix} 2 \\ 1 \end{pmatrix} + t\begin{pmatrix} 1 \\ 2 \end{pmatrix}$
   (b) $\mathbf{r} = \begin{pmatrix} 3 \\ 5 \end{pmatrix} + t\begin{pmatrix} -1 \\ 1 \end{pmatrix}$
   (c) $\mathbf{r} = \begin{pmatrix} -6 \\ -6 \end{pmatrix} + t\begin{pmatrix} 1 \\ 1 \end{pmatrix}$
   (d) $\mathbf{r} = \begin{pmatrix} 5 \\ 3 \end{pmatrix} + t\begin{pmatrix} 1 \\ 1 \end{pmatrix}$
   (e) $\mathbf{r} = t\begin{pmatrix} 2 \\ 1 \end{pmatrix}$
   (f) $\mathbf{r} = t\begin{pmatrix} -1 \\ 4 \end{pmatrix}$
   (g) $\mathbf{r} = t\begin{pmatrix} -1 \\ 4 \end{pmatrix}$
   (h) $\mathbf{r} = \begin{pmatrix} 3 \\ -12 \end{pmatrix} + t\begin{pmatrix} -1 \\ 4 \end{pmatrix}$

2. (a) $y = 3x - 1$
   (b) $y = \frac{1}{2}x + 1$
   (c) $y = x - 1$
   (d) $y = x - 1$
   (e) $y = 5$ (x may take any value)

3. (a) $\mathbf{r} = \begin{pmatrix} 0 \\ 3 \end{pmatrix} + t\begin{pmatrix} 1 \\ 2 \end{pmatrix}$
   (b) $\mathbf{r} = \begin{pmatrix} 0 \\ -4 \end{pmatrix} + t\begin{pmatrix} 1 \\ 1 \end{pmatrix}$
   (c) $\mathbf{r} = \begin{pmatrix} 0 \\ -1 \end{pmatrix} + t\begin{pmatrix} 2 \\ 1 \end{pmatrix}$
   (d) $\mathbf{r} = t\begin{pmatrix} -4 \\ 1 \end{pmatrix}$
   (e) $\mathbf{r} = \begin{pmatrix} 0 \\ 4 \end{pmatrix} + t\begin{pmatrix} -2 \\ 1 \end{pmatrix}$

4. (a) $\mathbf{r} = \begin{pmatrix} 2 \\ 4 \\ -1 \end{pmatrix} + t\begin{pmatrix} 3 \\ 6 \\ 4 \end{pmatrix}$
   (b) $\mathbf{r} = \begin{pmatrix} 1 \\ 0 \\ -1 \end{pmatrix} + t\begin{pmatrix} 1 \\ 0 \\ 0 \end{pmatrix}$
   (c) $\mathbf{r} = \begin{pmatrix} 1 \\ 0 \\ 4 \end{pmatrix} + t\begin{pmatrix} 5 \\ 3 \\ -6 \end{pmatrix}$
   (d) $\mathbf{r} = \begin{pmatrix} 0 \\ 0 \\ 1 \end{pmatrix} + t\begin{pmatrix} 2 \\ 1 \\ 3 \end{pmatrix}$
   (e) $\mathbf{r} = t\begin{pmatrix} 1 \\ 2 \\ 3 \end{pmatrix}$

5. (a) $\dfrac{x - 2}{3} = \dfrac{y - 4}{6} = \dfrac{z + 1}{4}$
   (b) $x - 1 = \dfrac{y}{3} = \dfrac{z + 1}{4}$
   (c) $x - 3 = \dfrac{z - 4}{2}$ and $y = 0$
   (d) $\dfrac{x}{2} = \dfrac{z - 1}{4}$ and $y = 4$
   (e) $x = -2$ and $z = 3$

6. (a) $\mathbf{r} = \begin{pmatrix} 3 \\ -2 \\ 1 \end{pmatrix} + t\begin{pmatrix} 5 \\ 3 \\ 4 \end{pmatrix}$
   (b) $\mathbf{r} = \begin{pmatrix} -6 \\ 0 \\ -4 \end{pmatrix} + t\begin{pmatrix} 6 \\ 2 \\ 3 \end{pmatrix}$
   (c) $\mathbf{r} = \begin{pmatrix} 0 \\ 0 \\ -1 \end{pmatrix} + t\begin{pmatrix} 1 \\ 2 \\ 3 \end{pmatrix}$
   (d) $\mathbf{r} = t\begin{pmatrix} 1 \\ 1 \\ 1 \end{pmatrix}$
   (e) $\mathbf{r} = t\begin{pmatrix} 2 \\ 0 \\ 0 \end{pmatrix} + t\begin{pmatrix} 0 \\ 1 \\ 1 \end{pmatrix}$

7. (a) $\begin{pmatrix} 7 \\ -4 \\ 4 \end{pmatrix}, \begin{pmatrix} 3 \\ -6 \\ 2 \end{pmatrix}, \begin{pmatrix} -7 \\ 4 \\ -4 \end{pmatrix}, \begin{pmatrix} 3 \\ -6 \\ 2 \end{pmatrix}$

(b) 9, 7, 9, 7

(c) $\mathbf{r} = \begin{pmatrix} 1 \\ 4 \\ -2 \end{pmatrix} + \alpha \begin{pmatrix} 7 \\ -4 \\ 4 \end{pmatrix}$, $\mathbf{r} = \begin{pmatrix} 8 \\ 0 \\ 2 \end{pmatrix} + \beta \begin{pmatrix} 3 \\ -6 \\ 2 \end{pmatrix}$

$\mathbf{r} = \begin{pmatrix} 11 \\ -6 \\ 4 \end{pmatrix} + \gamma \begin{pmatrix} -7 \\ 4 \\ -4 \end{pmatrix}$, $\mathbf{r} = \begin{pmatrix} 1 \\ 4 \\ -2 \end{pmatrix} + \delta \begin{pmatrix} 3 \\ -6 \\ 2 \end{pmatrix}$

(d) parallelogram

8 (a) $\mathbf{r} = \begin{pmatrix} 4 \\ -1 \\ 6 \end{pmatrix} + t \begin{pmatrix} 1 \\ -2 \\ 4 \end{pmatrix}$

(b) $\begin{pmatrix} 2\frac{1}{2} \\ 2 \\ 0 \end{pmatrix}$

(c) No

(d) $p = -5$, $q = 14$

EXERCISE 11F (Page 340)

1 (a) $\begin{pmatrix} 4 \\ 1 \end{pmatrix}$

(b) $\begin{pmatrix} 5 \\ 5 \end{pmatrix}$

(c) $\begin{pmatrix} 12 \\ 17 \end{pmatrix}$

(d) $\begin{pmatrix} -5 \\ 6 \end{pmatrix}$

(e) $\begin{pmatrix} 6 \\ 3 \end{pmatrix}$

2 (a) 12.8 km
  (b) 20 km h⁻¹, 5 km h⁻¹
  (c) After 40 minutes there is a collision

3 (a) L(10, 4.5); M(7, 3.5); N(4, 1)

  (b) AL: $\mathbf{r} = \begin{pmatrix} 1 \\ 0 \end{pmatrix} + \lambda \begin{pmatrix} 2 \\ 1 \end{pmatrix}$; BM: $\mathbf{r} = \begin{pmatrix} 7 \\ 2 \end{pmatrix} + \mu \begin{pmatrix} 0 \\ 1 \end{pmatrix}$;

  CN: $\mathbf{r} = \begin{pmatrix} 13 \\ 7 \end{pmatrix} + \nu \begin{pmatrix} 3 \\ 2 \end{pmatrix}$

  (c) (i) (7, 3)
      (ii) (7, 3)

  (d) The lines AL, BM and CN are concurrent. (They are the medians of the triangle, and this result holds for the medians of any triangle.)

4 (a) parallel
  (b) not parallel
  (c) parallel
  (d) not parallel
  (e) not parallel
  (f) parallel

5 (a) intersect at (3, 7, 0)
  (b) intersect at (−1, 1, 1)
  (c) skew
  (d) skew
  (e) intersect at (−7, 3, 11)
  (f) intersect at $(7, -1\frac{1}{2}, 5\frac{1}{2})$

EXERCISE 11G (Page 348)

1 (a) 42.3°
  (b) 90°
  (c) 18.4°
  (d) 31.0°
  (e) 90°
  (f) 180°

2 (a) $\mathbf{r} = \begin{pmatrix} 1 \\ 0 \end{pmatrix} + \lambda \begin{pmatrix} 2 \\ 1 \end{pmatrix}$

  (b) $\mathbf{r} = \begin{pmatrix} 6 \\ 1 \end{pmatrix} + \mu \begin{pmatrix} 1 \\ 2 \end{pmatrix}$

  (c) $\mathbf{r} = \begin{pmatrix} 7 \\ 3 \end{pmatrix}$

  (d) 36.9°

3 (a) parallelogram: AB//CD, BC//DA
  (b) A(5, 2); B(1, 1); C(2,4); D(6, 5)
  (c) 57.5°, 122.5°

4 (a) $\begin{pmatrix} 3 \\ 1 \end{pmatrix}, \begin{pmatrix} -1 \\ 3 \end{pmatrix}$

  (b) $\overrightarrow{BA}.\overrightarrow{BC} = 0$

  (c) $|\overrightarrow{AB}| = |\overrightarrow{BC}| = \sqrt{10}$

  (d) (2, 5)

5 (a) 29.0°
  (b) 76.2°
  (c) 162.0°

6 (a) 53.6°
  (b) 81.8°

7 (a) (0, 4, 3)

  (b) $\begin{pmatrix} -5 \\ 4 \\ 3 \end{pmatrix}$, $\sqrt{50}$

  (c) $\mathbf{r} = \begin{pmatrix} 5 \\ 0 \\ 0 \end{pmatrix} + \lambda \begin{pmatrix} -5 \\ 4 \\ 3 \end{pmatrix}$

  (d) $\begin{pmatrix} 3\frac{3}{4} \\ 1 \\ \frac{3}{4} \end{pmatrix}$, 63.4°

  (e) Spider is then at P(2.5, 2, 1.5) and $\overrightarrow{OP}.\overrightarrow{AG} = 0$, $|\overrightarrow{OP}| = 3.54$

Pure Mathematics: Core 4 — ANSWERS

8 (a) A(4, 0, 0), F(4, 0, 3)
(b) 114.1°, 109.5°
(c) They touch but are not perpendicular

9 (a) $\begin{pmatrix} -10 \\ 9 \\ 7 \end{pmatrix}$

(b) $\begin{pmatrix} -2 \\ 0 \\ 1 \end{pmatrix}$

10 (a) AB: $\mathbf{r} = \begin{pmatrix} 1 \\ 4 \\ -2 \end{pmatrix} + s\begin{pmatrix} 7 \\ -4 \\ 4 \end{pmatrix}$

BC: $\mathbf{r} = \begin{pmatrix} 8 \\ 0 \\ 2 \end{pmatrix} + t\begin{pmatrix} 3 \\ -6 \\ 2 \end{pmatrix}$

AC: $\mathbf{r} = \begin{pmatrix} 1 \\ 4 \\ -2 \end{pmatrix} + u\begin{pmatrix} 10 \\ -10 \\ 6 \end{pmatrix}$

(b) A = 14.3°, B = 147.3°, C = 18.5°

Exercise 11H (Page 351)

1 (a) 48.2°
(b) 14.5

2 (c) 150.8° or 29.2°

3 (b) (9, –4, 14)

(c) $\begin{pmatrix} -1 \\ 1 \\ 1 \end{pmatrix}$

(d) $\mathbf{r} = \begin{pmatrix} 9 \\ -4 \\ 14 \end{pmatrix} + \lambda\begin{pmatrix} -1 \\ 1 \\ 1 \end{pmatrix}$

4 (a) $\begin{pmatrix} 4 \\ -2 \\ 4 \end{pmatrix}$

(b) $\mathbf{r} = \lambda\begin{pmatrix} 2 \\ -1 \\ 2 \end{pmatrix}$

(c) 90°

5 (a) $\overrightarrow{AB} = \begin{pmatrix} 3 \\ 2 \\ -1 \end{pmatrix}$, $\overrightarrow{BC} = \begin{pmatrix} 4 \\ -5 \\ 2 \end{pmatrix}$

(b) $\sqrt{14}$, $\sqrt{45}$
(c) 60.8°
(d) 12.5 square units

6 (b) 1 : 3
(c) 7, 18
(d) 6
(e) 13.6 square units
(f) 40.8 square units

7 (a) $\mathbf{r} = \begin{pmatrix} 1 \\ -2 \\ 6 \end{pmatrix} + \mu\begin{pmatrix} 2 \\ 1 \\ -3 \end{pmatrix}$

(c) 81.8°

(d) $\begin{pmatrix} 0 \\ 0 \\ 1 \end{pmatrix}$

8 (a) 6
(b) 64.9°
(c) 70.5°
(d) $p = 2$

9 (a) $\mathbf{r} = \begin{pmatrix} 24 \\ 6 \\ 0 \end{pmatrix} + t\begin{pmatrix} 1 \\ 1 \\ 2 \end{pmatrix}$

(c) (35, 17, 22)
(d) 181 square units

10 (a) $\begin{pmatrix} 4 \\ -1 \\ -3 \end{pmatrix}$

(b) 71.4°

11 (a) $\mathbf{r} = \begin{pmatrix} 2 \\ 1 \\ 1 \end{pmatrix} + t\begin{pmatrix} -1 \\ 2 \\ 1 \end{pmatrix}$

(d) 18.7°

12 (a) $\begin{pmatrix} 1 \\ -3 \\ -1 \end{pmatrix}$

(c) 95.2°

Index

INDEX

accuracy of an answer 141
addition
  of fractions 3–4
  of vectors 317–18
algebra 162–74
algebraic fraction 2
angle, between vectors 347–50, 355
answers, accuracy 141
approximate values 131, 193, 298
arcos 56
arcsin 56
arctan 56
  see also trigonometric functions
area under the curve 184, 234
asymptote 82

binomial expansions 163, 195–205, 208
binomial theorem 196, 197
Bürgi, Jolst 87

Cartesian equations
  of curves 175–8, 192
  of a straight line 331–3
centre of mass 267
chain rule 105–11, 132, 133, 139, 209, 210, 215, 233
change of sign method 141–7, 160
change of variable see integration, by substitution
circle, parametric equations 180–1, 192
co-domain of a mapping 14, 49
cobweb diagram 150–1
coefficients, equating 165
common factors 4
compass bearing 311
composite functions 20–5
  differentiating 105–7

compound angle formulae 63–6, 71, 80, 130
compound functions, expanding 113–14
constant of integration 288, 309
convergence 147
coordinate geometry 175–92
  using vectors 326–42
$a\cos\theta + b\sin\theta$ 71–6, 80
cosecant (cosec) 53, 79
cosine (cos) 51
  differentiating 129–30
  see also trigonometric functions
cotangent (cot) 53, 78
curves
  area under 184, 234
  normal 97
  parametric and Cartesian equations 175–8, 180–4, 192

decimal search 142–4
definite integration by parts 268–71
derivatives
  of a^x 222–5
  of $(ax + b)^n$ 107–8
  dx/dy and dy/dx 123–8
  of e^{ax+b} 108
  of $e^{f(x)}$ 108–9
  of e^x 100–1
  of a function 97
  of $\ln(f(x))$ 109–10
  of $\ln x$ 101–2
differential equations 225–30, 233, 236
  general solution 288, 309
  particular solutions 290–5
  solving 287–95
differentiation 97–139, 209–33
  from first principles 130–1

functions and inverses 209, 233
implicit functions 209–15, 233
parametric functions 215–21, 233
and partial fractions 170–1
with respect to different variables 110–13
direction vector 329, 355
distance between two points 328–9
division
  algebraic 11–13
  of fractions 3
domain of a mapping 14, 49
double-angle formulae 67–9, 80, 275–7

e see exponential constant (e)
elimination of parameter 175–8, 192
ellipse
  parametric equations 181–2, 192
  standard equation 179
equal vectors 314
equations
  $e^{(ax+b)} = p$ 88–91, 96
  involving fractions 7–10
  $\ln(ax+b) = q$ 88–91, 96
  numerical solutions 140–1
  rearrangements 151–3
  of a straight line 329–33
  of the tangent at the general point 216
  $y = f(|x|)$ 33–4
  $y = \sin x$ 129–30
  $y = |f(x)|$ 32–3
Euler, Leonhard 82
ex see exponential function (ex)
exponential constant (e) 82
exponential function (ex) 81–4, 96
  derivative 100–1, 139
  integrating 239
exponential growth/decay 82, 222–5

Index

first-order differential equation 225, 233
fixed point, of an iteration 151
fractions
 addition 3–4
 algebraic 2
 division 3
 multiplication 3
 proper and improper 164, 168–9
 simplification 2, 49
 subtraction 3–4
functions 16–18
 composite 20–5, 49
 differentiation 233
 gradient 97
 graph 26–7
 impossible to expand 113–14
 inverse 25–31, 49
 notation 21
 order 22

general solution, of the differential equation 288, 309
geometrical figures, expressed as vectors 320
gradient, of a function 97
graphs
 exponential function 81, 82, 96
 of a function 26–7
 transforming 36–48
 of trigonometric functions 51
 of $y = f(|x|)$ 33–4
 of $y = \sin x$ 129–30
 of $y = |f(x)|$ 32–3

half-angle formulae 69–71
Hari Prasad, M. 193

identity symbol 271
implicit functions, differentiating 209–15, 233
integrals
 of $1/(ax + b)$ 247–9
 of $1/x$ 242–5
 approximate values 298
 of $(ax + b)^n$ 245–6
 by inspection 308
 definite and indefinite 235, 237
 of e^x 239
 of e^{ax+b} 246–7

 of $f'(x)/f(x)$ 249–50
 standard 296
integration 234–309
 by parts 262–71, 308
 by substitution 253–62, 308
 choosing a method 296–8
 constant 288, 309
 of parametrically defined functions 184–7, 192
 as reverse of differentiation 239–42
 trigonometric functions 251–3
 trigonometric identities 275–9
 volumes of revolution 279–84
intersection of two lines
 in three dimensions 338–42
 in two dimensions 337–8
interval, notation 141
inverse functions 96
 algebraic form 27–31
 differentiation 123–38, 139, 233
 gradients 124, 125
 integration 242–5
inverse trigonometric functions 56–9, 80
iterative methods 147–53, 160

lines
 angle between 347–50, 355
 in three-dimensional space 338–42
 in two-dimensional space 337–8
logarithmic functions ($\ln x$) 85–8, 96, 139
 derivative 101–2
logarithms, discovery 87
lowest common multiple 4

mappings 14–20
 domain and co-domain 14, 49
 inverse 57
 range 14, 49
maximum 97–8, 233
 see also stationary points
mental arithmetic 193
minimum 97–8
 see also stationary points
modulus
 function 32–5, 50
 of a vector 311, 354

moments of inertia 267
multiplying
 fractions 3
 vectors, by a scalar 316–17

Napier, John 87
natural logarithms 85, 87
normal, of a curve 97
numerical integration 298–302, 309
numerical methods 140–60

parabola
 equation 176
 parametric equations 182, 192
parameter, eliminating 175–8, 192
parametric equations 175–8, 192
 of standard curves 180–4, 192
 volumes of revolution 285–7
parametric functions, differentiation 215–21, 233
parametric integration 184–7, 192
partial fractions 163–71
 binomial expansion 203–5, 208
 in differentiation 170–1
 in integration 271–5, 308
Pascal's triangle 194–5, 196
point of inflection 98, 233
 see also stationary points
points
 coordinates 323–4
 distance between 328–9
polar coordinates 314
position vectors 314–15, 354
probability density function 267
product rule 113–15, 139, 209, 210, 233
proof 5

quotient 12
quotient rule 116–17, 131, 132, 139, 209, 233

range of a mapping 14, 49
rates of change 225–6
rational expressions 2–4, 49
rational functions 2, 49, 162–3
 expressed in terms of partial fractions 208
reciprocal trigonometric functions 53–6, 79
 derivatives 132

rectangular coordinates 314
rectangular hyperbola, parametric equations 182, 192
reflection 36, 37, 50
remainder 12
revolution *see* solids of revolution; volumes of revolution
roots
  close together 145
  decimal search 142–4
  finding 140–1
    see also numerical methods

scalar, definition 310, 354
scalar product 342–6
secant (sec) 53, 79
  see also trigonometric functions
second-order differential equation 225
separation of variables 288–90, 309
series, infinite 196
series expansions 193–208
sine (sin) 51
  differentiating 129–30
  see also trigonometric functions
small-angle approximation 131
solids of revolution 279
staircase diagram 150, 151
stationary points 211–14, 233

straight line
  Cartesian equation 331–3
  in two dimensions 334–6
  vector equation 329–31, 355
stretch 36, 37, 50
subtraction
  fractions 3–4
  vectors 318–19

tangent (tan) 51
  derivative 131–2
  see also trigonometric functions
transformations
  combinations 36–48
  order 42
translation 36, 50
trapezium rule 298–302, 309
triangles, standard 59
trigonometric functions 51–9
  differentiating 129–34, 139
  fundamental identities 59–62, 80, 275–9
  integration 251–3, 308
  inverse 56–9, 80
  reciprocal 53–6, 79
trigonometric parametric equations 178–80
trigonometry 51–80
turning points

maximum and minimum 97–8
  see also stationary points

unit vectors 321, 325, 354

variance 267
vector equation of a straight line 329–31, 355
vectors
  addition 317–18
  components 311, 354
  in coordinate geometry 326–42
  definition 310, 354
  equal 314
  geometrical figures 320
  joining two points 326–7
  magnitude 310, 324
  magnitude–direction form 310–11, 354
  multiplication by a scalar 316–17
  negative 317
  notation 311, 354
  subtraction 318–19
  in three dimensions 323–5
  in two dimensions 310–23
volumes of revolution 308
  finding by integration 279–84
  parametric equations 285–7